T0295155

HOMOTOPICAL QUANTUM FIELD THEORY

HOMOTOPICAL QUANTUM FIELD THEORY

Donald Yau

The Ohio State University at Newark, USA

 World Scientific

NEW JERSEY · LONDON · SINGAPORE · BEIJING · SHANGHAI · HONG KONG · TAIPEI · CHENNAI · TOKYO

Published by

World Scientific Publishing Co. Pte. Ltd.

5 Toh Tuck Link, Singapore 596224

USA office: 27 Warren Street, Suite 401-402, Hackensack, NJ 07601

UK office: 57 Shelton Street, Covent Garden, London WC2H 9HE

Library of Congress Control Number: 2019049784

British Library Cataloguing-in-Publication Data
A catalogue record for this book is available from the British Library.

HOMOTOPICAL QUANTUM FIELD THEORY

ISBN 978-981-121-285-7

For any available supplementary material, please visit
https://www.worldscientific.com/worldscibooks/10.1142/11626#t=suppl

To Eun Soo and Jacqueline

Preface

Algebraic quantum field theory and prefactorization algebra are two mathematical approaches to quantum field theory. In this monograph, using a new coend definition of the Boardman-Vogt construction of a colored operad, we define homotopy algebraic quantum field theories and homotopy prefactorization algebras, and investigate their homotopy coherent structures. Homotopy coherent diagrams, homotopy inverses, A_∞-algebras, E_∞-algebras, and E_∞-modules arise naturally in this context. Each homotopy algebraic quantum field theory has the structure of a homotopy coherent diagram of A_∞-algebras and satisfies a homotopy coherent version of the causality axiom. When the time-slice axiom is defined for algebraic quantum field theory, a homotopy coherent version of the time-slice axiom is satisfied by each homotopy algebraic quantum field theory. Over each topological space, every homotopy prefactorization algebra has the structure of a homotopy coherent diagram of E_∞-modules over an E_∞-algebra. To compare the two approaches, we construct a comparison morphism from the colored operad for (homotopy) prefactorization algebras to the colored operad for (homotopy) algebraic quantum field theories, and study the induced adjunctions on algebras.

Contents

Chapter 1

Introduction

Algebraic quantum field theory and prefactorization algebra are two mathematical approaches to quantum field theory. One of the main aims of this book is to provide robust definitions of homotopy algebraic quantum field theories and homotopy prefactorization algebras using a new definition of the Boardman-Vogt construction of a colored operad. To compare the two mathematical approaches to quantum field theory as well as their homotopy coherent analogues, we work within the framework of operads. This approach allows us to employ the powerful machinery from operad theory to quantum field theory. In the rest of this introduction, we briefly introduce each of these topics without going into too much details.

1.1 Algebraic Quantum Field Theory

Algebraic quantum field theory as introduced by Haag and Kastler [Haag and Kastler (1964)] provides one mathematical approach to quantum field theory that takes into account both quantum features and the theory of relativity. An algebraic quantum field theory $\mathfrak{A}$ assigns to each suitable spacetime region U in a fixed Lorentzian spacetime X an algebra $\mathfrak{A}(U)$. To each inclusion $i_U^V : U \subset V$, it assigns an algebra morphism

$$\mathfrak{A}(i_U^V) : \mathfrak{A}(U) \longrightarrow \mathfrak{A}(V)$$

in a functorial way. In other words, the morphism assigned to $i_U^U : U = U$ is the identity morphism of $\mathfrak{A}(U)$, and if $U \subset V \subset W$ then there is an equality

$$\mathfrak{A}(i_V^W) \circ \mathfrak{A}(i_U^V) = \mathfrak{A}(i_U^W). \tag{1.1.1}$$

This is just another way of saying that $\mathfrak{A}$ is a functor from the category of spacetime regions in X to the category of algebras. Physically $\mathfrak{A}(U)$ is the algebra of quantum observables in the region U. Each algebra $\mathfrak{A}(U)$ is only required to be associative, not commutative as in the classical case. The morphism $\mathfrak{A}(i_U^V)$ sends observables in U to observables in V.

An algebraic quantum field theory is more than just a functor from the category of spacetime regions in X to algebras. It is required to satisfy Einstein's *causality axiom*. It states that if U and V are causally disjoint regions in $W \subset X$, then the

1

images of $\mathfrak{A}(U)$ and $\mathfrak{A}(V)$ in $\mathfrak{A}(W)$ commute. The causality axiom, also known as causal locality or just locality, is a precise way of saying that physical influences cannot propagate faster than the speed of light. So causally disjoint regions are independent systems. An algebraic quantum field theory is also required to satisfy the *time-slice axiom*. It states that if $U \subset V$ contains a Cauchy surface of V, then the morphism

$$\mathfrak{A}(i_U^V) : \mathfrak{A}(U) \overset{\cong}{\longrightarrow} \mathfrak{A}(V)$$

is an isomorphism of algebras. Physically this means that all the observables in a spacetime region V are already determined by observables in a small time interval.

Traditionally one also asks that $\mathfrak{A}$ satisfy the *isotony axiom*, which states that each $\mathfrak{A}(i_U^V)$ is an injective morphism of algebras. However, various models of quantum gauge theories do not satisfy the isotony axiom; see, e.g., [Becker *et. al.* (2017b); Benini *et. al.* (2014,b)]. Therefore, recent literature on algebraic quantum field theory does not always include the isotony axiom. We follow this practice and only ask that each $\mathfrak{A}(i_U^V)$ be a morphism of algebras.

The Haag-Kastler framework is flexible in the sense that one can replace the domain category of spacetime regions in a fixed spacetime by another category C of spacetimes to obtain other versions of quantum field theories. One example is the category of all oriented, time-oriented, and globally hyperbolic Lorentzian manifolds of a fixed dimension. The resulting algebraic quantum field theories are locally covariant quantum field theories [Brunetti *et. al.* (2003); Fewster (2013); Fewster and Verch (2015)]. Similarly, to obtain chiral conformal quantum field theories [Bartels *et al.* (2015)] and Euclidean quantum field theories [Schlingemann (1999)], one uses the domain category of oriented manifolds and oriented Riemannian manifolds of a fixed dimension. One can also consider the category of spacetimes with extra geometric structures, such as principal bundles, connections, and spin structure [Benini and Schenkel (2017)], and the category of spacetimes with timelike boundaries [Benini *et. al.* (2018)].

To implement the causality axiom, one asks for a small category C that has a chosen subset $\perp$ of pairs of morphisms $\{ U_1 \longrightarrow V \longleftarrow U_2 \}$ with a common codomain. Such a pair formalizes the idea that U_1 and U_2 are disjoint in V. The pair $(\mathsf{C}, \perp)$ is called an *orthogonal category* [Benini *et. al.* (2017, 2019,b)]. To implement the time-slice axiom, one chooses a subset S of morphisms in C, which in the Lorentzian case is the set of Cauchy morphisms. An algebraic quantum field theory satisfies the time-slice axiom if the structure morphisms corresponding to morphisms in S are isomorphisms.

Furthermore, the target category of algebras over a field $\mathbb{K}$ can also be replaced by other categories of algebras. For example, instead of $\mathbb{K}$-algebras, which are monoids in the category $\mathsf{Vect}_{\mathbb{K}}$ of $\mathbb{K}$-vector spaces, one can use differential graded $\mathbb{K}$-algebras, which are monoids in the category $\mathsf{Chain}_{\mathbb{K}}$ of chain complexes of $\mathbb{K}$-vector spaces. In fact, conceptually it is easier to consider the category $\mathsf{Mon}(\mathsf{M})$ of monoids in a symmetric monoidal category M and then specify M as $\mathsf{Vect}_{\mathbb{K}}$, $\mathsf{Chain}_{\mathbb{K}}$, or whatever setting one wishes to work in, later if necessary.

In short, an algebraic quantum field theory on an orthogonal category $(\mathsf{C}, \perp)$ is a functor

$$\mathfrak{A} : \mathsf{C} \longrightarrow \mathsf{Mon}(\mathsf{M})$$

that satisfies the causality axiom and, if a set S of morphisms in C is chosen, the time-slice axiom with respect to S. This definition of an algebraic quantum field theory was introduced in [Benini *et. al.* (2017)], and we adopt it in this book.

1.2 Homotopy Algebraic Quantum Field Theory

Homotopy theory enters the picture with:

- recent toy examples of quantum gauge theories in [Benini and Schenkel (2017)] that are algebraic quantum field theories up to homotopy;
- the program in [Benini *et. al.* (2018b)] to study quantum Yang-Mills theory using the homotopy theory of stacks [Hollander (2007, 2008,b)];
- the use of homotopy (co)limits in [Benini *et. al.* (2015)] to study local-to-global extensions of field configurations and observables for abelian gauge theory from contractible manifolds to non-contractible manifolds.

At the most elementary level, this means that the strict equalities in algebraic quantum field theories are replaced by homotopies. For example, the equality (1.1.1) that expresses functoriality is replaced by the homotopy relation

$$\mathfrak{A}(i_V^W) \circ \mathfrak{A}(i_U^V) \sim \mathfrak{A}(i_U^W).$$

In other words, the composition on the left is chain homotopic to the morphism on the right. The causality axiom is replaced by a similar homotopical analogue that expresses commutativity up to chain homotopy. The homotopical version of the time-slice axiom says that, if $s \in S$ is one of the chosen morphisms, then $\mathfrak{A}(s)$ is a chain homotopy equivalence.

An important lesson from homotopy theory is that homotopies are only the first layer of a much richer homotopy coherent structure. This means that we do not simply ask for two things to be homotopic. Instead we ask for specific homotopies as part of the algebraic structure itself, and these homotopies satisfy higher homotopy relations via further structure morphisms, and so forth. For instance, for monoids and commutative monoids, the higher homotopical analogues are called A_∞-algebras and E_∞-algebras, respectively. Set theoretically, in an A_∞-algebra, instead of strict associativity $(ab)c = a(bc)$, it has a specific structure morphism that is a homotopy $(ab)c \sim a(bc)$. Instead of a strict two-sided unit 1, it has specific structure morphisms that are homotopies $1a \sim a \sim a1$. There is an infinite family of higher structure morphisms that relate these homotopies.

Operad theory is a powerful framework originating from homotopy theory that allows one to keep track of the enormous amount of data in higher homotopical structures in a manageable way. An operad O has a set of objects $\mathfrak{C}$, like a small category, but the domain of a morphism

$$(c_1, \ldots, c_n) \xrightarrow{f} d$$

is a finite, possibly empty, sequence of objects. Just like in a category, one can compose these morphisms. Moreover, the domain objects can be permuted. These morphisms should be thought of as models of n-ary operations, and they satisfy some reasonable unity, equivariance, and associativity axioms. To emphasize the set $\mathfrak{C}$ of objects, we call it a $\mathfrak{C}$-colored operad.

Similar to an algebra or a monoid, an operad O can act on objects, called O-algebras. For example, there is an associative operad As whose algebras are monoids, and there is a commutative operad Com whose algebras are commutative monoids. As a general rule, if there is an operad for a certain type of structure, then there is a colored operad for C-diagrams of such structure. So there is a colored operad whose algebras are C-diagrams of monoids in M, i.e., functors $\mathsf{C} \longrightarrow \mathsf{Mon}(\mathsf{M})$. With more work, one can even write down a colored operad $\mathsf{O}^{\mathsf{M}}_{\overline{\mathsf{C}}}$ whose algebras are algebraic quantum field theories on an orthogonal category $\overline{\mathsf{C}} = (\mathsf{C}, \perp)$. In other words, it is possible to incorporate the causality axiom, which is a kind of commutativity, and the time-slice axiom, which is a kind of invertibility, into the colored operad itself. The construction of the colored operad $\mathsf{O}^{\mathsf{M}}_{\overline{\mathsf{C}}}$ for algebraic quantum field theories was made explicit in [Benini *et. al.* (2017)].

To capture homotopy algebraic quantum field theories with all of the higher homotopical structure, we once again follow an established principle in homotopy theory. If O is an operad for a certain kind of algebras, then the homotopy coherent versions of these algebras are obtained as algebras over a suitable resolution $\mathsf{P} \xrightarrow{\ \sim\ } \mathsf{O}$ of O. This is analogous to replacing a module by a projective resolution in homological algebra, so we want P to be nice in some way. In the terminology of model category theory, we ask P to be a cofibrant resolution of O in the model category of operads. In homological algebra we learned that projective resolutions of a given module are not unique, and which projective resolution to use depends on one's intended applications. For instance, corresponding to monoids and commutative monoids, there are different versions of A_∞-algebras and E_∞-algebras, depending on which resolutions one chooses for the associative operad and the commutative operad. A good choice of a resolution of the colored operad $\mathsf{O}^{\mathsf{M}}_{\overline{\mathsf{C}}}$ for algebraic quantum field theories is its Boardman-Vogt resolution, which was originally defined for topological operads in [Boardman and Vogt (1972)] to study homotopy invariant algebraic structures.

In [Berger and Moerdijk (2006, 2007)] the Boardman-Vogt construction of a colored operad was extended from the category of topological spaces to a general symmetric monoidal category equipped with a segment, which provides a concept of length. For example, for topological spaces, a segment is given by the unit interval $[0, 1]$. For chain complexes over $\mathbb{K}$, a segment is given by the two-stage complex

$\mathbb{K} \xrightarrow{(+,-)} \mathbb{K} \oplus \mathbb{K}$ concentrated in degrees 1 and 0. In [Berger and Moerdijk (2006)] the Boardman-Vogt construction of an operad O is entrywise defined inductively as a sequential colimit, with each morphism in the sequence defined as a pushout that takes input from the previous inductive stage. To effectively apply the machinery to quantum field theory, we need a more direct construction. So we will introduce a new definition of the Boardman-Vogt construction of a colored operad that is entrywise defined in one step as a coend indexed by a category of trees, called a *substitution category*.

Given a flavor of spacetimes, i.e., a choice of an orthogonal category $\overline{\mathsf{C}} = (\mathsf{C}, \perp)$, we will define homotopy algebraic quantum field theories on $\overline{\mathsf{C}}$ as algebras over the Boardman-Vogt construction $\mathsf{WO}_{\overline{\mathsf{C}}}^{\mathsf{M}}$ of the colored operad $\mathsf{O}_{\overline{\mathsf{C}}}^{\mathsf{M}}$. All of the higher homotopy relations in homotopy algebraic quantum field theories are parametrized by the substitution categories in the coends. An algebraic quantum field theory is an $\mathsf{O}_{\overline{\mathsf{C}}}^{\mathsf{M}}$-algebra, which in turn is a C-diagram of monoids $\mathsf{C} \longrightarrow \mathsf{Mon}(\mathsf{M})$ that satisfies the causality axiom and possibly the time-slice axiom if a set S of morphisms in C is given. Replacing everything by their higher homotopical analogues, we will show that every homotopy algebraic quantum field theory on $\overline{\mathsf{C}}$, i.e., $\mathsf{WO}_{\overline{\mathsf{C}}}^{\mathsf{M}}$-algebra, has the structure of a homotopy coherent C-diagram of A_∞-algebras that satisfies a homotopy coherent version of the causality axiom. Furthermore, if a set S of morphisms in C is given, then it also satisfies a homotopy coherent version of the time-slice axiom.

An important point here is that all of the higher homotopies, such as the ones expressing homotopy functoriality, homotopy causality, and homotopy time-slice, are specific structure morphisms of a $\mathsf{WO}_{\overline{\mathsf{C}}}^{\mathsf{M}}$-algebra. In other words, all of the higher homotopies are already encoded in the Boardman-Vogt construction $\mathsf{WO}_{\overline{\mathsf{C}}}^{\mathsf{M}}$ itself. Our coend definition of the Boardman-Vogt construction plays a crucial role in our understanding of the structure in homotopy algebraic quantum field theories. In future work, using our Boardman-Vogt construction and the approach in this book, it would be interesting to develop a higher homotopical version of algebraic field theories as developed in [Bruinsma and Schenkel (2018)].

1.3 Homotopy Prefactorization Algebra

Prefactorization algebras were introduced in [Costello and Gwilliam (2017)] to provide another mathematical framework for quantum field theory that is analogous to the deformation quantization approach to quantum mechanics. For a given topological space X, to each open subset $U \subset X$, a prefactorization algebra $\mathcal{F}$ on X assigns a chain complex $\mathcal{F}(U)$. To each finite sequence $U_1, \ldots, U_n$ of pairwise disjoint open subsets in $V \subset X$, $\mathcal{F}$ assigns a chain map

$$\mathcal{F}(U_1) \otimes \cdots \otimes \mathcal{F}(U_n) \xrightarrow{\mathcal{F}_{U_1,\ldots,U_n}^V} \mathcal{F}(V) .$$

In particular, for an inclusion $U \subset V$ of open subsets in X,

$$\mathcal{F}_U^V : \mathcal{F}(U) \longrightarrow \mathcal{F}(V)$$

is a chain map. This data is required to satisfy some reasonable unity, equivariance, and associativity conditions. For example, for open subsets $U \subset V \subset W$ in X, a part of the associativity condition is the equality

$$\mathcal{F}_V^W \circ \mathcal{F}_U^V = \mathcal{F}_U^W : \mathcal{F}(U) \longrightarrow \mathcal{F}(W).$$

In particular, a prefactorization algebra $\mathcal{F}$ on X has the structure of a functor

$$\mathsf{Open}(X) \longrightarrow \mathsf{Chain}_{\mathbb{K}}$$

from the category of open subsets in X with inclusions as morphisms. There is also a time-slice axiom in this setting, called local constancy in [Costello and Gwilliam (2017)]. If S is a chosen set of morphisms in $\mathsf{Open}(X)$, then one asks that each structure morphism $\mathcal{F}_U^V$ with $(U \subset V) \in S$ be an isomorphism.

Physically $\mathcal{F}(U)$ is the collection of quantum observables in U. The chain map $\mathcal{F}_{U_1,\dots,U_n}^V$ means that if the U_i's are pairwise disjoint in V, then their observables can be multiplied in $\mathcal{F}(V)$. This is the main difference between a prefactorization algebra and an algebraic quantum field theory. In an algebraic quantum field theory, every object $\mathfrak{A}(U)$ is a monoid, so observables in a spacetime region U can always be multiplied. On the other hand, in a prefactorization algebra $\mathcal{F}$, only observables from pairwise disjoint regions can be multiplied. Furthermore, part of the equivariance condition says that $\mathcal{F}(\varnothing_X)$, where $\varnothing_X \subset X$ denotes the empty subset, is a commutative differential graded algebra. For each open subset $V \subset X$, since $\varnothing_X$ is disjoint from V, there is a structure morphism

$$\overbrace{\mathcal{F}(\varnothing_X) \otimes \cdots \otimes \mathcal{F}(V) \otimes \cdots \otimes \mathcal{F}(\varnothing_X)}^{\text{only one } \mathcal{F}(V)} \xrightarrow{\mathcal{F}_{\varnothing_X,\dots,V,\dots,\varnothing_X}^V} \mathcal{F}(V)$$

that gives each $\mathcal{F}(V)$ the structure of an $\mathcal{F}(\varnothing_X)$-module. These objectwise $\mathcal{F}(\varnothing_X)$-modules are compatible with the structure morphisms $\mathcal{F}_U^V$.

To facilitate the comparison between the above two mathematical approaches to quantum field theory, we will take a slightly more abstract approach to prefactorization algebras. To define a prefactorization algebra, what one really needs is a small category C, whose objects are thought of as spacetime regions, that has a suitable notion of pairwise disjointness. In other words, one chooses a set $\triangle$ of finite sequences of morphisms $\{f_i : U_i \to V\}_{i=1}^n$ in C with a common codomain. Such a finite sequence, called a *configuration*, formalizes the idea that the U_i's are pairwise disjoint in V. These configurations are required to satisfy some natural axioms, such as closure under composition and permutation. The pair $\widehat{\mathsf{C}} = (\mathsf{C}, \triangle)$ is called a *configured category*.

As in the case of algebraic quantum field theory, we allow the base category to be a general symmetric monoidal category M instead of just $\mathsf{Chain}_{\mathbb{K}}$. A prefactorization algebra on a configured category $\widehat{\mathsf{C}}$ is defined as an algebra over the colored operad

$\mathsf{O}^{\mathsf{M}}_{\widehat{\mathsf{C}}}$ whose entries are coproducts $\coprod_{\Delta'} \mathbb{1}$, with Δ' a suitable subset of configurations and $\mathbb{1}$ the monoidal unit in M. To implement the time-slice axiom with respect to a set S of morphisms in C, we replace the colored operad $\mathsf{O}^{\mathsf{M}}_{\widehat{\mathsf{C}}}$ by a suitable localization.

Proceeding as in the story of algebraic quantum field theory, we define homotopy prefactorization algebras on a configured category $\widehat{\mathsf{C}}$ as algebras over the Boardman-Vogt construction $\mathsf{WO}^{\mathsf{M}}_{\widehat{\mathsf{C}}}$ of the colored operad $\mathsf{O}^{\mathsf{M}}_{\widehat{\mathsf{C}}}$. Once again due to our one-step coend definition of the Boardman-Vogt construction, we are able to make explicit the structure of homotopy prefactorization algebras. Let us take as an example the configured category associated to the category $\mathsf{Open}(X)$ of open subsets of a topological space X with configurations defined by pairwise disjointedness. In this setting, we will show that a homotopy prefactorization algebra Y has, first of all, an E_∞-algebra structure in the entry $Y_{\varnothing_X}$ corresponding to the empty subset of X. It also has the structure of a homotopy coherent $\mathsf{Open}(X)$-diagram and satisfies a homotopy coherent version of the time-slice axiom if a set S of open subset inclusions is given. Furthermore, for each open subset $V \subset X$, the entry Y_V admits the structure of an E_∞-module over the E_∞-algebra $Y_{\varnothing_X}$. These objectwise E_∞-modules are homotopy coherently compatible with the homotopy coherent $\mathsf{Open}(X)$-diagram structure of Y. Once again, all of the higher homotopical structure is already encoded in the Boardman-Vogt construction $\mathsf{WO}^{\mathsf{M}}_{\widehat{\mathsf{C}}}$ itself.

1.4 Comparison

Given that both algebraic quantum field theory and prefactorization algebra are mathematical approaches to quantum field theory, a natural question is how they are related. The two approaches certainly have something in common. In both settings, we consider functors from some category C, whose objects are thought of as spacetime regions, to some target category, such as $\mathsf{Vect}_{\mathbb{K}}$ or $\mathsf{Chain}_{\mathbb{K}}$. Moreover, in each setting there is a time-slice axiom that says that some chosen structure morphisms are invertible. Our comparison of the two approaches happens at two levels.

We first compare orthogonal categories, on which (homotopy) algebraic quantum field theories are defined, and configured categories, on which (homotopy) prefactorization algebras are defined. Informally, every orthogonal category generates a configured category, in which a configuration is a finite sequence of pairwise orthogonal morphisms. Conversely, every configured category restricts to an orthogonal category, in which the orthogonal pairs are the binary configurations. The precise version says that the category of orthogonal categories embeds as a full reflective subcategory in the category of configured categories. This is not an adjoint equivalence, so the two categories are genuinely different.

Next we compare prefactorization algebras on a configured category $\widehat{\mathsf{C}}$ and algebraic quantum field theories on the associated orthogonal category $\overline{\mathsf{C}}$. We construct a comparison morphism

$$\mathsf{O}^{\mathsf{M}}_{\widehat{\mathsf{C}}} \longrightarrow \mathsf{O}^{\mathsf{M}}_{\overline{\mathsf{C}}}$$

from the colored operad $\mathsf{O}^{\mathsf{M}}_{\widehat{\mathsf{C}}}$ defining prefactorization algebras on $\widehat{\mathsf{C}}$ to the colored operad $\mathsf{O}^{\mathsf{M}}_{\overline{\mathsf{C}}}$ defining algebraic quantum field theories on $\overline{\mathsf{C}}$. Since our Boardman-Vogt construction is natural, there is an induced comparison morphism

$$\mathsf{WO}^{\mathsf{M}}_{\widehat{\mathsf{C}}} \longrightarrow \mathsf{WO}^{\mathsf{M}}_{\overline{\mathsf{C}}}$$

from the colored operad $\mathsf{WO}^{\mathsf{M}}_{\widehat{\mathsf{C}}}$ defining homotopy prefactorization algebras to the colored operad $\mathsf{WO}^{\mathsf{M}}_{\overline{\mathsf{C}}}$ defining homotopy algebraic quantum field theories. These comparison morphisms induce various comparison adjunctions between (homotopy) prefactorization algebras and (homotopy) algebraic quantum field theories, with or without the time-slice axiom.

Although prefactorization algebras and algebraic quantum field theories are different in general, there is one important case when they are equal. This situation corresponds to the maximal configured category and the maximal orthogonal category for a given small category C. In this case, both the category of prefactorization algebras and the category of algebraic quantum field theories are isomorphic to the category of C-diagrams of commutative monoids. We interpret this situation as saying that the two mathematical approaches to quantum field theory both reduce to the classical case, where observables form commutative algebras. Furthermore, since E_∞-algebras are homotopy coherent versions of commutative algebras, in this case both the category of homotopy prefactorization algebras and the category of homotopy algebraic quantum field theories are isomorphic to the category of homotopy coherent C-diagrams of E_∞-algebras.

1.5   Organization

This book is divided into two parts. The first part is about operads, with special emphasize on our version of the Boardman-Vogt construction in terms of coends. The second part is the application of the machinery in the first part to algebraic quantum field theory, prefactorization algebra, and their homotopy coherent analogues. Each chapter has its own introduction. A brief description of each chapter follows.

To keep this book relatively self-contained, Part 1 begins with Chapter 2 in which we review basic concepts of category theory, including colimits, coends, adjoint functors, monoidal categories, monads, and localization. Our coend definition of the Boardman-Vogt construction uses the language of trees. In Chapter 3 we review the basic combinatorics of trees and their composition, called tree substitution.

Colored operads are defined in Chapter 4. We give four equivalent definitions of colored operads. We first define colored operads as monoids with respect to the colored circle product. Then we give three more equivalent descriptions in terms of generating operations, partial compositions, and trees. As soon as we start discussing colored operads in this chapter, we will work over a general symmetric monoidal category M. The reader who is interested in a specific base category, such as $\mathsf{Chain}_{\mathbb{K}}$, should feel free to take M as this category throughout.

In Chapter 5 we discuss further properties of operads, including change-of-operad adjunctions and change-of-category functors. We briefly discuss the model category structure on the category of algebras over a colored operad. This chapter ends with the discussion of a localization of a colored operad, which is analogous to the localization of a category. A localized colored operad is a colored operad in which some unary elements have been inverted. We need localized colored operads when we discuss the time-slice axiom in (homotopy) prefactorization algebras.

In Chapter 6 we define the Boardman-Vogt construction of a colored operad using a coend indexed by a category of trees and discuss its naturality properties. Each colored operad O has a Boardman-Vogt construction WO together with an augmentation $\eta : \mathsf{WO} \longrightarrow \mathsf{O}$. In favorable situations, such as when the underlying category is $\mathsf{Chain}_{\mathbb{K}}$ with $\mathbb{K}$ a field of characteristic zero, the augmentation is a weak equivalence, and the induced adjunction between the categories of algebras is a Quillen equivalence. However, to understand the structure of homotopy algebraic quantum field theories and homotopy prefactorization algebras, we only need the Boardman-Vogt construction itself, not its homotopical properties.

In Chapter 7 we study the Boardman-Vogt construction of various colored operads of interest. Due to our one-step coend definition of the Boardman-Vogt construction, we are able to write down a coherence theorem for their algebras. As examples, we discuss in details homotopy coherent diagrams, homotopy inverses in homotopy coherent diagrams, specific models of A_∞-algebras and E_∞-algebras, and homotopy coherent diagrams of A_∞-algebras and of E_∞-algebras. All of these homotopy coherent algebraic structures are relevant in our study of homotopy algebraic quantum field theories and homotopy prefactorization algebras. This finishes Part 1.

Part 2 begins with Chapter 8 in which we discuss the colored operad for algebraic quantum field theories following [Benini *et. al.* (2017)]. Examples include diagrams of (commutative) monoids, quantum field theories on (equivariant) topological spaces, chiral conformal quantum field theories, Euclidean quantum field theories, locally covariant quantum field theories, and quantum field theories on structured spacetimes and on spacetimes with timelike boundary.

In Chapter 9 we define homotopy algebraic quantum field theories as algebras over the Boardman-Vogt construction $\mathsf{WO}_{\overline{\mathsf{C}}}^{\mathsf{M}}$ applied to the colored operad $\mathsf{O}_{\overline{\mathsf{C}}}^{\mathsf{M}}$ for algebraic quantum field theories. We record a coherence theorem for homotopy algebraic quantum field theories. Each homotopy algebraic quantum field theory is shown to have the structure of a homotopy coherent diagram of A_∞-algebras and to satisfy a homotopy coherent version of the causality axiom. When a set of morphisms in C is given, each homotopy algebraic quantum field theory also satisfies a homotopy coherent version of the time-slice axiom.

In Chapter 10 we define configured categories and prefactorization algebras on them. We will see that commutative monoids and their modules feature prominently in prefactorization algebras. In Chapter 11 we define homotopy prefactorization algebras on a configured category $\widehat{\mathsf{C}}$ as algebras over the Boardman-Vogt construction $\mathsf{WO}^{\mathsf{M}}_{\widehat{\mathsf{C}}}$ of the colored operad $\mathsf{O}^{\mathsf{M}}_{\widehat{\mathsf{C}}}$ for prefactorization algebras. We record a coherence theorem for homotopy prefactorization algebras. In addition to a homotopy coherent diagram structure, we will see that E_∞-algebras and their E_∞-modules play prominent roles in homotopy prefactorization algebras.

In Chapter 12 we compare the two mathematical approaches to quantum field theory featured in this book. We show that the category of orthogonal categories embeds in the category of configured categories as a full reflective subcategory. Then we construct a comparison morphism $\mathsf{O}_{\widehat{\mathsf{C}}} \longrightarrow \mathsf{O}_{\overline{\mathsf{C}}}$ that we use to compare (homotopy) prefactorization algebras and (homotopy) algebraic quantum field theories. We discuss examples of prefactorization algebras that come from algebraic quantum field theories and those that do not. This concludes Part 2.

Audience

This book is intended for graduate students, mathematicians, and physicists. Throughout this book, we include many examples and a lot of motivation and interpretation of results both mathematically and physically. Since we actually review the basics of categories and operads, an ambitious advanced undergraduate should be able to follow this book.

Chapter 2

Category Theory

In this chapter, we recall some basic concepts of category theory and some relevant examples. The reader who is familiar with basic category theory can just read the examples in Section 2.2 and skip the rest of this chapter. Our references for category theory are [Borceux (1994,b); Mac Lane (1998)]. Categories were originally introduced by Eilenberg and Mac Lane [Eilenberg and Mac Lane (1945)].

In Section 2.1 we review categories, functors, natural transformations, and equivalences. A long list of examples of categories that will be used in later chapters are given in Section 2.2. In Section 2.3 we review limits, colimits, and coends, which will play a crucial role in our definition of the Boardman-Vogt construction of a colored operad in Chapter 6. In Section 2.4 we discuss adjoint functors. In Section 2.5 we review symmetric monoidal categories, which are the most natural setting to discuss colored operads. In Section 2.6 and Section 2.7 we review monoids and monads, which are important because algebras over a colored operad are defined as algebras over the associated monad. In Section 2.8 we review localization of categories, which will be needed to discuss the time-slice axiom in algebraic quantum field theories in Chapter 8.

2.1 Basics of Categories

Definition 2.1.1. A *category* C consists of the following data:

- a class $\mathsf{Ob}(\mathsf{C})$ of *objects*;
- for any two objects $a, b \in \mathsf{Ob}(\mathsf{C})$, a set $\mathsf{C}(a,b)$ of *morphisms* with *domain a* and *codomain b*;
- for each object $a \in \mathsf{Ob}(\mathsf{C})$, an *identity morphism* $\mathrm{Id}_a \in \mathsf{C}(a,a)$;
- for any objects $a, b, c \in \mathsf{Ob}(\mathsf{C})$, a function called the *composition*

$$\mathsf{C}(b,c) \times \mathsf{C}(a,b) \xrightarrow{\;\circ\;} \mathsf{C}(a,c)$$

sending (g, f) to $g \circ f = gf$, called the *composition* of g and f.

The above data is required to satisfy the following two axioms.

Associativity Suppose $(h, g, f) \in C(c, d) \times C(b, c) \times C(a, b)$. Then there is an equality

$$h \circ (g \circ f) = (h \circ g) \circ f \quad \text{in} \quad C(a, d).$$

Unity For any objects $a, b \in \text{Ob}(C)$ and morphism $f \in C(a, b)$, there are equalities

$$f \circ \text{Id}_a = f = \text{Id}_b \circ f \quad \text{in} \quad C(a, b).$$

The collection of all morphisms in C is written as $\text{Mor}(C)$.

Definition 2.1.2. Suppose C is a category.

(1) The *opposite category* C^{op} is the category with the same objects as C and with morphism sets $C^{\text{op}}(a, b) = C(b, a)$. Its identity morphisms and composition are defined by those in C.

(2) A *subcategory* of C is a category D such that:

  (a) There is an inclusion $\text{Ob}(D) \subseteq \text{Ob}(C)$ on objects.
  (b) For any objects $a, b \in \text{Ob}(D)$, there is a subset inclusion $D(a, b) \subseteq C(a, b)$ on morphisms.
  (c) For each object $a \in \text{Ob}(D)$, the identity morphism $\text{Id}_a \in D(a, a)$ is the identity morphism of $a \in \text{Ob}(C)$.
  (d) Suppose $(g, f) \in D(b, c) \times D(a, b)$. Then their composition $g \circ f \in D(a, c)$ in D is equal to the composition $g \circ f \in C(a, c)$ in C.

(3) A subcategory D of C is a *full subcategory* if, for any objects $a, b \in \text{Ob}(D)$, there is an equality $D(a, b) = C(a, b)$ of morphism sets.

(4) C is called a *small category* if $\text{Ob}(C)$ is a set.

(5) An *isomorphism* $f \in C(a, b)$ is a morphism such that there exists an *inverse* $f^{-1} \in C(b, a)$ satisfying

$$f \circ f^{-1} = \text{Id}_b \quad \text{and} \quad f^{-1} \circ f = \text{Id}_a.$$

An inverse is unique if it exists. An isomorphism is also denoted by $\cong$.

(6) A *groupoid* is a category in which all morphisms are isomorphisms.

(7) A *discrete category* is a category whose only morphisms are the identity morphisms.

Definition 2.1.3. A *functor* $F : C \longrightarrow D$ from a category C to a category D consists of

- an *assignment on objects*

$$\text{Ob}(C) \longrightarrow \text{Ob}(D), \qquad a \longmapsto Fa;$$

- for any objects $a, b \in \text{Ob}(C)$, a *function on morphism sets*

$$C(a, b) \longrightarrow D(Fa, Fb), \qquad f \longmapsto Ff. \qquad (2.1.4)$$

The above data is required to satisfy the following two axioms.

Preservation of Identity For each object $a \in \mathsf{Ob}(\mathsf{C})$, there is an equality

$$F(\mathrm{Id}_a) = \mathrm{Id}_{Fa} \quad \text{in} \quad \mathsf{D}(Fa, Fa).$$

Preservation of Composition For any morphisms $(g, f) \in \mathsf{C}(b, c) \times \mathsf{C}(a, b)$, there is an equality

$$F(g \circ f) = Fg \circ Ff \quad \text{in} \quad \mathsf{D}(Fa, Fc).$$

If C is a small category, we also call a functor $\mathsf{C} \longrightarrow \mathsf{D}$ a C-*diagram in* D. The *identity functor* $\mathrm{Id}_{\mathsf{C}} : \mathsf{C} \longrightarrow \mathsf{C}$ is the functor given by the identity functions on both objects and morphisms.

Definition 2.1.5. A functor $F : \mathsf{C} \longrightarrow \mathsf{D}$ is called:

(1) *full* (resp., *faithful*) if for any objects $a, b \in \mathsf{C}$, the function on morphism sets in (2.1.4) is surjective (resp., injective);
(2) *essentially surjective* if for each object $d \in \mathsf{D}$, there exist an object $c \in \mathsf{C}$ and an isomorphism $Fc \xrightarrow{\cong} d$.

An object $d \in \mathsf{D}$ is in the *essential image of* F if there exist an object $c \in \mathsf{C}$ and an isomorphism $Fc \xrightarrow{\cong} d$.

Definition 2.1.6. Suppose $F : \mathsf{C} \longrightarrow \mathsf{D}$ and $G : \mathsf{D} \longrightarrow \mathsf{E}$ are functors.

(1) The *composition* of functors

$$GF = G \circ F : \mathsf{C} \longrightarrow \mathsf{E}$$

is defined by composing the assignments on objects and the functions on morphism sets.
(2) We call F an *isomorphism* if there exists a functor $F^{-1} : \mathsf{D} \longrightarrow \mathsf{C}$ such that

$$FF^{-1} = \mathrm{Id}_{\mathsf{D}} \quad \text{and} \quad F^{-1}F = \mathrm{Id}_{\mathsf{C}}.$$

An inverse F^{-1} is unique if it exists.

Definition 2.1.7. Suppose $F, G, H : \mathsf{C} \longrightarrow \mathsf{D}$ are functors from C to D.

(1) A *natural transformation* $\theta : F \longrightarrow G$ consists of a *structure morphism* $\theta_a \in \mathsf{D}(Fa, Ga)$ for each $a \in \mathsf{Ob}(\mathsf{C})$ such that, if $f \in \mathsf{C}(a, b)$ is a morphism for some object $b \in \mathsf{C}$, then

$$Gf \circ \theta_a = \theta_b \circ Ff \quad \text{in} \quad \mathsf{D}(Fa, Gb).$$

(2) If $\theta : F \longrightarrow G$ and $\eta : G \longrightarrow H$ are natural transformations, their *composition* is the natural transformation $\eta\theta : F \longrightarrow H$ with structure morphisms $(\eta\theta)_a = \eta_a \circ \theta_a$ for $a \in \mathsf{Ob}(\mathsf{C})$.
(3) A *natural isomorphism* is a natural transformation in which every structure morphism is an isomorphism.

Definition 2.1.8. An *equivalence* between categories C and D consists of a pair of functors $F : \mathsf{C} \longrightarrow \mathsf{D}$ and $G : \mathsf{D} \longrightarrow \mathsf{C}$ and a pair of natural isomorphisms

$$\mathrm{Id}_\mathsf{C} \overset{\cong}{\longrightarrow} GF \quad \text{and} \quad \mathrm{Id}_\mathsf{D} \overset{\cong}{\longrightarrow} FG \,.$$

In this setting, we say that F is an *equivalence of categories* and that the categories C and D are *equivalent* via the functors F and G. A category is said to be *essentially small* if it is equivalent to a small category.

Notation 2.1.9. The following notations and conventions will be used.

- If x is an object or a morphism in a category C, we will often write $x \in \mathsf{C}$ instead of $x \in \mathsf{Ob}(\mathsf{C})$ or $x \in \mathsf{Mor}(\mathsf{C})$.

- A morphism $f \in \mathsf{C}(a, b)$ is also written as $f : a \longrightarrow b$ or $a \overset{f}{\longrightarrow} b$.

- A functor $F : \mathsf{C} \longrightarrow \mathsf{D}$ is also written as $\mathsf{C} \overset{F}{\longrightarrow} \mathsf{D}$.

- A natural transformation $\theta : F \longrightarrow G$ is also written as $F \overset{\theta}{\longrightarrow} G$.

2.2 Examples of Categories

In this section, we list some relevant examples of categories that we will use later.

Example 2.2.1. The *empty category*, with no objects and no morphisms, is denoted by ∅. ◇

Example 2.2.2 (Functor categories). Given any categories C and D, there is a *functor category* $\mathsf{Fun}(\mathsf{C}, \mathsf{D})$ with functors $\mathsf{C} \longrightarrow \mathsf{D}$ as objects and natural transformations between them as morphisms. If C is a small category, we also call the functor category $\mathsf{Fun}(\mathsf{C}, \mathsf{D})$ a *diagram category* and denote it by D^C. ◇

Example 2.2.3 (Product categories). Given any categories C and D, there is a *product category* $\mathsf{C} \times \mathsf{D}$ with objects $\mathsf{Ob}(\mathsf{C}) \times \mathsf{Ob}(\mathsf{D})$ and morphism sets

$$(\mathsf{C} \times \mathsf{D})\big((c, d), (c', d')\big) = \mathsf{C}(c, c') \times \mathsf{D}(d, d')$$

for $c, c' \in \mathsf{C}$ and $d, d' \in \mathsf{D}$. ◇

Example 2.2.4 (Under categories). Suppose a is an object in a category C. The *under category* $a \downarrow \mathsf{C}$ is the category whose objects are morphisms in C of the form $a \longrightarrow b$. A morphism $f : (a \to b) \longrightarrow (a \to c)$ in the under category is a morphism $f : b \longrightarrow c$ in C such that the triangle

in C is commutative. The identity morphisms and composition are defined by those in C. ◇

Example 2.2.5 (Sets). There is a category Set with sets as objects and functions as morphisms. ◇

Example 2.2.6 (Vector spaces). For a field $\mathbb{K}$, there is a category $\mathsf{Vect}_{\mathbb{K}}$ with $\mathbb{K}$-vector spaces as objects and linear maps as morphisms. ◇

Example 2.2.7 (Chain complexes). There is a category $\mathsf{Chain}_{\mathbb{K}}$ with chain complexes of $\mathbb{K}$-vector spaces as objects and chain maps as morphisms. Via the reindexing $X_n \longmapsto X^{-n}$, one can also regard $\mathsf{Chain}_{\mathbb{K}}$ as the category of cochain complexes of $\mathbb{K}$-vector spaces. With this in mind, everything below about chain complexes also holds for cochain complexes. ◇

Example 2.2.8 (Topological spaces). There is a category Top whose objects are compactly generated weak Hausdorff spaces and whose morphisms are continuous maps. ◇

Example 2.2.9 (Simplex category). The *simplex category* Δ has objects the finite totally ordered sets

$$[n] = \{0 < 1 < \cdots < n\}$$

for $n \geq 0$. A morphism is a weakly order-preserving map, i.e., $f(i) \leq f(j)$ if $i < j$. ◇

Example 2.2.10 (Simplicial sets). For a category C, the diagram category

$$\mathsf{C}^{\Delta^{\mathrm{op}}} = \mathsf{Fun}(\Delta^{\mathrm{op}}, \mathsf{C})$$

is called the category of *simplicial objects in* C. If C is the category of sets, then $\mathsf{Set}^{\Delta^{\mathrm{op}}}$ is also written as SSet, and its objects are called simplicial sets. ◇

Example 2.2.11 (Small categories). There is a category Cat whose objects are small categories and whose morphisms are functors. ◇

Example 2.2.12 (Partially ordered sets and lattices). Each partially ordered set $(S, \leq)$ becomes a small category with object set S, and there is a morphism $a \longrightarrow b$ if and only if $a \leq b$. We will denote this category by S. For $a, b \in S$, the morphism set $S(a, b)$ is either empty or a one-element set. A *lattice* is a partially ordered set such that every pair of distinct elements $\{a, b\}$ has both a least upper bound $a \vee b$ and a greatest lower bound $a \wedge b$. A *bounded lattice* is a lattice with a least element 0 and a greatest element 1. ◇

Example 2.2.13 (Open subsets of a topological space). For each topological space X, there is a partially ordered set $(\mathsf{Open}(X), \subset)$ consisting of open subsets of X in which $U \subset V$ if and only if U is a subset of V. By Example 2.2.12 we will also consider $\mathsf{Open}(X)$ as a category with open subsets of X as objects and subset inclusions as morphisms. Note that $\mathsf{Open}(X)$ is a bounded lattice. For open subsets $U, V \subset X$, their least upper bound is the union $U \cup V$, and their greatest lower bound is the intersection $U \cap V$. The least element is the empty subset of X, and the greatest element is X. ◇

Example 2.2.14 (Equivariant topological spaces). Suppose G is a group, and X is a topological space in which G acts on the left by homeomorphisms. Suppose $\mathsf{Open}(X)_G$ is the category obtained from $\mathsf{Open}(X)$ in Example 2.2.13 by adjoining the isomorphisms

$$g : U \xrightarrow{\;\cong\;} gU$$

for each open subset $U \subset X$ and each $g \in G$, subject to the following three relations:

(1) $e : U \longrightarrow eU = U$ is Id_U, where e is the multiplicative unit in G.
(2) The composition of $g : U \longrightarrow gU$ and $h : gU \longrightarrow hgU$ is $hg : U \longrightarrow hgU$.
(3) The diagram

$$
\begin{array}{ccc}
U & \xrightarrow{\text{inclusion}} & V \\
{\scriptstyle g}\downarrow & & \downarrow{\scriptstyle g} \\
gU & \xrightarrow{\text{inclusion}} & gV
\end{array}
$$

is commutative for all open subsets $U \subset V$ in X and $g \in G$.

Each morphism in $\mathsf{Open}(X)_G$ decomposes as

$$U \xrightarrow[\cong]{\;g\;} gU \xrightarrow{\text{inclusion}} gV$$

for some $g \in G$. If G is the trivial group, then $\mathsf{Open}(X)_G$ is the category $\mathsf{Open}(X)$ in Example 2.2.13. ◇

Example 2.2.15 (Oriented manifolds). For each integer $d \geq 1$, there is a category Man^d with d-dimensional oriented manifolds as objects and orientation-preserving open embeddings as morphisms. The reader may consult [O'Neill (1983)] for discussion of manifolds. We always assume that a manifold is Hausdorff and second-countable. By Whitney Embedding Theorem, Man^d is essentially small. In what follows, we will tacitly replace Man^d by an equivalent small category. ◇

Example 2.2.16 (Discs). There is a full subcategory

$$i : \mathsf{Disc}^d \longrightarrow \mathsf{Man}^d$$

whose objects are oriented manifolds diffeomorphic to $\mathbb{R}^d$, where $\mathbb{R}$ is the field of real numbers. ◇

Example 2.2.17 (Oriented Riemannian manifolds). For each integer $d \geq 1$, there is a category Riem^d with d-dimensional oriented Riemannian manifolds as objects and orientation-preserving isometric open embeddings as morphisms. As in the Example 2.2.15, Riem^d is essentially small, and we will tacitly replace it by an equivalent small category. ◇

Example 2.2.18 (Lorentzian manifolds). The reader is referred to [Bär *et al.* (2007); Beem (1996); O'Neill (1983)] for detailed discussion of Lorentzian geometry. A *Lorentzian manifold* is a manifold X equipped with a pseudo-Riemannian metric g of signature $(+, -, \ldots, -)$. A tangent vector v in a Lorentzian manifold (X, g) is *timelike* (resp., *causal*) if $g(v, v) > 0$ (resp., $g(v, v) \geq 0$). A smooth curve $f : [0, 1] \longrightarrow X$ is a *timelike/causal curve* if its tangent vectors are all timelike/causal. A *time-orientation* t on an oriented Lorentzian manifold (X, g, o) is a smooth vector field t on X such that the vector t_x is timelike at each point $x \in X$.

Suppose (X, g, o, t) is an oriented and time-oriented Lorentzian manifold. A causal curve f is *future-directed* if $g(t_x, \dot{f}_x) > 0$ and *past-directed* if $g(t_x, \dot{f}_x) < 0$ at each $x \in X$, where $\dot{f}_x$ is the tangent vector of f at x. The *causal future/past* of a point $x \in X$ is the set $J_X^+(x)$ (resp., $J_X^-(x)$) consisting of x and points in X that can be reached from x by a future/past-directed causal curve. A subset $A \subset X$ is *causally compatible* if for each $a \in A$, $J_X^\pm(a) \cap A = J_A^\pm(a)$. Two subsets A and B in X are *causally disjoint* if for each point $a \in \overline{A}$, $J_X^\pm(a) \cap \overline{B} = \varnothing$. A *Cauchy surface* in (X, g, o, t) is a smooth hypersurface that intersects every inextensible timelike curve exactly once. We call (X, g, o, t) *globally hyperbolic* if it contains a Cauchy surface.

For each integer $d \geq 1$, there is a category Loc^d with d-dimensional oriented, time-oriented, and globally hyperbolic Lorentzian manifolds as objects. A morphism in Loc^d is an isometric embedding that preserves the orientations and time-orientations whose image is causally compatible and open. As in Example 2.2.15, Loc^d is essentially small, and we will tacitly replace it by an equivalent small category. ◇

Example 2.2.19 (Globally hyperbolic open subsets). Similar to Example 2.2.13, for a fixed Lorentzian manifold $X \in \mathsf{Loc}^d$, there is a category $\mathsf{Gh}(X)$ with globally hyperbolic open subsets of X as objects and subset inclusions as morphisms. There is a functor

$$i : \mathsf{Gh}(X) \longrightarrow \mathsf{Loc}^d$$

given by restricting the structures of X to globally hyperbolic open subsets. ◇

Example 2.2.20 (Lorentzian manifolds with bundles). Suppose G is a Lie group. There is a category Loc_G^d in which an object is a pair (X, P) with $X \in \mathsf{Loc}^d$ and P a principal G-bundle over X. A morphism $f : (X, P) \longrightarrow (Y, Q)$ in Loc_G^d is a principal G-bundle morphism $f : P \longrightarrow Q$ covering a morphism $f' : X \longrightarrow Y$. There is a forgetful functor

$$\pi : \mathsf{Loc}_G^d \longrightarrow \mathsf{Loc}^d$$

that forgets about the bundle. ◇

Example 2.2.21 (Lorentzian manifolds with bundles and connections). Suppose G is a Lie group. There is a category $\mathsf{Loc}_{G,\mathsf{con}}^d$ in which an object is a triple (X, P, C) with $(X, P) \in \mathsf{Loc}_G^d$ and C a connection on P. A morphism in $\mathsf{Loc}_{G,\mathsf{con}}^d$ is a morphism in Loc_G^d that preserves the connections. There is a forgetful functor

$$p : \mathsf{Loc}_{G,\mathsf{con}}^d \longrightarrow \mathsf{Loc}_G^d$$

that forgets about the connection. Composing with the forgetful functor in Example 2.2.20, there is a forgetful functor

$$\pi p : \mathsf{Loc}^d_{G,\mathrm{con}} \longrightarrow \mathsf{Loc}^d$$

that forgets about both the bundle and the connection. ◇

Example 2.2.22 (Lorentzian manifolds with spin structures). Suppose $d \geq 4$. There is a category SLoc^d with d-dimensional oriented, time-oriented, and globally hyperbolic Lorentzian spin manifolds as objects. To be more precise, an object is a triple (X, P, ψ) with $X \in \mathsf{Loc}^d$, P a principal $\mathrm{Spin}_0(1, d-1)$-bundle over X, and $\psi : P \longrightarrow FX$ a $\mathrm{Spin}_0(1, d-1)$-equivariant bundle map over Id_X to the pseudo-orthonormal oriented and time-oriented frame bundle FX over X. A morphism $f : (X, P, \psi) \longrightarrow (Y, Q, \phi)$ in SLoc^d is a principal $\mathrm{Spin}_0(1, d-1)$-bundle morphism $f : P \longrightarrow Q$ covering a morphism $f' : X \longrightarrow Y$ such that $\phi f = f'_* \psi$. Here $f'_* : FX \longrightarrow FY$ is the pseudo-orthonormal oriented and time-oriented frame bundle morphism induced by f'.

There is a forgetful functor

$$\pi : \mathsf{SLoc}^d \longrightarrow \mathsf{Loc}^d$$

that forgets the spin structure such that the fiber $\pi^{-1}(X)$ is a groupoid for each $X \in \mathsf{Loc}^d$. Here $\pi^{-1}(X)$ is the subcategory of SLoc^d whose objects are sent to X and whose morphisms are sent to Id_X by π. ◇

Example 2.2.23 (Regions in spacetime with timelike boundary). Following [Benini et. al. (2018)] we define a *spacetime with timelike boundary* as an oriented and time-oriented Lorentzian manifold X with boundary [Lee (2013)] such that the pullback of the Lorentzian metric along the boundary inclusion $\partial X \longrightarrow X$ defines a Lorentzian metric on the boundary ∂X. There is a *category of regions* in X, denoted $\mathsf{Reg}(X)$, in which an object is a causally convex open subset in X with inclusions as morphisms. It contains a full subcategory $\mathsf{Reg}(X_0)$ whose objects are causally convex open subsets contained in the interior X_0 of X. ◇

2.3   Limits and Colimits

Definition 2.3.1. Suppose $F : \mathsf{D} \longrightarrow \mathsf{C}$ is a functor

(1) For an object $c \in \mathsf{C}$, the *constant functor* $\Delta_c : \mathsf{D} \longrightarrow \mathsf{C}$ is the functor that sends every object in D to c and every morphism in D to Id_c.
(2) A *limit of F* is a pair $(\lim F, \theta)$ consisting of
   - an object $\lim F \in \mathsf{C}$ and
   - a natural transformation $\theta : \Delta_{\lim F} \longrightarrow F$

   that satisfies the following universal property: If (y, ϕ) is another such pair, then there exists a unique morphism

$$f : y \longrightarrow \lim F \in \mathsf{C} \quad \text{such that} \quad \phi = \theta \circ \Delta_f,$$

where $\Delta_f : \Delta_y \longrightarrow \Delta_{\lim F}$ is the obvious natural transformation induced by f. Omitting Δ we may represent a limit of F as follows.

If a limit of F exists, then it is unique up to a unique isomorphism in C.

(3) A *colimit of* F is a pair $(\text{colim} F, \theta)$ consisting of

- an object $\text{colim} F \in \mathsf{C}$ and
- a natural transformation $\theta : F \longrightarrow \Delta_{\text{colim} F}$

that satisfies the following universal property: If (z, ψ) is another such pair, then there exists a unique morphism

$$f : \text{colim} F \longrightarrow z \in \mathsf{C} \quad \text{such that} \quad \psi = \Delta_f \circ \theta.$$

We may represent a colimit of F as follows.

$$F \xrightarrow{\ \theta\ } \text{colim} F$$
$$\forall \psi \searrow \quad \vdots \exists! f$$
$$z$$

If a colimit of F exists, then it is unique up to a unique isomorphism in C. We will often write a colimit of F as either $\text{colim} F$ or $\text{colim}_\mathsf{D} F$ if we wish to emphasize its domain category.

(4) C is *(co)complete* if every functor from a small category to C has a (co)limit.

Example 2.3.2. The categories Set, Top, SSet, Cat, $\text{Vect}_\mathbb{K}$, and $\text{Chain}_\mathbb{K}$ are complete and cocomplete. ◇

Example 2.3.3 (Initial and Terminal Objects). Taking D to be the empty category $\varnothing$ and $F : \varnothing \longrightarrow \mathsf{C}$ the trivial functor, a limit of F (i.e., of the empty diagram) is called a *terminal object* in C. More explicitly, a terminal object $*$ in C is an object such that, for each object $c \in \mathsf{C}$, there exists a unique morphism $c \longrightarrow *$. Dually, a colimit of the empty diagram is called an *initial object* in C. An initial object $i \in \mathsf{C}$ is characterized by the universal property that for each object $c \in \mathsf{C}$, there exists a unique morphism $i \longrightarrow c$. In what follows, we will often use the symbol $\varnothing$ to denote an initial object in C, and the reader should not confuse it with the empty category.

For instance:

(1) In the category Set, the empty set is an initial object, and a terminal object is exactly a one-element set.
(2) In $\text{Vect}_\mathbb{K}$ and $\text{Chain}_\mathbb{K}$, the 0 vector space (or chain complex) is both an initial object and a terminal object.
(3) For a topological space X, the category $\text{Open}(X)$ has the empty subset of X as an initial object and X as a terminal object. ◇

Example 2.3.4 (Coproducts and Products). Taking D to be a small discrete category, a functor $F : \mathsf{D} \longrightarrow \mathsf{C}$ is determined by the set of objects $\{Fd : d \in \mathsf{D}\}$. A (co)limit of F is called a *(co)product* of the set of objects $\{Fd : d \in \mathsf{D}\}$, denoted by $\prod_{d \in \mathsf{D}} Fd$ (resp., $\coprod_{d \in \mathsf{D}} Fd$). For instance, in $\mathsf{Vect}_{\mathbb{K}}$ and $\mathsf{Chain}_{\mathbb{K}}$, coproducts and finite products are both given by direct sums. ◇

Example 2.3.5 (Pushouts and Pullbacks). Suppose D is the category

$$1 \longleftarrow 0 \longrightarrow 2$$

with three objects and two non-identity morphisms as indicated. A colimit of $F : \mathsf{D} \longrightarrow \mathsf{C}$ is called a *pushout* of the diagram $F1 \longleftarrow F0 \longrightarrow F2$. A limit of $F : \mathsf{D}^{\mathrm{op}} \longrightarrow \mathsf{C}$ is called a *pullback* of the diagram $F1 \longrightarrow F0 \longleftarrow F2$. ◇

Example 2.3.6 (Coequalizers and Equalizers). A *coequalizer* of a pair of parallel morphisms $f, g : a \longrightarrow b$ in C is a pair (c, u) consisting of an object $c \in \mathsf{C}$ and a morphism $u : b \longrightarrow c$ such that $uf = ug$ and that is initial among such pairs. In other words, for every other such pair (d, v), there exists a unique morphism $v' : c \longrightarrow d$ such that $v = v'u$.

$$
\begin{array}{ccc}
a \underset{g}{\overset{f}{\rightrightarrows}} b \xrightarrow{\;u\;} c & & uf = ug \\
\qquad\quad \downarrow^{\forall v} \;\;\nearrow_{\exists! v'} & & \\
\qquad\quad d & & vf = vg
\end{array}
$$

A *reflexive pair* is a pair of morphisms $f, g : a \longrightarrow b$ with a common section $s : b \longrightarrow a$ in the sense that $fs = gs = \mathrm{Id}_b$. A *reflexive coequalizer* is a coequalizer of a reflexive pair. Note that a coequalizer of f and g is the same as a pushout of the diagram

$$
\begin{array}{ccc}
a \amalg b & \xrightarrow{(g,\mathrm{Id}_b)} & b \\
{\scriptstyle (f,\mathrm{Id}_b)} \downarrow & & \\
b & &
\end{array}
$$

and hence is a particular kind of colimit. The dual concept is called an *equalizer* of f and g. ◇

Definition 2.3.7 (Coends). Suppose $F : \mathsf{C}^{\mathrm{op}} \times \mathsf{C} \longrightarrow \mathsf{M}$ is a functor.

(1) A *wedge* of F is a pair (X, ζ) consisting of
  - an object $X \in \mathsf{M}$ and
  - morphisms $\zeta_c : F(c, c) \longrightarrow X$ for $c \in \mathsf{C}$

  such that the diagram

$$
\begin{array}{ccc}
F(d, c) & \xrightarrow{F(d,g)} & F(d, d) \\
{\scriptstyle F(g,c)} \downarrow & & \downarrow {\scriptstyle \zeta_d} \\
F(c, c) & \xrightarrow{\;\zeta_c\;} & X
\end{array}
$$

is commutative for each morphism $g : c \longrightarrow d \in \mathsf{C}$.

(2) A *coend* of F is an initial wedge $\left(\int^{c \in \mathsf{C}} F(c,c), \omega \right)$.

In other words, a coend of F is a wedge of F such that given any wedge (X, ζ) of F, there exists a unique morphism

$$h : \int^{c \in \mathsf{C}} F(c,c) \longrightarrow X \in \mathsf{M}$$

such that the diagram

$$F(c,c) \xrightarrow{\ \omega_c\ } \int^{c \in \mathsf{C}} F(c,c)$$

with ζ_c and h mapping to X

is commutative for each object $c \in \mathsf{C}$. The dual concept of a coend is called an end, which is originally due to Yoneda [Yoneda (1960)]. We will not need to use ends in this book.

The proof of the following result is a simple exercise in checking the definitions of a coend and of a coequalizer.

Proposition 2.3.8. *Suppose given a functor* $F : \mathsf{C}^{\mathrm{op}} \times \mathsf{C} \longrightarrow \mathsf{M}$ *with* C *a small category and* M *a cocomplete category. Then a coend of* F *exists and is given by a coequalizer*

$$\int^{c \in \mathsf{C}} F(c,c) = \mathrm{coequal}\left(\coprod_{g \in \mathrm{Mor}(\mathsf{C})} F(d,c) \underset{i_c \circ F(g,c)}{\overset{i_d \circ F(d,g)}{\rightrightarrows}} \coprod_{c \in \mathsf{C}} F(c,c) \right)$$

in which $g : c \longrightarrow d$ *runs through all the morphisms in* C *and*

$$i_c : F(c,c) \longrightarrow \coprod_{c \in \mathsf{C}} F(c,c)$$

is the natural inclusion. The natural morphism ω_c *is the composition*

$$F(c,c) \xrightarrow{\ \omega_c\ } \int^{c \in \mathsf{C}} F(c,c)$$

with i_c to $\coprod_{c \in \mathsf{C}} F(c,c)$ and natural

for each object $c \in \mathsf{C}$.

In particular, in the above setting, a coend is a particular kind of colimit.

2.4   Adjoint Functors

Adjoint functors will provide us with ways to compare (i) algebraic quantum field theories of various flavors, (ii) prefactorization algebras of various flavors, and (iii) algebraic quantum field theories with prefactorization algebras. The concept of an adjunction is due to Kan [Kan (1958)].

Definition 2.4.1. Suppose $F : \mathsf{C} \longrightarrow \mathsf{D}$ and $G : \mathsf{D} \longrightarrow \mathsf{C}$ are functors. We call the pair (F, G) an *adjoint pair*, or an *adjunction*, if for each object $c \in \mathsf{C}$ and each object $d \in \mathsf{D}$, there exist a bijection

$$\theta_{c,d} : \mathsf{D}(Fc, d) \cong \mathsf{C}(c, Gd)$$

that is natural in both c and d. In this case:

(1) We call F a *left adjoint of* G and G a *right adjoint of* F and write $F \dashv G$.
(2) For each $c \in \mathsf{C}$, the morphism $\eta_c : c \longrightarrow GFc$ corresponding under $\theta_{c,Fc}$ to Id_{Fc} is called the *unit of c*.
(3) For each $d \in \mathsf{D}$, the morphism $\epsilon_d : FGd \longrightarrow d$ corresponding under $\theta_{Gd,d}$ to Id_{Gd} is called the *counit of d*.

Convention 2.4.2. We will always write the left adjoint on top (if displayed horizontally) or on the left (if displayed vertically).

Definition 2.4.3. Suppose $i : \mathsf{D} \longrightarrow \mathsf{C}$ is the inclusion functor of a full subcategory. Then D is called a *reflective subcategory* of C if i admits a left adjoint.

The unit and the counit actually characterize an adjoint pair; the proof of the following result can be found in [Borceux (1994)] Section 3.1.

Theorem 2.4.4. *Suppose $F : \mathsf{C} \longrightarrow \mathsf{D}$ and $G : \mathsf{D} \longrightarrow \mathsf{C}$ are functors. The following statements are equivalent.*

(1) (F, G) is an adjoint pair.
(2) There exist natural transformations $\eta : \mathrm{Id}_{\mathsf{C}} \longrightarrow GF$, called the unit, and $\epsilon : FG \longrightarrow \mathrm{Id}_{\mathsf{D}}$, called the counit, such that the diagrams

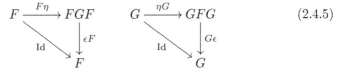

(2.4.5)

are commutative.
(3) There exists a natural transformation $\eta : \mathrm{Id}_{\mathsf{C}} \longrightarrow GF$ such that, given any morphism $f \in \mathsf{C}(c, Gd)$ with $c \in \mathsf{C}$ and $d \in \mathsf{D}$, there exists a unique morphism $\overline{f} \in \mathsf{D}(Fc, d)$ such that the diagram

is commutative.

The two commutative diagrams in (2.4.5) are called the *triangle identities*. Adjoint functors are unique up to isomorphisms.

Example 2.4.6. The full subcategory inclusion from the category of abelian groups Ab to the category of groups Grp admits a left adjoint, namely, the abelianization functor that sends a group G to the quotient $G/[G,G]$. So Ab is a full reflective subcategory of Grp. ◇

Example 2.4.7. In the context of Examples 2.2.5 and 2.2.6, there is a free-forgetful adjunction

$$\mathsf{Set} \underset{U}{\overset{F}{\rightleftarrows}} \mathsf{Vect}_{\mathbb{K}}$$

in which the right adjoint U sends a vector space to its underlying set. The left adjoint F sends a set X to the vector space $\oplus_X \mathbb{K}$ freely generated by X. ◇

Recall the concept of an equivalence in Definition 2.1.8.

Definition 2.4.8. An adjoint pair $F : \mathsf{C} \rightleftarrows \mathsf{D} : G$ is an *adjoint equivalence* if the categories C and D are equivalent via the functors F and G.

The following characterizations of an adjoint equivalence is [Mac Lane (1998)] IV.4 Theorem 1.

Theorem 2.4.9. *The following properties of a functor $G : \mathsf{D} \longrightarrow \mathsf{C}$ are equivalent.*

(1) G is an equivalence of categories.
(2) G admits a left adjoint $F : \mathsf{C} \longrightarrow \mathsf{D}$ such that $F \dashv G$ is an adjoint equivalence.
(3) G is full, faithful, and essentially surjective.

An important example of an adjunction is Kan extension.

Definition 2.4.10. Suppose $F : \mathsf{C} \longrightarrow \mathsf{D}$ is a functor, and M is a category. If the induced functor

$$F^* = \mathsf{Fun}(F, \mathsf{M}) : \mathsf{Fun}(\mathsf{D}, \mathsf{M}) \longrightarrow \mathsf{Fun}(\mathsf{C}, \mathsf{M}), \quad F^*(G) = GF$$

on functor categories admits a left adjoint $F_!$, then for a functor $H \in \mathsf{Fun}(\mathsf{C}, \mathsf{M})$, the image $F_! H \in \mathsf{Fun}(\mathsf{D}, \mathsf{M})$ is called a *left Kan extension* of H along F and is written as $\mathsf{Lan}_F H$ or $\mathsf{Lan} H$.

The following existence result is the dual of [Mac Lane (1998)] (p.239 Corollary 2). If D is also small, then the following result can be obtained as a special case of the change-of-operad Theorem 5.1.8; see Example 5.1.12.

Theorem 2.4.11. *Suppose $F : \mathsf{C} \longrightarrow \mathsf{D}$ is a functor with C a small category, and M is a cocomplete category. Then the induced functor $\mathsf{Fun}(F, \mathsf{M})$ admits a left adjoint. In particular, every functor $H : \mathsf{C} \longrightarrow \mathsf{M}$ admits a left Kan extension along F.*

Example 2.4.12 (Left Kan Extensions as Coends). In the setting of Theorem 2.4.11, a left Kan extension of $H : C \longrightarrow M$ along $F : C \longrightarrow D$ is given objectwise by the coend (Definition 2.3.7)

$$\left(\mathsf{Lan}_F H\right)(d) = \int^{c \in C} \mathsf{D}(Fc, d) \cdot Hc \qquad (2.4.13)$$

for each object $d \in \mathsf{D}$. In this coend formula, the integrand is the *copower* defined by

$$S \cdot X = \coprod_{s \in S} X \qquad (2.4.14)$$

for a set S and an object $X \in \mathsf{M}$. For a proof that the coend formula (2.4.13) actually yields a left Kan extension, see [Mac Lane (1998)] (p.240 Theorem 1) or [Loregian (2015)] (p.23).   ◇

An important property of a general left adjoint is that it preserves colimits. Similarly, a right adjoint preserves limits. For a proof of the following two results, see [Borceux (1994)] Section 3.2.

Theorem 2.4.15 (Left Adjoints Preserve Colimits). *Suppose $F : C \longrightarrow D$ admits a right adjoint, and $H : E \longrightarrow C$ has a colimit $(\mathrm{colim} H, \theta : H \longrightarrow \Delta_{\mathrm{colim} H})$. Then the pair*

$$\left(F \mathrm{colim} H, F\theta : FH \longrightarrow F\Delta_{\mathrm{colim} H} = \Delta_{F \mathrm{colim} H}\right)$$

is a colimit of $FH : E \longrightarrow D$.

Theorem 2.4.16 (Right Adjoints Preserve Limits). *Suppose $G : D \longrightarrow C$ admits a left adjoint, and $H : E \longrightarrow D$ has a limit $(\lim H, \theta : \Delta_{\lim H} \longrightarrow H)$. Then the pair*

$$\left(G \lim H, G\theta : G\Delta_{\lim H} = \Delta_{G \lim H} \longrightarrow GH\right)$$

is a limit of $GH : E \longrightarrow C$.

2.5   Symmetric Monoidal Categories

A symmetric monoidal category is the most natural setting to discuss operads and their algebras.

Definition 2.5.1. A *monoidal category* is a tuple

$$(\mathsf{M}, \otimes, \mathbb{1}, \alpha, \lambda, \rho)$$

consisting of the following data.

- M is a category.
- $\otimes : \mathsf{M} \times \mathsf{M} \longrightarrow \mathsf{M}$ is a functor, called the *monoidal product*.
- $\mathbb{1}$ is an object in M, called the *monoidal unit*.

- α is a natural isomorphism

$$(X \otimes Y) \otimes Z \xrightarrow[\cong]{\alpha} X \otimes (Y \otimes Z) \tag{2.5.2}$$

  for all objects $X, Y, Z \in \mathsf{M}$, called the *associativity isomorphism*.
- λ and ρ are natural isomorphisms

$$\mathbb{1} \otimes X \xrightarrow[\cong]{\lambda} X \quad \text{and} \quad X \otimes \mathbb{1} \xrightarrow[\cong]{\rho} X \tag{2.5.3}$$

  for all objects $X \in \mathsf{M}$, called the *left unit* and the *right unit*, respectively.

This data is required to satisfy the following two axioms.

Unit Axioms The diagram

$$(X \otimes \mathbb{1}) \otimes Y \xrightarrow[\cong]{\alpha} X \otimes (\mathbb{1} \otimes Y) \tag{2.5.4}$$

$$\rho \otimes \mathrm{Id} \downarrow \qquad\qquad \downarrow \mathrm{Id} \otimes \lambda$$

$$X \otimes Y \xrightarrow{=} X \otimes Y$$

is commutative for all objects $X, Y \in \mathsf{M}$, and

$$\lambda = \rho : \mathbb{1} \otimes \mathbb{1} \xrightarrow{\cong} \mathbb{1}. \tag{2.5.5}$$

Pentagon Axiom The pentagon

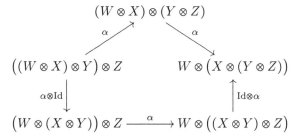

is commutative for all objects $W, X, Y, Z \in \mathsf{M}$.

A *strict monoidal category* is a monoidal category in which the natural isomorphisms α, λ, and ρ are all identity morphisms.

So in a strict monoidal category, an iterated monoidal product $a_1 \otimes \cdots \otimes a_n$ without any parentheses has an unambiguous meaning.

Convention 2.5.6. In a monoidal category, an *empty tensor product*, written as $X^{\otimes 0}$ or $X^{\otimes \varnothing}$, means the monoidal unit $\mathbb{1}$.

Definition 2.5.7. A *symmetric monoidal category* is a pair (M, ξ) in which:

- $\mathsf{M} = (\mathsf{M}, \otimes, \mathbb{1}, \alpha, \lambda, \rho)$ is a monoidal category.
- ξ is a natural isomorphism

$$X \otimes Y \xrightarrow[\cong]{\xi_{X,Y}} Y \otimes X \tag{2.5.8}$$

  for objects $X, Y \in \mathsf{M}$, called the *symmetry isomorphism*.

This data is required to satisfy the following three axioms.

Symmetry Axiom The diagram

$$X \otimes Y \xrightarrow{\xi_{X,Y}} Y \otimes X \qquad (2.5.9)$$

$$\downarrow{\xi_{Y,X}}$$

$$X \otimes Y$$

is commutative for all objects $X, Y \in \mathsf{M}$.

Compatibility with Units The diagram

$$X \otimes 1 \xrightarrow{\xi_{X,1}} 1 \otimes X$$

$$\rho \downarrow \qquad \qquad \downarrow \lambda$$

$$X \xrightarrow{\quad = \quad} X$$

is commutative for all objects $X \in \mathsf{M}$.

Hexagon Axiom The diagram

$$X \otimes (Z \otimes Y) \xrightarrow{\mathrm{Id} \otimes \xi_{Z,Y}} X \otimes (Y \otimes Z) \qquad (2.5.10)$$

$$\nearrow \alpha \qquad \qquad \alpha^{-1} \searrow$$

$$(X \otimes Z) \otimes Y \qquad \qquad (X \otimes Y) \otimes Z$$

$$\xi_{X \otimes Z, Y} \searrow \qquad \qquad \nearrow \xi_{Y,X} \otimes \mathrm{Id}$$

$$Y \otimes (X \otimes Z) \xrightarrow{\alpha^{-1}} (Y \otimes X) \otimes Z$$

is commutative for all objects $X, Y, Z \in \mathsf{M}$.

Definition 2.5.11. A *symmetric monoidal closed category* is a symmetric monoidal category M in which for each object Y, the functor

$$- \otimes Y : \mathsf{M} \longrightarrow \mathsf{M},$$

admits a right adjoint

$$\mathrm{Hom}_{\mathsf{M}}(Y, -) : \mathsf{M} \longrightarrow \mathsf{M},$$

called the *internal hom*. In other words, for any objects $X, Y, Z \in \mathsf{M}$, there is a specified bijection, called the $\otimes$-$\mathrm{Hom}_{\mathsf{M}}$ *adjunction*,

$$\mathsf{M}\big(X \otimes Y, Z\big) \xrightarrow[\cong]{\phi} \mathsf{M}\big(X, \mathrm{Hom}_{\mathsf{M}}(Y, Z)\big) \qquad (2.5.12)$$

that is natural in X, Y, and Z.

Example 2.5.13. The categories Set, Top, and SSet are symmetric monoidal closed categories via the Cartesian product. The category $\mathsf{Vect}_{\mathbb{K}}$ is a symmetric monoidal closed category with the usual tensor product of vector spaces. The category $\mathsf{Chain}_{\mathbb{K}}$ is a symmetric monoidal closed category via the monoidal product $X \otimes Y$ with

$$(X \otimes Y)_n = \bigoplus_{k \in \mathbb{Z}} X_k \otimes_{\mathbb{K}} Y_{n-k}$$

and differential

$$d(x \otimes y) = (dx) \otimes y + (-1)^{|x|} x \otimes (dy).$$

For the internal homs, the reader is referred to [Hovey (1999)] Chapters 2 and 3. ◇

Definition 2.5.14. Suppose M and N are monoidal categories. A *monoidal functor*

$$(F, F_2, F_0) : \mathsf{M} \longrightarrow \mathsf{N}$$

consists of the following data:

- a functor $F : \mathsf{M} \longrightarrow \mathsf{N}$;
- a natural transformation

$$F(X) \otimes F(Y) \xrightarrow{\ F_2\ } F(X \otimes Y) \in \mathsf{N}, \tag{2.5.15}$$

  where X and Y are objects in M;
- a morphism

$$\mathbb{1}_{\mathsf{N}} \xrightarrow{\ F_0\ } F(\mathbb{1}_{\mathsf{M}}) \in \mathsf{N}, \tag{2.5.16}$$

  where $\mathbb{1}_{\mathsf{N}}$ and $\mathbb{1}_{\mathsf{M}}$ are the monoidal units in N and M, respectively.

This data is required to satisfy the following three axioms.

Compatibility with the Associativity Isomorphisms The diagram

$$\big(F(X) \otimes F(Y)\big) \otimes F(Z) \xrightarrow[\cong]{\ \alpha_{\mathsf{N}}\ } F(X) \otimes \big(F(Y) \otimes F(Z)\big) \tag{2.5.17}$$

$$
\begin{array}{ccc}
& F_2 \otimes \mathrm{Id} \downarrow & \qquad\qquad \downarrow \mathrm{Id} \otimes F_2 \\
F(X \otimes Y) \otimes F(Z) & & F(X) \otimes F(Y \otimes Z) \\
F_2 \downarrow & & \uparrow F_2 \\
F\big((X \otimes Y) \otimes Z\big) & \xrightarrow[\cong]{\ F(\alpha_{\mathsf{M}})\ } & F\big(X \otimes (Y \otimes Z)\big)
\end{array}
$$

is commutative for all objects $X, Y, Z \in \mathsf{M}$.

Compatibility with the Left Units The diagram

$$\mathbb{1}_{\mathsf{N}} \otimes F(X) \xrightarrow[\cong]{\ \lambda_{\mathsf{N}}\ } F(X) \tag{2.5.18}$$

$$
\begin{array}{ccc}
F_0 \otimes \mathrm{Id} \downarrow & & \cong \uparrow F(\lambda_{\mathsf{M}}) \\
F(\mathbb{1}_{\mathsf{M}}) \otimes F(X) & \xrightarrow{\ F_2\ } & F(\mathbb{1}_{\mathsf{M}} \otimes X)
\end{array}
$$

is commutative for all objects $X \in \mathsf{M}$.

Compatibility with the Right Units The diagram

$$
\begin{array}{ccc}
F(X) \otimes \mathbb{1}_N & \xrightarrow[\cong]{\ \rho_N\ } & F(X) \\
{\scriptstyle \mathrm{Id} \otimes F_0} \downarrow & & \uparrow {\scriptstyle \cong}\ \Big| {\scriptstyle F(\rho_M)} \\
F(X) \otimes F(\mathbb{1}_M) & \xrightarrow[\ ]{\ F_2\ } & F(X \otimes \mathbb{1}_M)
\end{array}
\tag{2.5.19}
$$

is commutative for all objects $X \in \mathsf{M}$.

A *strong monoidal functor* is a monoidal functor in which the morphisms F_0 and F_2 are all isomorphisms.

Definition 2.5.20. A *monoidal natural transformation*

$$
\theta : (F, F_2, F_0) \longrightarrow (G, G_2, G_0)
$$

between monoidal functors $F, G : \mathsf{M} \longrightarrow \mathsf{N}$ is a natural transformation of the underlying functors $\theta : F \longrightarrow G$ that is compatible with the structure morphisms in the sense that the diagrams

$$
\begin{array}{ccc}
F(X) \otimes F(Y) & \xrightarrow{(\theta_X, \theta_Y)} & G(X) \otimes G(Y) \\
{\scriptstyle F_2} \downarrow & & \downarrow {\scriptstyle G_2} \\
F(X \otimes Y) & \xrightarrow[\ \theta_{X \otimes Y}\ ]{} & G(X \otimes Y)
\end{array}
\qquad
\begin{array}{ccc}
\mathbb{1}_N & \xrightarrow{\ F_0\ } & F(\mathbb{1}_M) \\
& {\scriptstyle G_0} \searrow & \downarrow {\scriptstyle \theta_{\mathbb{1}_M}} \\
& & G(\mathbb{1}_M)
\end{array}
$$

are commutative for all objects $X, Y \in \mathsf{M}$.

The proof of the following result can be found in [Mac Lane (1998)] (XI.3).

Theorem 2.5.21 (Mac Lane's Coherence Theorem). *Suppose* M *is a monoidal category. Then there exist a strict monoidal category* $\overline{\mathsf{M}}$ *and an adjoint equivalence*

$$
\overline{\mathsf{M}} \underset{G}{\overset{F}{\rightleftarrows}} \mathsf{M}
$$

such that both F *and* G *are strong monoidal functors.*

Convention 2.5.22. Following common practice, using Mac Lane's Coherence Theorem 2.5.21, we will omit parentheses for monoidal products of multiple objects in a monoidal category, replacing it by an adjoint equivalent strict monoidal category, via strong monoidal functors, if necessary. In the rest of this book, Mac Lane's Coherence Theorem will be used without further comment.

Definition 2.5.23. Suppose M and N are symmetric monoidal categories. A *symmetric monoidal functor* $(F, F_2, F_0) : \mathsf{M} \longrightarrow \mathsf{N}$ is a monoidal functor that is compatible with the symmetry isomorphisms, in the sense that the diagram

$$
\begin{array}{ccc}
F(X) \otimes F(Y) & \xrightarrow[\cong]{\ \xi_{FX, FY}\ } & F(Y) \otimes F(X) \\
{\scriptstyle F_2} \downarrow & & \downarrow {\scriptstyle F_2} \\
F(X \otimes Y) & \xrightarrow[\cong]{\ F\xi_{X,Y}\ } & F(Y \otimes X)
\end{array}
\tag{2.5.24}
$$

is commutative for all objects $X, Y \in \mathsf{M}$.

Example 2.5.25. Suppose $(M, \otimes, \mathbb{1})$ is a symmetric monoidal category with all set-indexed coproducts. Then the functor

$$\mathsf{Set} \xrightarrow{\;F\;} \mathsf{M}, \quad FX = \coprod_{x \in X} \mathbb{1}$$

is a strong symmetric monoidal functor. ◇

Example 2.5.26. The singular chain functor $C : \mathsf{Top} \longrightarrow \mathsf{Chain}_{\mathbb{Z}}$ is a symmetric monoidal functor [Massey (1991)] (XI.3). ◇

Example 2.5.27. Given two monoidal categories M and N, there is a category

$$\mathsf{MFun}(\mathsf{M}, \mathsf{N})$$

whose objects are monoidal functors $\mathsf{M} \longrightarrow \mathsf{N}$ and whose morphisms are monoidal natural transformations between such monoidal functors. If M and N are furthermore symmetric monoidal categories, then there is a category

$$\mathsf{SMFun}(\mathsf{M}, \mathsf{N})$$

whose objects are symmetric monoidal functors $\mathsf{M} \longrightarrow \mathsf{N}$ and whose morphisms are monoidal natural transformations between such symmetric monoidal functors. ◇

Example 2.5.28. Suppose $*$ is a category with one object and only the identity morphism. It has an obvious symmetric strict monoidal structure. ◇

Example 2.5.29. For each category C, $\big(\mathsf{Fun}(\mathsf{C}, \mathsf{C}), \circ, \mathsf{Id}_\mathsf{C}\big)$ is a strict monoidal category, where $\circ$ is composition of functors. ◇

An important property of a symmetric monoidal closed category is that its monoidal product preserves colimits in each variable. Indeed, a left adjoint preserves colimits (Theorem 2.4.15). So by symmetry each side of a symmetric monoidal product preserves colimits. The following observation is a special case of the dual of [Mac Lane (1998)] (p.231 Corollary)

Theorem 2.5.30. *Suppose* M *is a symmetric monoidal closed category, and* $F : \mathsf{C} \longrightarrow \mathsf{M}$ *and* $G : \mathsf{D} \longrightarrow \mathsf{M}$ *are functors with* C *and* D *small categories that admit colimits. Then there is a canonical isomorphism*

$$\operatorname*{colim}_{\mathsf{C} \times \mathsf{D}} F \otimes G \xrightarrow{\;\cong\;} \Big(\operatorname*{colim}_{\mathsf{C}} F\Big) \otimes \Big(\operatorname*{colim}_{\mathsf{D}} G\Big)$$

in which $F \otimes G$ *is the composition of the functors*

$$\mathsf{C} \times \mathsf{D} \xrightarrow{\;(F,G)\;} \mathsf{M} \times \mathsf{M} \xrightarrow{\;\otimes\;} \mathsf{M}.$$

Example 2.5.31. Suppose M is a symmetric monoidal closed category with all set-indexed coproducts. Then there is a canonical isomorphism

$$\Big(\coprod_{a \in A} X_a\Big) \otimes \Big(\coprod_{b \in B} Y_b\Big) \cong \coprod_{(a,b) \in A \times B} X_a \otimes Y_b$$

for any sets A and B with $X_a, Y_b \in \mathsf{M}$. ◇

2.6 Monoids

Below $*$ denotes the category with one object and only the identity morphism.

Definition 2.6.1 (Monoids). Suppose M is a monoidal category.

(1) Define the category
$$\mathsf{Mon}(\mathsf{M}) = \mathsf{MFun}(*, \mathsf{M})$$
of monoidal functors $* \longrightarrow \mathsf{M}$, whose objects are called *monoids in* M.

(2) Suppose M is also symmetric. Define the category
$$\mathsf{Com}(\mathsf{M}) = \mathsf{SMFun}(*, \mathsf{M})$$
of symmetric monoidal functors $* \longrightarrow \mathsf{M}$, whose objects are called *commutative monoids in* M.

A simple exercise in unwrapping the definitions yields the following more explicit description of a (commutative) monoid.

Proposition 2.6.2. *Suppose* M *is a monoidal category.*

(1) A monoid in M *is exactly a triple* (A, μ, ε) *consisting of*

- *an object* $A \in \mathsf{M}$,
- *a multiplication morphism* $\mu : A \otimes A \longrightarrow A$, *and*
- *a unit* $\varepsilon : \mathbb{1} \longrightarrow A$

such that the associativity and unity diagrams

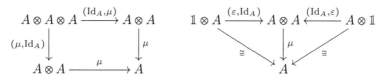

are commutative. A morphism of monoids is a morphism of the underlying objects that is compatible with the multiplications and the units.

(2) Suppose M *is a symmetric monoidal category. Then a commutative monoid in* M *is exactly a monoid whose multiplication is commutative in the sense that the diagram*

$$
\begin{array}{ccc}
A \otimes A & \xrightarrow{\text{permute}} & A \otimes A \\
{\scriptstyle\mu}\downarrow & & \downarrow{\scriptstyle\mu} \\
A & \xrightarrow{\text{Id}_A} & A
\end{array}
$$

is commutative. A morphism of commutative monoids is a morphism of the underlying objects that is compatible with the multiplications and the units.

Example 2.6.3. A monoid in Set is a monoid in the usual sense. A monoid in Top is a topological monoid.   ◇

Example 2.6.4. In $\mathsf{Vect}_{\mathbb{K}}$ a (commutative) monoid is exactly a (commutative) $\mathbb{K}$-algebra. In $\mathsf{Chain}_{\mathbb{K}}$ a (commutative) monoid is exactly a (commutative) differential graded $\mathbb{K}$-algebra. ◇

The following result is a slight extension of Definition 2.6.1 of (commutative) monoids as (symmetric) monoidal functors.

Proposition 2.6.5. *Suppose* C *is a small category with all finite coproducts, regarded as a symmetric monoidal category* $(\mathsf{C}, \sqcup, \varnothing_{\mathsf{C}})$ *under coproducts. Suppose* M *is a monoidal category.*

(1) Then there is a canonical isomorphism

$$\mathsf{Mon}(\mathsf{M})^{\mathsf{C}} \xrightarrow{\cong} \mathsf{MFun}(\mathsf{C}, \mathsf{M})$$

between the category of C*-diagrams of monoids in* M *and the category of monoidal functors from* C *to* M.

(2) Suppose M *is a symmetric monoidal category. Then there is a canonical isomorphism*

$$\mathsf{Com}(\mathsf{M})^{\mathsf{C}} \xrightarrow{\cong} \mathsf{SMFun}(\mathsf{C}, \mathsf{M})$$

between the category of C*-diagrams of commutative monoids in* M *and the category of symmetric monoidal functors from* C *to* M.

Proof. A (symmetric) monoidal functor $F : \mathsf{C} \longrightarrow \mathsf{M}$ is equipped with a morphism (2.5.16)

$$\mathbb{1} \xrightarrow{F_0} F(\varnothing_{\mathsf{C}}) \in \mathsf{M}$$

with $\varnothing_{\mathsf{C}}$ an initial object in C, which exists by our assumption on C. For each object $c \in \mathsf{C}$, the unique morphism $0_c : \varnothing_{\mathsf{C}} \longrightarrow c$ and F_0 yield the composition

$$
\begin{array}{ccc}
\mathbb{1} & \xrightarrow{\;1_c\;} & F(c) \in \mathsf{M}. \\
{\scriptstyle F_0}\downarrow & & \| \\
F(\varnothing_{\mathsf{C}}) & \xrightarrow{F(0_c)} & F(c)
\end{array}
$$

The monoidal functor F is also equipped with a morphism (2.5.15)

$$F(c) \otimes F(d) \xrightarrow{F_2} F(c \sqcup d) \in \mathsf{M}$$

that is natural in $c, d \in \mathsf{C}$. The morphism

$$(\mathrm{Id}_c, \mathrm{Id}_c) : c \sqcup c \longrightarrow c \in \mathsf{C}$$

and F_2 yield the composition

$$
\begin{array}{ccc}
F(c) \otimes F(c) & \xrightarrow{\;\mu_c\;} & F(c) \in \mathsf{M} \\
{\scriptstyle F_2}\downarrow & & \| \\
F(c \sqcup c) & \xrightarrow{F(\mathrm{Id}_c, \mathrm{Id}_c)} & F(c)
\end{array}
$$

for each object $c \in$ C.

Now one checks that the associativity diagram (2.5.17) corresponds to the associativity of the morphism μ_c, while the unity diagrams (2.5.18) and (2.5.19) correspond to the property that 1_c is a two-sided unit of μ_c as in Proposition 2.6.2. Furthermore, the symmetry diagram (2.5.24) corresponds to the commutativity of μ_c. Therefore, $(F(c), \mu_c, 1_c)$ is a (commutative) monoid for each $c \in$ C. That we have a C-diagram of (commutative) monoids in M corresponds to the functoriality of F and F_2.

Conversely, given $F \in \mathsf{Mon(M)}^{\mathsf{C}}$, we will write $(F(c), \mu_c, 1_c) \in \mathsf{Mon(M)}^{\mathsf{C}}$ for its value at $c \in$ C. The monoidal structure on F is defined as follows. The monoid unit for $F(\varnothing_{\mathsf{C}})$ is a morphism $F_0 : \mathbb{1} \longrightarrow F(\varnothing_{\mathsf{C}})$. The composition

$$
\begin{array}{ccc}
F(c) \otimes F(d) & \xrightarrow{\quad F_2 \quad} & F(c \sqcup d) \\
{\scriptstyle (F\iota_c, F\iota_d)} \downarrow & & \| \\
F(c \sqcup d) \otimes F(c \sqcup d) & \xrightarrow{\mu_{c \sqcup d}} & F(c \sqcup d)
\end{array}
$$

is natural in $c, d \in$ C, where

$$\iota_c : c \longrightarrow c \sqcup d \in \mathsf{C}$$

is the natural morphism. To simplify the typography below, we will write coproducts in C as concatenation, so ab means $a \sqcup b$. The desired associativity diagram (2.5.17) of F is the outermost diagram in

$$
\begin{array}{ccccc}
F(a) \otimes F(b) \otimes F(c) & \xrightarrow{(\mathrm{Id}, F\iota_b, F\iota_c)} & F(a) \otimes F(bc)^{\otimes 2} & \xrightarrow{(\mathrm{Id}, \mu_{bc})} & F(a) \otimes F(bc) \\
{\scriptstyle (F\iota_a, F\iota_b, \mathrm{Id})} \downarrow \quad {\scriptstyle (F\iota_a, F\iota_b, F\iota_c) \quad (2)} & & {\scriptstyle (F\iota_a, (F\iota_{bc})^{\otimes 2})} \downarrow & & \downarrow {\scriptstyle (F\iota_a, F\iota_{bc})} \\
{\scriptstyle (1)} & & & & \\
F(ab)^{\otimes 2} \otimes F(c) & \xrightarrow{((F\iota_{ab})^{\otimes 2}, F\iota_c)} & F(abc)^{\otimes 3} & \xrightarrow{(\mathrm{Id}, \mu_{abc})} & F(abc)^{\otimes 2} \\
{\scriptstyle (\mu_{ab}, \mathrm{Id})} \downarrow & & \downarrow {\scriptstyle (\mu_{abc}, \mathrm{Id})} & & \downarrow {\scriptstyle \mu_{abc}} \\
F(ab) \otimes F(c) & \xrightarrow{(F\iota_{ab}, F\iota_c)} & F(abc)^{\otimes 2} & \xrightarrow{\mu_{abc}} & F(abc)
\end{array}
$$

for $a, b, c \in$ C. The sub-diagrams (1) and (2) are commutative by the functoriality of F. The lower left and upper right rectangles are commutative because $F\iota_{ab}$ and $F\iota_{bc}$ are morphisms of monoids, hence compatible with the multiplication. The lower right rectangle is commutative by the associativity of the monoid multiplication μ_{abc}.

The compatibility with the left unit (2.5.18) is the outer diagram in

$$
\begin{array}{ccc}
\mathbb{1} \otimes F(c) & \xrightarrow{\quad \cong \quad} & F(c) \\
{\scriptstyle (1_{\varnothing_{\mathsf{C}}}, \mathrm{Id})} \downarrow \quad {\scriptstyle (1_c, \mathrm{Id})} & & \uparrow {\scriptstyle \mu_c} \\
F(\varnothing_{\mathsf{C}}) \otimes F(c) & \xrightarrow{(F0_c, \mathrm{Id})} & F(c) \otimes F(c)
\end{array}
$$

for $c \in \mathsf{C}$. The lower left triangle is commutative because $F0_c$ preserves the monoid units. The upper right triangle is commutative by part of the unity condition of the monoid $F(c)$ in Proposition 2.6.2. The compatibility with the right unit (2.5.19) is proved similarly.

Finally, one can check that, under the above correspondence, natural transformations in $\mathsf{Mon}(\mathsf{M})^\mathsf{C}$ and $\mathsf{Com}(\mathsf{M})^\mathsf{C}$ correspond to monoidal natural transformations in $\mathsf{MFun}(\mathsf{C}, \mathsf{M})$ and $\mathsf{SMFun}(\mathsf{C}, \mathsf{M})$, respectively. □

As one would expect, monoids can act on objects.

Definition 2.6.6 (Modules over a Monoid). Suppose (A, μ, ε) is a monoid in a monoidal category M.

(1) A *left A-module* is a pair (X, m) consisting of

- an object $X \in \mathsf{M}$ and
- a left A-action $m : A \otimes X \longrightarrow X \in \mathsf{M}$

such that the associativity and unity diagrams

$$
\begin{array}{ccc}
A \otimes A \otimes X & \xrightarrow{(\mathrm{Id}_A, m)} & A \otimes X \\
{\scriptstyle (\mu, \mathrm{Id}_X)} \downarrow & & \downarrow {\scriptstyle m} \\
A \otimes X & \xrightarrow{\quad m \quad} & X
\end{array}
\qquad
\begin{array}{ccc}
\mathbb{1} \otimes X & \xrightarrow{(\varepsilon, \mathrm{Id}_X)} & A \otimes X \\
& {\scriptstyle \cong} \searrow & \downarrow {\scriptstyle m} \\
& & X
\end{array}
$$

are commutative. A morphism of left A-modules is a morphism of the underlying objects that is compatible with the left A-actions in the obvious sense.
(2) The category of left A-modules is denoted by $\mathsf{Mod}(A)$.

Example 2.6.7. In $\mathsf{Vect}_\mathbb{K}$ and $\mathsf{Chain}_\mathbb{K}$, this concept of a left module coincides with the usual one. ◇

2.7 Monads

Definition 2.7.1. For a category C, a *monad on* C is defined as a monoid in the strict monoidal category $(\mathsf{Fun}(\mathsf{C}, \mathsf{C}), \circ, \mathrm{Id}_\mathsf{C})$, where $\circ$ is composition of functors.

Unwrapping this definition using Proposition 2.6.2, a monad can be described more explicitly as follows.

Proposition 2.7.2. *Given a category* C, *a monad on* C *is exactly a triple* (T, μ, ε) *consisting of*

- *a functor* $T : \mathsf{C} \longrightarrow \mathsf{C}$,
- *a natural transformation* $\mu : TT \longrightarrow T$ *called the* multiplication, *and*
- *a natural transformation* $\varepsilon : \mathrm{Id}_\mathsf{C} \longrightarrow T$ *called the* unit,

such that the following associativity and unity diagrams are commutative.

$$
\begin{array}{ccc}
TTT & \xrightarrow{\ T\mu\ } & TT \\
{\scriptstyle \mu T}\downarrow & & \downarrow{\scriptstyle \mu} \\
TT & \xrightarrow{\ \mu\ } & T
\end{array}
\qquad\qquad
\begin{array}{ccc}
T \xrightarrow{\ \varepsilon T\ } & TT & \xleftarrow{\ T\varepsilon\ } T \\
{\scriptstyle \mathrm{Id}}\searrow & \downarrow{\scriptstyle \mu} & \swarrow{\scriptstyle \mathrm{Id}} \\
& T &
\end{array}
$$

Example 2.7.3. Suppose $F : \mathsf{C} \rightleftarrows \mathsf{D} : G$ is an adjunction. Then

$$T = GF : \mathsf{C} \longrightarrow \mathsf{C}$$

is the functor of a monad on C whose unit is the unit of the adjunction $\eta :$ $\mathrm{Id}_{\mathsf{C}} \longrightarrow GF$. The multiplication is

$$\mu = G\varepsilon F : TT = GFGF \longrightarrow GF = T,$$

where $\varepsilon : FG \longrightarrow \mathrm{Id}_{\mathsf{D}}$ is the counit of the adjunction.   ◇

We defined monads as monoids in the functor category $\mathsf{Fun}(\mathsf{C}, \mathsf{C})$. Conversely, the next example shows that each monoid in a monoidal category yields a monad.

Example 2.7.4. Suppose $(\mathsf{M}, \otimes, \mathbb{1})$ is a monoidal category, and (A, μ, ε) is a monoid in M as in Proposition 2.6.2. Then there is a monad on M with the functor $T = A \otimes -$, whose multiplication and unit are induced by those of A. The monadic associativity and unity diagrams are exactly those of the monoid A in Proposition 2.6.2.   ◇

Definition 2.7.5. Suppose (T, μ, ε) is a monad on a category C. A *T-algebra* is a pair (X, λ) consisting of

- an object $X \in \mathsf{C}$ and
- a structure morphism $\lambda : TX \longrightarrow X$

such that the following associativity and unity diagrams are commutative.

$$
\begin{array}{ccc}
TTX & \xrightarrow{\ T\lambda\ } & TX \\
{\scriptstyle \mu_X}\downarrow & & \downarrow{\scriptstyle \lambda} \\
TX & \xrightarrow{\ \lambda\ } & X
\end{array}
\qquad\qquad
\begin{array}{ccc}
X & \xrightarrow{\ \varepsilon_X\ } & TX \\
{\scriptstyle \mathrm{Id}}\searrow & & \downarrow{\scriptstyle \lambda} \\
& & X
\end{array}
$$

A morphism of T-algebras $f : (X, \lambda) \longrightarrow (Y, \pi)$ is a morphism $f : X \longrightarrow Y$ in M such that the diagram

$$
\begin{array}{ccc}
TX & \xrightarrow{\ Tf\ } & TY \\
{\scriptstyle \lambda}\downarrow & & \downarrow{\scriptstyle \pi} \\
X & \xrightarrow{\ f\ } & Y
\end{array}
$$

is commutative. The category of T-algebras and their morphisms is denoted by $\mathsf{Alg}_{\mathsf{C}}(T)$.

Example 2.7.3 says that each adjunction yields a monad on the domain category of the left adjoint. The next example is the converse.

Example 2.7.6. Suppose (T, μ, ε) is a monad on a category C. For each object $X \in \mathsf{C}$, the pair

$$\big(TX, \mu_X : TTX \longrightarrow TX\big)$$

is a T-algebra, called the *free T-algebra of X*. There is a free-forgetful adjunction

$$\mathsf{C} \underset{U}{\overset{T}{\rightleftarrows}} \mathsf{Alg}_{\mathsf{C}}(T) \qquad (2.7.7)$$

in which the right adjoint U forgets about the T-algebra structure and remembers only the underlying object. The left adjoint sends an object to its free T-algebra. This adjunction is known as the *Eilenberg-Moore adjunction*.   ◇

Example 2.7.8. In the setting of Example 2.7.4, a T-algebra is a pair (X, λ) consisting of an object $X \in \mathsf{M}$ and a structure morphism $\lambda : A \otimes X \longrightarrow X$ such that the following associativity and unity diagrams are commutative.

$$
\begin{array}{ccc}
A \otimes A \otimes X & \xrightarrow{(A, \lambda)} & A \otimes X \\
{\scriptstyle (\mu, \mathrm{Id}_X)} \downarrow & & \downarrow {\scriptstyle \lambda} \\
A \otimes X & \xrightarrow{\quad \lambda \quad} & X
\end{array}
\qquad
\begin{array}{ccc}
X \cong \mathbb{1} \otimes X & \xrightarrow{(\varepsilon, X)} & A \otimes X \\
& {\scriptstyle \mathrm{Id}} \searrow & \downarrow {\scriptstyle \lambda} \\
& & X
\end{array}
$$

This is exactly a left A-module.   ◇

The following coequalizer characterization of an algebra over a monad is [Borceux (1994b)] Lemma 4.3.3.

Proposition 2.7.9. *Suppose (T, μ, ε) is a monad on a category* C*, and (X, λ) is a T-algebra. Then the diagram*

$$\big(TTX, \mu_{TX}\big) \underset{T\lambda}{\overset{\mu_X}{\rightrightarrows}} \big(TX, \mu_X\big) \xrightarrow{\ \lambda\ } (X, \lambda)$$

is a coequalizer in $\mathsf{Alg}_{\mathsf{C}}(T)$.

Interpretation 2.7.10. Proposition 2.7.9 says that every algebra over a monad T is a quotient of the free T-algebra on its underlying object, with relations given by the monad multiplication and the T-algebra structure morphism.   ◇

2.8   Localization

Localization of categories will play an important role in encoding the time-slice axiom in (homotopy) algebraic quantum field theory. Here we recall its definition and construction. The idea of localization is to formally invert some morphisms and make them into isomorphisms. The process is similar to the construction of the rational numbers from the integers. Later we will also need the operad version of localization.

Definition 2.8.1. Suppose C is a category, and $S \subseteq \mathsf{Mor}(\mathsf{C})$. An *$S$-localization of* C, if it exists, is a pair $(\mathsf{C}[S^{-1}], \ell)$ consisting of

- a category $C[S^{-1}]$ and
- a functor $\ell : C \longrightarrow C[S^{-1}]$

that satisfies the following two conditions:

(1) $\ell(f)$ is an isomorphism for each $f \in S$.
(2) $(C[S^{-1}], \ell)$ is initial with respect to the previous property. In other words, if $F : C \longrightarrow D$ is a functor such that $F(f)$ is an isomorphism for each $f \in S$, then there exists a unique functor

$$F' : C[S^{-1}] \longrightarrow D \quad \text{such that} \quad F = F'\ell.$$

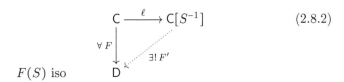

$$(2.8.2)$$

In this setting, ℓ is called the *S-localization functor*.

By the universal property of an S-localization, $C[S^{-1}]$ is unique up to a unique isomorphism if it exists. The following observation says that when S is small enough, the localization always exists.

Theorem 2.8.3. *Suppose* C *is a category, and* S *is a set of morphisms in* C. *Then the* S-*localization* $C[S^{-1}]$ *exists such that* $\mathsf{Ob}(C) = \mathsf{Ob}(C[S^{-1}])$ *and that the localization functor* ℓ *is the identity function on objects.*

Proof. The proof can be found in [Borceux (1994)] Section 5.2. Since we will need the operad version later, we provide a sketch of the proof here. Without loss of generality, we may assume that S is closed under composition. For each $f \in S$, suppose f^{-1} is a symbol such that the sets S and $S^{-1} = \{ f^{-1} : f \in S \}$ are disjoint. Define a category $C[S^{-1}]$ by setting $\mathsf{Ob}(C) = \mathsf{Ob}(C[S^{-1}])$. For objects $a, b \in C$, the morphisms in $C[S^{-1}](a, b)$ are the equivalence classes of finite alternating sequences

$$\varphi = \left(g_{n+1}, f_n^{-1}, g_n, \cdots, f_2^{-1}, g_2, f_1^{-1}, g_1 \right)$$

with each $g_i \in \mathsf{Mor}(C)$ and each $f_i \in S$, which we visualize as follows.

$$a \xrightarrow{\;g_1\;} \bullet \underset{f_1}{\overset{f_1^{-1}}{\rightleftarrows}} \bullet \xrightarrow{\;g_2\;} \cdots \xrightarrow{\;g_n\;} \bullet \underset{f_n}{\overset{f_n^{-1}}{\rightleftarrows}} \bullet \xrightarrow{\;g_{n+1}\;} b$$

Such a sequence is not required to start or end with some g_i. For $1 \le i \le n$, the domain of f_i is the domain of g_{i+1} (if it exists in the sequence), and the codomain of f_i is the codomain of g_i (again if it exists in the sequence). If g_1 is part of φ, then its domain is a. Otherwise, the codomain of f_1 is a. If g_{n+1} is part of φ, then its codomain is b. Otherwise, the domain of f_n is b.

The equivalence relation is generated by the following three identifications:

(1) If g_i is the identity morphism, then φ is identified with the sequence ϕ obtained by replacing the subsequence $(f_i^{-1}, g_i, f_{i-1}^{-1})$ with the entry $(f_{i-1}f_i)^{-1}$. If this g_i happens to be the first or the last entry of φ, then it is omitted in ϕ.

(2) If f_i is the identity morphism, then φ is identified with the sequence ϕ obtained by replacing the subsequence (g_{i+1}, f_i^{-1}, g_i) with the entry $g_{i+1}g_i$. If this f_i happens to be the first or the last entry of φ, then it is omitted in ϕ.

(3) If $g_i = f_i$ (resp., $f_i = g_{i+1}$), then the sequence φ is identified with the subsequence in which g_i and f_i^{-1} (resp., f_i^{-1} and g_{i+1}) are omitted.

The assumption that S be a set implies that $\mathsf{C}[S^{-1}](a, b)$ is a set. For each object $a \in \mathsf{C}[S^{-1}]$, its identity morphism is the equivalence class of the empty sequence. Composition in $\mathsf{C}[S^{-1}]$ is induced by concatenation of sequences and composition in C, with (h_1^{-1}, f_n^{-1}) identified with $(f_n h_1)^{-1}$ if one sequence ends with f_n^{-1} and the next sequence starts with h_1^{-1} for $f_n, h_1 \in S$. One checks that this composition is well-defined (i.e., respects the three identifications above) and that $\mathsf{C}[S^{-1}]$ is indeed a category.

The localization functor $\ell : \mathsf{C} \longrightarrow \mathsf{C}[S^{-1}]$ is defined as the identity function on objects. For a morphism $g \in \mathsf{C}(a, b)$, we define $\ell(g) \in \mathsf{C}[S^{-1}](a, b)$ to be the sequence (g). This defines a functor ℓ that sends each morphism $f \in S$ to an isomorphism in $\mathsf{C}[S^{-1}]$. Suppose $F : \mathsf{C} \longrightarrow \mathsf{D}$ is a functor such that $F(f)$ is an isomorphism for each $f \in S$. The requirement that $F = F'\ell$ (2.8.2) forces us to define the functor $F' : \mathsf{C}[S^{-1}] \longrightarrow \mathsf{D}$ by defining it to be the same as F on objects and

$$F'(\varphi) = (Fg_{n+1})(Ff_n)^{-1}(Fg_n)\cdots(Ff_1)^{-1}(Fg_1)$$

on morphisms. One checks that this F' is well-defined (i.e., respects the three identifications above). So $(\mathsf{C}[S^{-1}], \ell)$ has the required universal property of the S-localization. $\qquad\qquad\Box$

Chapter 3

Trees

One of the important descriptions of operads uses the language of trees, which we discuss in this chapter. The definitions of the Boardman-Vogt construction of a colored operad, homotopy algebraic quantum field theories, and homotopy prefactorization algebras also use trees. The following material on graphs and trees are adapted from [Yau and Johnson (2015)] Part 1, where much more details and many more examples can be found. In practice, since we mostly work with isomorphism classes of trees, it is sufficient to work pictorially as in the examples below.

3.1 Graphs

An *involution* is a self-map τ such that $\tau^2 = \text{Id}$; it is *free* if it has no fixed points.

Definition 3.1.1. Fix an infinite set $\mathfrak{F}$ once and for all. A *graph* is a tuple

$$G = \left(\mathsf{Flag}(G), \lambda_G, \iota_G, \pi_G\right)$$

consisting of:

- a finite set $\mathsf{Flag}(G) \subset \mathfrak{F}$ of *flags*;
- a partition λ_G of $\mathsf{Flag}(G)$ into finitely many possibly empty subsets, called *cells*, together with a distinguished cell G_0, called the *exceptional cell*;
- an involution ι_G on $\mathsf{Flag}(G)$ such that $\iota_G(G_0) = G_0$;
- a free involution π_G on the set of ι_G-fixed points in G_0.

An *isomorphism* of graphs is a bijection on flags that preserves the partition and both involutions. For graphs with any further structure as we will introduce later, an isomorphism is required to preserve that structure as well.

Definition 3.1.2. Suppose G is a graph.

- Flags not in the exceptional cell G_0 are called *ordinary flags*. Flags in G_0 are called *exceptional flags*.
- G is said to be an *ordinary graph* if the exceptional cell is empty.

- A *vertex* is a cell that is not the exceptional cell. A flag in a vertex v is said to be *adjacent to* v. An *isolated vertex* is a vertex that is empty. The cardinality of a vertex v is denoted by $|v|$. The set of vertices is denoted by $\mathsf{Vt}(G)$.
- Two distinct vertices u and v are *adjacent* if there exist flags $a \in u$ and $b \in v$ such that $\iota_G(a) = b$.
- The fixed points of ι_G are celled *legs*. The set of legs is denoted by $\mathsf{Leg}(G)$. A leg in a vertex is called an *ordinary leg*. A leg in the exceptional cell is called an *exceptional leg*.
- The orbits of π_G and of ι_G away from its fixed points in G_0 are called *edges*. The set of edges is denoted by $\mathsf{Ed}(G)$.
- The non-trivial orbits of ι_G are called *internal edges*. The set of internal edges in G is denoted by $|G|$. The non-trivial orbits of ι_G within the vertices are called *ordinary internal edges*. Those within the exceptional cell are called *exceptional loops* and denoted by $\bigcirc$.
- An orbit of π_G is called an *exceptional edge* and denoted by $|$.
- If $f = \{f_\pm\}$ is an ordinary internal edge with $f_+ \in u$ and $f_- \in v$, then we say that f is *adjacent* to u and v.

Example 3.1.3. The *empty graph* $\varnothing$ has an empty set of flags, hence an empty exceptional cell, and no vertices.   ◇

Example 3.1.4. The graph $\bullet$ with an empty set of flags, hence an empty exceptional cell, and a single empty vertex is an isolated vertex.   ◇

Example 3.1.5. The graph with no vertices and with only two exceptional legs $f_\pm$, which must be paired by the involution π, is the exceptional edge $|$.   ◇

Example 3.1.6. The graph with no vertices and with only two exceptional flags $e_\pm$ paired by ι is the exceptional loop $\bigcirc$.   ◇

Definition 3.1.7. Suppose G is a graph.

(1) A *path* of length $r \geq 0$ is a pair

$$P = \left(\{e^i\}_{i=1}^r, \{v_i\}_{i=0}^r \right)$$

in which:

- the v_i's are distinct vertices, except possibly for v_0 and v_r;
- each e^i is an ordinary internal edge adjacent to both v_{i-1} and v_i;

(2) A *cycle* is a path of length $r \geq 1$ with $v_0 = v_r$.

Definition 3.1.8. A non-empty graph G is:

(1) *connected* if it satisfies one of the following two conditions:

(a) It is an isolated vertex $\bullet$, the exceptional edge $|$, or the exceptional loop $\bigcirc$.

(b) It is an ordinary graph that has no isolated vertices such that, for each pair of distinct flags $\{f_1, f_2\}$, there exists a path $P = (\{e^i\}, \{v_i\})$ with f_1 adjacent to some v_k and f_2 adjacent to some v_l.

(2) *simply-connected* if it (i) is connected; (ii) is not the exceptional loop; (iii) contains no cycles.

We will need to consider the following extra structures on graphs.

Definition 3.1.9. For a non-empty set $\mathfrak{C}$, whose elements are called colors, a $\mathfrak{C}$-*coloring* of a graph G is a function

$$\kappa : \mathsf{Flag}(G) \longrightarrow \mathfrak{C}$$

that is constant on each orbit of the involutions ι_G and π_G. In other words, a $\mathfrak{C}$-coloring assigns to each edge a color.

Definition 3.1.10. A *direction* of a graph G is a function

$$\delta : \mathsf{Flag}(G) \longrightarrow \{1, -1\}$$

such that:

- If $(f, \iota_G(f))$ is an internal edge, then $\delta(\iota_G(f)) = -\delta(f)$.
- If $(f, \pi_G(f))$ is an exceptional edge, then $\delta(\pi_G(f)) = -\delta(f)$.

Definition 3.1.11. Suppose G is a graph equipped with a direction δ.

- A leg f with $\delta(f) = 1$ is an *input* of G.
- A leg f with $\delta(f) = -1$ is an *output* of G.
- If v is a vertex with $f \in v$ and $\delta(f) = 1$, then f is an *input* of v.
- If v is a vertex with $f \in v$ and $\delta(f) = -1$, then f is an *output* of v.
- For $z \in \{G\} \sqcup \mathsf{Vt}(G)$, the set of inputs of z is denoted by $\mathsf{in}(z)$, and the set of outputs of z is denoted by $\mathsf{out}(z)$.
- An internal edge is regarded as oriented from the flag with $\delta = -1$ to the flag with $\delta = 1$.
- For an ordinary internal edge $f = \{f_\pm\}$ with $\delta(f_\pm) = \pm 1$, the vertex containing f_- (resp., f_+) is the *initial vertex* (resp., *terminal vertex*) of f.
- A *directed path* is a path P as in Definition 3.1.7 such that each e^i has initial vertex v_{i-1} and terminal vertex v_i. We call v_0 (resp., v_r) the *initial vertex* (resp., *terminal vertex*) of P.

Definition 3.1.12. An *unordered tree* is a pair (T, δ) consisting of

- a simply-connected graph T and
- a direction δ

such that $|\mathsf{out}(v)| = 1$ for each $v \in \mathsf{Vt}(T)$. In an unordered tree T that is not isomorphic to an exceptional edge, the unique vertex containing the output of T is called the *root vertex*.

Definition 3.1.13. Suppose (T, δ) is an unordered tree.

(1) An *ordering at a vertex v* is a bijection

$$\zeta_v : \{1, \ldots, |\mathsf{in}(v)|\} \xrightarrow{\;\cong\;} \mathsf{in}(v).$$

(2) An *ordering of T* is a bijection

$$\zeta_T : \{1, \ldots, |\mathsf{in}(T)|\} \xrightarrow{\;\cong\;} \mathsf{in}(T).$$

A *listing* of (T, δ) is a choice of an ordering for each $z \in \{T\} \sqcup \mathsf{Vt}(T)$. Given a listing, we will regard each $\mathsf{in}(z)$ as an ordered set.

Definition 3.1.14. Suppose $\mathfrak{C}$ is a non-empty set, whose elements are called *colors*. A $\mathfrak{C}$-*profile* is a finite sequence of elements in $\mathfrak{C}$.

- If $\mathfrak{C}$ is clear from the context, then we simply say *profile*.
- The empty $\mathfrak{C}$-profile is denoted by $\varnothing$.
- We write $|\underline{c}| = m$ for the *length* of a profile $\underline{c} = (c_1, \ldots, c_m)$.
- The set of $\mathfrak{C}$-profiles is denoted by $\mathsf{Prof}(\mathfrak{C})$.
- An element in $\mathsf{Prof}(\mathfrak{C}) \times \mathfrak{C}$ is written either horizontally as $(\underline{c}; d)$ or vertically as $\binom{d}{\underline{c}}$.

Definition 3.1.15. A $\mathfrak{C}$-*colored tree* is a tuple $(T, \delta, \kappa, \zeta)$ consisting of an unordered tree (T, δ), a $\mathfrak{C}$-coloring κ, and a listing ζ. Given such a $\mathfrak{C}$-colored tree, using the $\mathfrak{C}$-coloring κ:

(1) For $z \in \{T\} \sqcup \mathsf{Vt}(T)$, we regard the ordered set $\mathsf{in}(z)$ as a $\mathfrak{C}$-profile, called the *input profile of z*, whose jth entry is denoted by $\mathsf{in}(z)_j$. Similarly, we regard the element $\mathsf{out}(z) \in \mathfrak{C}$, called the *output color of z*.

(2) For $z \in \{T\} \sqcup \mathsf{Vt}(T)$, the *profile of z* is the pair

$$\mathsf{Prof}(z) = \big(\mathsf{in}(z); \mathsf{out}(z)\big) = \binom{\mathsf{out}(z)}{\mathsf{in}(z)} \in \mathsf{Prof}(\mathfrak{C}) \times \mathfrak{C}.$$

The set of isomorphism classes of $\mathfrak{C}$-colored trees with profile $(\underline{c}; d) \in \mathsf{Prof}(\mathfrak{C}) \times \mathfrak{C}$ is denoted by $\mathsf{Tree}^{\mathfrak{C}}(\underline{c}; d)$ or $\mathsf{Tree}^{\mathfrak{C}}\binom{d}{\underline{c}}$. We will omit mentioning $\mathfrak{C}$ if it is clear from the context.

Convention 3.1.16. For a vertex v in a $\mathfrak{C}$-colored tree, to simplify the typography, we will often abbreviate $\mathsf{Prof}(v)$ to just (v). From now on, the single word *tree* will mean a $\mathfrak{C}$-colored tree, unless otherwise specified.

Definition 3.1.17. A $\mathfrak{C}$-*colored linear graph* is a $\mathfrak{C}$-colored tree T such that $|\mathsf{in}(v)| = 1$ for each $v \in \mathsf{Vt}(T)$. The set of isomorphism classes of $\mathfrak{C}$-colored linear graphs with profile $(c; d) \in \mathfrak{C}^{\times 2}$ is denoted by $\mathsf{Linear}^{\mathfrak{C}}(c; d)$ or $\mathsf{Linear}^{\mathfrak{C}}\binom{d}{c}$.

Example 3.1.18 (Exceptional edges). The exceptional edge | in Example 3.1.5 can be given a direction δ with $\delta(f_\pm) = \pm 1$. For each color $c \in \mathfrak{C}$, it becomes a c-colored linear graph $\uparrow_c$ with profile $(c; c)$, called the $\mathfrak{C}$-*colored exceptional edge*, in which the bottom (resp., top) flag is f_+ (resp., f_-), with coloring $\kappa(f_\pm) = c$ and with a trivial listing. These colored exceptional edges are the only colored trees with exceptional flags. As we will see later, the c-colored exceptional edge corresponds to the c-colored unit of a $\mathfrak{C}$-colored operad. ◇

Example 3.1.19 (Linear graphs). For each $\mathfrak{C}$-profile $\underline{c} = (c = c_0, \ldots, c_n = d)$ with $n \geq 0$, there is a $\mathfrak{C}$-colored *linear graph*

$$\mathrm{Lin}_{\underline{c}} \in \mathsf{Linear}^{\mathfrak{C}}\left(\begin{smallmatrix} d \\ \underline{c} \end{smallmatrix}\right)$$

defined as follows.

- $\mathsf{Flag}(\mathrm{Lin}_{\underline{c}}) = \{i, e_\pm^1, \ldots, e_\pm^{n-1}, o\}$, all of which are ordinary flags. Note that the flags $e_\pm^j$ are only in $\mathsf{Flag}(\mathrm{Lin}_{\underline{c}})$ if $n \geq 2$.
- The involution ι fixes i and o, and $\iota(e_\pm^j) = e_\mp^j$ for $1 \leq j \leq n - 1$.
- There are n vertices $v_j = \{e_+^{j-1}, e_-^j\}$ for $1 \leq j \leq n$, where $e_+^0 = i$ and $e_-^n = o$.
- $\kappa(i) = c_0$, $\kappa(e_\pm^j) = c_j$ for $1 \leq j \leq n - 1$, and $\kappa(o) = c_n$.
- $\delta(i) = 1$, $\delta(e_\pm^j) = \pm 1$ for $1 \leq j \leq n - 1$, and $\delta(o) = -1$.

If $n = 0$, then $\mathrm{Lin}_{\underline{c}}$ is the c-colored exceptional edge $\uparrow_c$ in Example 3.1.18. If $n \geq 1$, then we depict the linear graph $\mathrm{Lin}_{\underline{c}}$ as follows.

It has $n - 1$ internal edges $e^j = \{e_\pm^j\}$ for $1 \leq j \leq n - 1$. The initial vertex of e^j is v_j, and its terminal vertex is v_{j+1}. The input flag of $\mathrm{Lin}_{\underline{c}}$ is i, and its output flag is o. ◇

Example 3.1.20 (Truncated linear graphs). For each $\mathfrak{C}$-profile $(c_1, \ldots, c_n)$ with $n \geq 1$, the *truncated linear graph*

$$\mathrm{lin}_{(c_1, \ldots, c_n)}$$

with profile $(\varnothing; c_n)$ has the same definition as the linear graph in Example 3.1.19 but without the input flag i. We visualize the truncated linear graph $\mathrm{lin}_{(c_1, \ldots, c_n)}$ as follows.

Note that it does not have any inputs. ◇

Example 3.1.21 (Corollas). For each pair $\left(\underline{c} = (c_1, \ldots, c_m); d\right) \in \mathsf{Prof}(\mathfrak{C}) \times \mathfrak{C}$, there is a $\mathfrak{C}$-colored tree

$$\mathrm{Cor}_{(\underline{c}; d)},$$

called the $(\underline{c}; d)$-*corolla*, with profile $(\underline{c}; d)$. It is the ordinary $\mathfrak{C}$-colored tree defined as follows.

- Flag$\left(\mathrm{Cor}_{(\underline{c};d)}\right) = \{i_1, \ldots, i_m, o\}$, all of which are ordinary legs at a unique vertex v.
- $\kappa(i_p) = c_p$ for $1 \le p \le m$ and $\kappa(o) = d$.
- $\delta(i_p) = 1$ for $1 \le p \le m$ and $\delta(o) = -1$.
- $\zeta_z(p) = i_p$ for $z \in \{v, \mathrm{Cor}_{(\underline{c};d)}\}$ and $1 \le p \le m$.

The corolla $\mathrm{Cor}_{(\underline{c};d)}$ is a linear graph if and only if $|\underline{c}| = 1$. We depict the $(\underline{c}; d)$-corolla as

in which the input legs are drawn from left to right according to their ordering. The jth input leg is i_j, and the output is o.   ◇

Example 3.1.22 (Permuted corollas). With the same setting as in Example 3.1.21, suppose given a permutation $\tau \in \Sigma_m$. There is a $\mathfrak{C}$-colored tree

$$\mathrm{Cor}_{(\underline{c};d)}\tau,$$

called the *permuted corolla*, with profile $(\underline{c}\tau; d)$. It is defined just like the corolla $\mathrm{Cor}_{(\underline{c};d)}$, except for the ordering of the whole graph:

$$\zeta_{\mathrm{Cor}_{(\underline{c};d)}\tau}(p) = i_{\tau(p)} \quad \text{for} \quad 1 \le p \le m.$$

Note that $\mathrm{Prof}(v) = (\underline{c}; d)$ for its unique vertex v, while $\mathrm{Prof}(\mathrm{Cor}_{(\underline{c};d)}\tau) = (\underline{c}\tau; d)$. For example, if $\underline{c} = (c_1, c_2)$ and $\tau = (1\ 2) \in \Sigma_2$, then we may visualize the permuted corolla $\mathrm{Cor}_{(\underline{c};d)}\tau$ as:

As we will see below, permuted corollas provide operads with their equivariant structure.   ◇

Example 3.1.23 (2-level trees). Suppose $d \in \mathfrak{C}$, $\underline{c} = (c_1, \ldots, c_m) \in \mathrm{Prof}(\mathfrak{C})$ with $m \ge 1$, $\underline{b}_j = \left(b_{j,1}, \ldots, b_{j,k_j}\right) \in \mathrm{Prof}(\mathfrak{C})$ for $1 \le j \le m$ with $|\underline{b}_j| = k_j$, and $\underline{b} = (\underline{b}_1, \ldots, \underline{b}_m)$. There is a $\mathfrak{C}$-colored tree

$$T\left(\{\underline{b}_j\}; \underline{c}; d\right)$$

with profile $(\underline{b}; d)$, called a 2-*level tree*, that can be pictorially represented as:

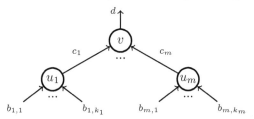

Formally $T = T\left(\{\underline{b}_j\}; \underline{c}; d\right)$ is defined as follows.

- $\mathsf{Flag}(T) = \left\{o, \{f_{\pm}^j\}_{1 \le j \le m}, \{g_{j,i}\}_{1 \le j \le m}^{1 \le i \le k_j}\right\}$, all of which are ordinary flags.
- $\iota(f_{\pm}^j) = f_{\mp}^j$, and ι fixes all other flags.
- There are $m + 1$ vertices:
$$v = \left\{o, f_+^1, \dots, f_+^m\right\} \quad \text{and} \quad u_j = \left\{f_-^j, g_{j,1}, \dots, g_{j,k_j}\right\} \quad \text{for} \quad 1 \le j \le m$$
- $\kappa(o) = d$, $\kappa\left(f_{\pm}^j\right) = c_j$, and $\kappa\left(g_{j,i}\right) = b_{j,i}$.
- $\delta(o) = -1 = \delta\left(f_-^j\right)$, and $\delta\left(g_{j,i}\right) = 1 = \delta\left(f_+^j\right)$.
- $\zeta_v\left(f_+^j\right) = j$, $\zeta_{u_j}\left(g_{j,i}\right) = i$, and $\zeta_T\left(g_{j,i}\right) = i + k_1 + \dots + k_{j-1}$.

There are m internal edges $f^j = \{f_{\pm}^j\}$. The unique output is the flag o, and the flags $g_{j,i}$ are the inputs of T. As we will see below, these $\mathfrak{C}$-colored trees correspond to the operadic composition γ. ◇

In what follows, we will often draw a colored tree without writing down its detailed definition. The reader can fill in the details using the examples above as a guide.

3.2 Tree Substitution

The main reason for considering trees is the operation called tree substitution. Suppose $\mathfrak{C}$ is a non-empty set. All the trees below are $\mathfrak{C}$-colored trees.

Definition 3.2.1 (Tree Substitution at a Vertex). Suppose T is a tree, and v is a vertex in T. Suppose H is a tree such that $\mathsf{Prof}(H) = \mathsf{Prof}(v)$. Define the tree $T(H)$ with $\mathsf{Prof}\big(T(H)\big) = \mathsf{Prof}(T)$, called the *tree substitution at* v, as follows.

(1) If H is not an exceptional edge, then we identify (i) the ordered sets $\mathsf{in}(v)$ and $\mathsf{in}(H)$ and (ii) the flags $\mathsf{out}(v)$ and $\mathsf{out}(H)$. We define
$$\mathsf{Flag}\big(T(H)\big) = \big(\mathsf{Flag}(T) \smallsetminus v\big) \coprod \mathsf{Flag}(H),$$
$$\mathsf{Vt}\big(T(H)\big) = \big[\mathsf{Vt}(T) \smallsetminus \{v\}\big] \coprod \mathsf{Vt}(H),$$
$$\iota_{T(H)}(f) = \begin{cases} \iota_H(f) & \text{if } f \in \mathsf{Flag}(H) \smallsetminus \mathsf{Leg}(H), \\ \iota_T(f) & \text{otherwise,} \end{cases}$$
with an empty exceptional cell. Its coloring, direction, and listing are induced from those of T and H.

(2) If H is the exceptional edge $\uparrow_c$ and if T is the corollary $\mathrm{Cor}_{(c;c)}$ with v its unique vertex, then $T(H) = \uparrow_c$.

(3) If H is the exceptional edge $\uparrow_c$ and if T is not the corollary $\mathrm{Cor}_{(c;c)}$, then we define

$$\mathsf{Flag}(T(H)) = \mathsf{Flag}(T) \smallsetminus v \quad \text{and} \quad \mathsf{Vt}(T(H)) = \mathsf{Vt}(T) \smallsetminus \{v\}$$

with an empty exceptional cell. Furthermore:

(a) If $\mathsf{out}(v) = \mathsf{out}(T)$, then we define

$$\iota_{T(H)}(f) = \begin{cases} f & \text{if } f = \iota_T(\mathsf{in}(v)), \\ \iota_T(f) & \text{otherwise.} \end{cases}$$

(b) If $\mathsf{in}(v)$ is an input of T, then we define

$$\iota_{T(H)}(f) = \begin{cases} f & \text{if } f = \iota_T(\mathsf{out}(v)), \\ \iota_T(f) & \text{otherwise.} \end{cases}$$

(c) If $\mathsf{out}(v) \neq \mathsf{out}(T)$ and if $\mathsf{in}(v)$ is not an input of T, then we define

$$\iota_{T(H)}(f) = \begin{cases} \iota_T(\mathsf{in}(v)) & \text{if } f = \iota_T(\mathsf{out}(v)), \\ \iota_T(\mathsf{out}(v)) & \text{if } f = \iota_T(\mathsf{in}(v)), \\ \iota_T(f) & \text{otherwise.} \end{cases}$$

Its coloring, direction, and listing are induced from those of T.

The following properties of tree substitution are proved by directly checking the definitions, so we omit the proofs here. The reader may consult [Yau and Johnson (2015)] Chapter 5 for proofs.

Proposition 3.2.2. *Consider the setting of Definition 3.2.1.*

(1) *If there are isomorphisms $T \cong T'$ and $H \cong H'$, then there is an isomorphism $T(H) \cong T'(H')$.*

(2) *$T(H)$ is a linear graph if and only if T and H are both linear graphs.*

(3) *Internal edges in H yield internal edges in $T(H)$.*

(4) *Suppose u is a vertex in H, and K is a tree such that $\mathsf{Prof}(K) = \mathsf{Prof}(u)$. Then there is a canonical isomorphism*

$$[T(H)](K) \cong T(H(K)). \tag{3.2.3}$$

(5) *Suppose $w \neq v$ is another vertex in T, and G is a tree such that $\mathsf{Prof}(G) = \mathsf{Prof}(w)$. Then there is a canonical isomorphism*

$$[T(H)](G) \cong [T(G)](H). \tag{3.2.4}$$

(6) *There are canonical isomorphisms*

$$T(\mathrm{Cor}_v) \cong T \quad \text{and} \quad \mathrm{Cor}_T(H) \cong H, \tag{3.2.5}$$

where Cor_z is the $(\mathsf{in}(z); \mathsf{out}(z))$-corolla for $z \in \{v, T\}$.

(7) If T is a permuted corolla $\mathsf{Cor}_{(\underline{c};d)}\tau$ as in Example 3.1.22 for some permutation $\tau \in \Sigma_{|\mathsf{in}(T)|}$, then $\big(\mathsf{Cor}_{(\underline{c};d)}\tau\big)(H)$ is canonically isomorphic to H except that its input profile is $\mathsf{in}(H)\tau$.

(8) If H is a permuted corolla $\mathsf{Cor}_{v'}\tau$ for some permutation τ, then $T\big(\mathsf{Cor}_{v'}\tau\big)$ is canonically isomorphic to T except that the input profile at v' is

$$\mathsf{Prof}(v') = \big(\mathsf{in}(v)\tau^{-1}; \mathsf{out}(v)\big).$$

Definition 3.2.6 (Tree Substitution). Suppose T is a tree, and H_v is a tree with $\mathsf{Prof}(H_v) = \mathsf{Prof}(v)$ for each vertex v in T. Suppose $\{v_1,\ldots,v_n\}$ is an ordering of the set $\mathsf{Vt}(T)$. Define the tree $T(H_v)_{v\in\mathsf{Vt}(T)}$ with the same profile as T, called the *tree substitution*, by

$$T(H_v)_{v\in\mathsf{Vt}(T)} = \big(\cdots(T(H_{v_1}))(H_{v_2})\cdots\big)(H_{v_n}).$$

Notation 3.2.7. To simplify the notation, we will often write $v \in T$ to mean $v \in \mathsf{Vt}(T)$. Furthermore, we will sometimes abbreviate $T(H_v)_{v\in\mathsf{Vt}(T)}$ to $T(H_v)$. We will say that H_v is *substituted into* v.

The following properties of tree substitution are consequences of Proposition 3.2.2.

Corollary 3.2.8. *Consider the setting of Definition 3.2.1.*

(1) *The isomorphism class of the tree substitution $T(H_v)_{v\in T}$ is independent of the choices of (i) an ordering of $\mathsf{Vt}(T)$, (ii) a representative in the isomorphism class of T, and (iii) a representative in the isomorphism class of each H_v.*

(2) *There is a decomposition*

$$\mathsf{Vt}\big(T(H_v)_{v\in T}\big) = \coprod_{v\in T}\mathsf{Vt}(H_v). \tag{3.2.9}$$

(3) *Up to isomorphisms, tree substitution is associative in the sense that, if I_u is a tree with $\mathsf{Prof}(I_u) = \mathsf{Prof}(u)$ for each $u \in \mathsf{Vt}(H_v)$ and each $v \in \mathsf{Vt}(T)$, then there is an isomorphism*

$$\Big[T(H_v)_{v\in T}\Big](I_u)_{u\in T(H_v)_{v\in T}} \cong T\Big(H_v(I_u)_{u\in H_v}\Big)_{v\in T}.$$

(4) *Up to isomorphisms, tree substitution is unital in the sense that there are isomorphisms*

$$T\big(\mathsf{Cor}_v\big)_{v\in T} \cong T \quad and \quad \mathsf{Cor}_T(T) \cong T.$$

Proof. The first assertion follows from the first assertion in Proposition 3.2.2 and (3.2.4). The decomposition 3.2.9 on vertex set follows from Definition 3.2.1. The associativity isomorphism follows from (3.2.3) and (3.2.4). The unity isomorphisms follow from (3.2.5). □

Convention 3.2.10. To simplify the presentation, in what follows we will minimize the distinction between a tree and its isomorphism class and will use the same symbol to denote both.

Definition 3.2.11 (Substitution Category). Define the *substitution category* $\underline{\text{Tree}}^{\mathfrak{C}}$ as the small category with:

- $\mathfrak{C}$-colored trees as objects;
- $\underline{\text{Tree}}^{\mathfrak{C}}(K, T)$ the set of finite sets $(H_v)_{v \in T}$ such that $K = T(H_v)_{v \in T}$;
- $(\text{Cor}_v)_{v \in T}$ as the identity morphism of T;
- composition given by tree substitution.

For a pair $(\underline{c}; d) \in \text{Prof}(\mathfrak{C}) \times \mathfrak{C}$, denote by $\underline{\text{Tree}}^{\mathfrak{C}}\binom{d}{\underline{c}}$ or $\underline{\text{Tree}}^{\mathfrak{C}}(\underline{c}; d)$, called the *substitution category* with profile $(\underline{c}; d)$, the full subcategory of $\underline{\text{Tree}}^{\mathfrak{C}}$ consisting of trees with profile $(\underline{c}; d)$. For a vertex v in a $\mathfrak{C}$-colored tree, we will also write $\underline{\text{Tree}}^{\mathfrak{C}}(v)$ for $\underline{\text{Tree}}^{\mathfrak{C}}\binom{\text{out}(v)}{\text{in}(v)}$. We similarly define the substitution categories $\underline{\text{Linear}}^{\mathfrak{C}}$ and $\underline{\text{Linear}}^{\mathfrak{C}}\binom{d}{\underline{c}}$ using linear graphs instead of trees.

Remark 3.2.12. Suppose

$$(H_v)_{v \in T} : K \longrightarrow T \quad \text{and} \quad (G_u)_{u \in K} : E \longrightarrow K$$

are morphisms in $\underline{\text{Tree}}^{\mathfrak{C}}$. Then

$$E = K(G_u)_{u \in K} = \big(T(H_v)_{v \in T}\big)(G_u)_{u \in H_v,\, v \in T} = T\Big(H_v(G_u)_{u \in H_v}\Big)_{v \in T}.$$

This defines the composition

$$E \xrightarrow{\ (G_u)_{u \in K}\ } K \xrightarrow{\ (H_v)_{v \in T}\ } T$$

with the top arc labeled $\big(H_v(G_u)_{u \in H_v}\big)_{v \in T}$

of $(G_u)_{u \in K}$ and $(H_v)_{v \in T}$ in the substitution category $\underline{\text{Tree}}^{\mathfrak{C}}$.   ◇

Example 3.2.13. Consider the morphism

$$(H_u, H_v, H_w) : K \longrightarrow T$$

indicated by the following picture.

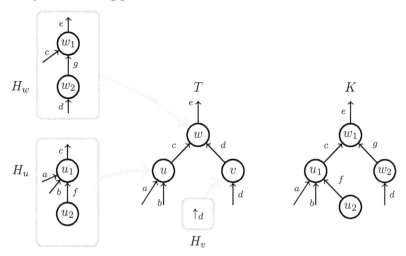

For simplicity, in each tree, all the orderings at the vertices and for the whole tree are from left to right as displayed. Each gray arrow indicates substituting the tree inside the originating gray box into the corresponding vertex in T. Observe that

$$\mathsf{Prof}(u) = \binom{c}{a,b} = \mathsf{Prof}(H_u), \quad \mathsf{Prof}(v) = \binom{d}{d} = \mathsf{Prof}(H_v),$$
$$\mathsf{Prof}(w) = \binom{e}{c,d} = \mathsf{Prof}(H_w), \quad \text{and} \quad \mathsf{Prof}(T) = \binom{e}{a,b,d} = \mathsf{Prof}(K).$$

Internal edges in the H's become internal edges in the tree substitution K. The d-colored internal edge in T is no longer an internal edge in K because H_v is the d-colored exceptional edge. ◇

3.3 Grafting

Grafting is a special kind of tree substitution, which can be pictorially interpreted as gluing the outputs of a finite family of trees with the inputs of another tree. In the next definition, we will use the 2-level tree in Example 3.1.23.

Definition 3.3.1 (Grafting of Trees). Suppose

- $\binom{d}{\underline{c}} = \binom{d}{c_1,\dots,c_m} \in \mathsf{Prof}(\mathfrak{C}) \times \mathfrak{C}$ with $m \geq 1$.
- $\underline{b}_j \in \mathsf{Prof}(\mathfrak{C})$ for $1 \leq j \leq m$ with $|\underline{b}_j| = k_j$, and $\underline{b} = (\underline{b}_1,\dots,\underline{b}_m)$.
- G is a $\mathfrak{C}$-colored tree with profile $(\underline{c};d)$.
- H_j is a $\mathfrak{C}$-colored tree with profile $(\underline{b}_j;c_j)$ for each $1 \leq j \leq m$.

The *grafting of G with $H_1,\dots,H_m$* is defined as the tree substitution

$$\mathsf{Graft}(G;H_1,\dots,H_m) = \left[T\left(\{\underline{b}_j\};\underline{c};d\right)\right](G,H_1,\dots,H_m),$$

where $T\left(\{\underline{b}_j\};\underline{c};d\right)$ is the 2-level tree with profile $(\underline{b};d)$ in Example 3.1.23, with G substituted into v and H_j substituted into u_j.

Note that the profile of the grafting $\mathsf{Graft}(G;\{H_j\})$ is $(\underline{b};d)$, which is the same as the profile of the 2-level tree $T\left(\{\underline{b}_j\};\underline{c};d\right)$.

Example 3.3.2. With G as in Definition 3.3.1, we have

$$\mathsf{Graft}(G;\uparrow_{c_1},\dots,\uparrow_{c_m}) = G = \mathsf{Graft}(\uparrow_d;G).$$

That is, grafting with an exceptional edge has no effect. ◇

Example 3.3.3. Suppose T is the tree in Example 3.2.13 with profile $\binom{e}{a,b,d}$. With the trees H_1, H_2, and H_3 as drawn below, the grafting

$$G = \mathsf{Graft}(T;H_1,H_2,H_3) = \left[T(\{(f),(b),\varnothing\};(a,b,d);e)\right](T,H_1,H_2,H_3)$$

is the tree on the right.

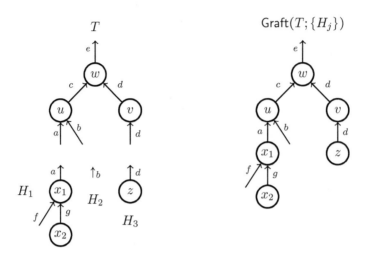

Observe that grafting with an exceptional edge, as with $H_2 = \;\uparrow_b$ above, has no effect. Internal edges in T and in the H's remain internal edges in the grafting $\mathsf{Graft}(T; \{H_j\})$. New internal edges are created by the grafting, unless it involves an exceptional edge.   ◇

Grafting allows us to construct bigger trees from smaller ones, as illustrated in Example 3.3.3. The following observation says that corollas are the building blocks of trees with respect to grafting. With slightly different terminology, the following result is [Yau (2016)] Theorem 5.7.3.

Theorem 3.3.4. *Suppose T is a $\mathfrak{C}$-colored tree with at least one vertex. Up to a reordering of its inputs, T is an iterated grafting of corollas and exceptional edges.*

Proof. If T has only one vertex, then it is a permuted corolla as in Example 3.1.22, which is a corolla with its inputs reordered. Inductively, suppose T has $n > 1$ vertices, and v is the root vertex in T. Suppose $u_1, \ldots, u_n$ are the vertices adjacent to v, with an internal edge $e^j = \{e^j_\pm\}$ adjacent to u_j and v for $1 \leq j \leq n$. For each j, suppose T_j is the tree consisting of the largest subset of flags in T with root vertex u_j, output e^j_-, and the induced direction, $\mathfrak{C}$-coloring, and listing. Suppose the inputs of the root vertex v are the flags $\{f_1, \ldots, f_r\}$, which contain $\{e^1_+, \ldots, e^n_+\}$. Up to a reordering of its inputs, T is the grafting

$$T = \mathsf{Graft}\big(\mathsf{Cor}_v; H_1, \ldots, H_r\big)$$

in which:

- Cor_v is the corolla with the same profile as v.
- H_i is:

  - the $\kappa(f_i)$-colored exceptional edge $\uparrow_{\kappa(f_i)}$ if $f_i \notin \{e^1_+, \ldots, e^n_+\}$.
  - the tree T_j if $f_i = e^j_+$.

We finish the proof by observing that the induction hypothesis applies to each T_j, since it has strictly fewer vertices than T. □

Example 3.3.5. In the setting of Example 3.3.3, we have

$$T = \mathsf{Graft}\big(\mathrm{Cor}_w; \mathrm{Cor}_u, \mathrm{Cor}_v\big) \quad \text{and} \quad H_1 = \mathsf{Graft}\big(\mathrm{Cor}_{x_1}; \uparrow_f, \mathrm{Cor}_{x_2}\big).$$

Moreover, $H_2 = \uparrow_b$ is an exceptional edge, and $H_3 = \mathrm{Cor}_{(\varnothing;d)}$ is a corolla. ◇

Chapter 4

Colored Operads

In this chapter, we define colored operads and their algebras. A colored operad is a generalization of a category in which the domain of each morphism is a finite sequence of objects. Colored operads provide an efficient way to encode operations with multiple inputs and one output. This efficient bookkeeping aspect of operad theory is especially important when we discuss homotopy algebraic quantum field theories and homotopy prefactorization algebras, in which the desired structures are too complicated to encode without colored operads.

Historically, colored non-symmetric operads in Set were defined by Lambek [Lambek (1969)], who called them *multicategories*. May [May (1972)] defined a one-colored topological operad and coined the term *operad*. In [Kelly (2005)] Kelly gave a more categorical construction of one-colored operads in symmetric monoidal categories in terms of coends and Day convolutions. For an introduction to colored operads, the reader may also consult [Benini *et. al.* (2017); White and Yau (2018); Yau (2016)].

In Section 4.1 we define a colored operad as a monoid in certain monoidal category. In Section 4.2 to Section 4.4, we provide three equivalent and more explicit descriptions of a colored operad. In particular, the definition in Section 4.2 in terms of generating structure morphisms and axioms and the definition in Section 4.4 in terms of trees will be used throughout the rest of this book. In Section 4.5, we define algebras over a colored operad and discuss some key examples.

Throughout this chapter, $(\mathsf{M}, \otimes, \mathbb{1})$ is a cocomplete symmetric monoidal closed category with an initial object $\varnothing$ and an internal hom $\mathsf{Hom_M}$.

4.1 Operads as Monoids

As we will see later, each operad yields a monad (Definition 2.7.1). We defined a monad as a monoid in the strict monoidal category of functors from a category to itself. In this section, we define operads analogously as monoids in a suitable monoidal category. We first recall from [Yau and Johnson (2015)] some notations regarding colors and profiles. The symmetric group on n letters is denoted by Σ_n, whose unit is id_n. Recall from Definition 3.1.14 that a $\mathfrak{C}$-profile is a finite sequence of elements in $\mathfrak{C}$, and $\mathsf{Prof}(\mathfrak{C})$ denotes the set of all $\mathfrak{C}$-profiles.

Definition 4.1.1. Fix a non-empty set $\mathfrak{C}$, whose elements are called *colors*.

(1) If $\underline{a} = (a_1, \ldots, a_m)$ and $\underline{b}$ are $\mathfrak{C}$-profiles, then a *left permutation* $\sigma : \underline{a} \longrightarrow \underline{b}$ is a permutation $\sigma \in \Sigma_{|\underline{a}|}$ such that

$$\sigma\underline{a} = (a_{\sigma^{-1}(1)}, \ldots, a_{\sigma^{-1}(m)}) = \underline{b}$$

(2) The *groupoid of $\mathfrak{C}$-profiles*, with left permutations as the isomorphisms, is denoted by $\Sigma_{\mathfrak{C}}$. The opposite groupoid $\Sigma_{\mathfrak{C}}^{\mathrm{op}}$ is regarded as the groupoid of $\mathfrak{C}$-profiles with *right permutations*

$$\underline{a}\sigma = (a_{\sigma(1)}, \ldots, a_{\sigma(m)})$$

as isomorphisms.

(3) The objects of the diagram category

$$\mathsf{SymSeq}^{\mathfrak{C}}(\mathsf{M}) = \mathsf{M}^{\Sigma_{\mathfrak{C}}^{\mathrm{op}} \times \mathfrak{C}}$$

are called $\mathfrak{C}$-*colored symmetric sequences* in M. For an object X in $\mathsf{M}^{\mathsf{Prof}(\mathfrak{C}) \times \mathfrak{C}}$ or $\mathsf{SymSeq}^{\mathfrak{C}}(\mathsf{M})$, we write

$$X(\underline{c}; d) = X(\tfrac{d}{\underline{c}}) \in \mathsf{M}$$

for the value of X at $(\underline{c}; d) \in \mathsf{Prof}(\mathfrak{C}) \times \mathfrak{C}$ and call it an *m-ary entry* of X if $|\underline{c}| = m$. We call $\underline{c}$ the *input profile*, c_i the *ith input color*, and d the *output color*.

(4) An object in the product category $\prod_{\mathfrak{C}} \mathsf{M} = \mathsf{M}^{\mathfrak{C}}$ is called a $\mathfrak{C}$-*colored object in* M, and similarly for a morphism of $\mathfrak{C}$-colored objects. A $\mathfrak{C}$-colored object X is also written as $\{X_c\}$ with $X_c \in \mathsf{M}$ for each color $c \in \mathfrak{C}$.

(5) A $\mathfrak{C}$-colored object $\{X_c\}_{c \in \mathfrak{C}}$ is also regarded as a $\mathfrak{C}$-colored symmetric sequence concentrated in 0-ary entries:

$$X(\tfrac{d}{\underline{c}}) = \begin{cases} X_d & \text{if } \underline{c} = \varnothing, \\ \varnothing & \text{if } \underline{c} \neq \varnothing. \end{cases} \tag{4.1.2}$$

Definition 4.1.3. Suppose $X, Y \in \mathsf{SymSeq}^{\mathfrak{C}}(\mathsf{M})$.

(1) For each $\underline{c} = (c_1, \ldots, c_m) \in \mathsf{Prof}(\mathfrak{C})$, define the object $Y^{\underline{c}} \in \mathsf{M}^{\Sigma_{\mathfrak{C}}^{\mathrm{op}}}$ entrywise as the coend

$$Y^{\underline{c}}(\underline{b}) = \int^{\{\underline{a}_j\} \in \prod_{j=1}^m \Sigma_{\mathfrak{C}}^{\mathrm{op}}} \Sigma_{\mathfrak{C}}^{\mathrm{op}}(\underline{a}_1, \ldots, \underline{a}_m; \underline{b}) \cdot \left[\bigotimes_{j=1}^m Y(\tfrac{c_j}{\underline{a}_j})\right] \in \mathsf{M} \tag{4.1.4}$$

for $\underline{b} \in \mathsf{Prof}(\mathfrak{C})$, in which $(\underline{a}_1, \ldots, \underline{a}_m)$ is the concatenation. Note that $Y^{\underline{c}}$ is natural in $\underline{c} \in \mathsf{Prof}(\mathfrak{C})$ via left permutations of the tensor factors in $\bigotimes_{j=1}^m Y(\tfrac{c_j}{\underline{a}_j})$.

(2) The $\mathfrak{C}$-*colored circle product*

$$X \circ Y \in \mathsf{SymSeq}^{\mathfrak{C}}(\mathsf{M})$$

is defined entrywise as the coend

$$(X \circ Y)(\tfrac{d}{\underline{b}}) = \int^{\underline{c} \in \Sigma_{\mathfrak{C}}} X(\tfrac{d}{\underline{c}}) \otimes Y^{\underline{c}}(\underline{b}) \tag{4.1.5}$$

for $(\underline{b}; d) \in \mathsf{Prof}(\mathfrak{C}) \times \mathfrak{C}$.

(3) Define the object $\mathsf{I} \in \mathsf{SymSeq}^{\mathfrak{C}}(\mathsf{M})$ by

$$\mathsf{I}\binom{d}{\underline{c}} = \begin{cases} \mathbb{1} & \text{if } \underline{c} = d, \\ \varnothing & \text{otherwise} \end{cases} \tag{4.1.6}$$

for $(\underline{c}; d) \in \mathsf{Prof}(\mathfrak{C}) \times \mathfrak{C}$.

The one-colored case of the following result is in [Kelly (2005)]. The general colored case is proved in [White and Yau (2018)], in which the colored circle product was written in terms of a left Kan extension. When the colored circle product is written as a coend as in (4.1.5), the proof below can be found in [Benini *et. al.* (2017)].

Proposition 4.1.7. $\big(\mathsf{SymSeq}^{\mathfrak{C}}(\mathsf{M}), \circ, \mathsf{I}\big)$ *is a monoidal category.*

Proof. Suppose $X, Y, Z \in \mathsf{SymSeq}^{\mathfrak{C}}(\mathsf{M})$. We will exhibit the associativity isomorphism. First note that, for $\underline{c} = (c_1, \ldots, c_m), \underline{b} \in \Sigma_{\mathfrak{C}}^{\mathsf{op}}$, there exist canonical isomorphisms:

$$
\begin{aligned}
(Y \circ Z)^{\underline{c}}(\underline{b}) &= \int^{\underline{a}_1, \ldots, \underline{a}_m} \Sigma_{\mathfrak{C}}^{\mathsf{op}}\big(\underline{a}_1, \ldots, \underline{a}_m; \underline{b}\big) \cdot \left[\bigotimes_{j=1}^{m} (Y \circ Z)\binom{c_j}{\underline{a}_j}\right] \\
&\cong \int^{\underline{a}_1, \ldots, \underline{a}_m} \int^{\underline{d}_1, \ldots, \underline{d}_m} \Sigma_{\mathfrak{C}}^{\mathsf{op}}(\underline{a}; \underline{b}) \cdot \bigotimes_{j=1}^{m} \left[Y\binom{c_j}{\underline{d}_j} \otimes Z^{\underline{d}_j}(\underline{a}_j)\right] \\
&\cong \int^{\underline{d}_1, \ldots, \underline{d}_m} \left[\bigotimes_{j=1}^{m} Y\binom{c_j}{\underline{d}_j}\right] \otimes Z^{\underline{d}}(\underline{b}) \\
&\cong \int^{\underline{d}_1, \ldots, \underline{d}_m} \int^{\underline{e}} \Sigma_{\mathfrak{C}}^{\mathsf{op}}(\underline{d}; \underline{e}) \cdot \left[\bigotimes_{j=1}^{m} Y\binom{c_j}{\underline{d}_j}\right] \otimes Z^{\underline{e}}(\underline{b}) \\
&\cong \int^{\underline{e}} Y^{\underline{c}}(\underline{e}) \otimes Z^{\underline{e}}(\underline{b}).
\end{aligned}
\tag{4.1.8}
$$

In the above calculation, we wrote $\underline{a} = (\underline{a}_1, \ldots, \underline{a}_m)$ and $\underline{d} = (\underline{d}_1, \ldots, \underline{d}_m)$ for the concatenations. Now we have the equalities and canonical isomorphism

$$
\begin{aligned}
\big((X \circ Y) \circ Z\big)\binom{d}{\underline{a}} &= \int^{\underline{c}} (X \circ Y)\binom{d}{\underline{c}} \otimes Z^{\underline{c}}(\underline{a}) \\
&= \int^{\underline{c}} \int^{\underline{b}} X\binom{d}{\underline{b}} \otimes Y^{\underline{b}}(\underline{c}) \otimes Z^{\underline{c}}(\underline{a}) \\
&\cong \int^{\underline{b}} X\binom{d}{\underline{b}} \otimes (Y \circ Z)^{\underline{b}}(\underline{a}) \\
&= \big(X \circ (Y \circ Z)\big)\binom{d}{\underline{a}}
\end{aligned}
$$

for $(\underline{a}; d) \in \mathsf{Prof}(\mathfrak{C}) \times \mathfrak{C}$, in which the isomorphism uses (4.1.8). The rest of the axioms of a monoidal category are straightforward to check. $\qquad\square$

Recall from Definition 2.6.1 that each monoidal category has a category of monoids.

Definition 4.1.9. Define the category
$$\mathsf{Operad}^{\mathfrak{C}}(\mathsf{M}) = \mathsf{Mon}\big(\mathsf{SymSeq}^{\mathfrak{C}}(\mathsf{M})\big)$$
of $\mathfrak{C}$-*colored operads in* M as the category of monoids in $\big(\mathsf{SymSeq}^{\mathfrak{C}}(\mathsf{M}), \circ, \mathsf{I}\big)$. If $\mathfrak{C}$ has $n < \infty$ elements, we also refer to objects in $\mathsf{Operad}^{\mathfrak{C}}(\mathsf{M})$ as n-*colored operads*.

Using Proposition 2.6.2 we may express a $\mathfrak{C}$-colored operad as follows.

Corollary 4.1.10. *A* $\mathfrak{C}$-*colored operad in* M *is exactly a triple* $(\mathsf{O}, \mu, \varepsilon)$ *consisting of*

- *an object* O,
- *a multiplication morphism* $\mu : \mathsf{O} \circ \mathsf{O} \longrightarrow \mathsf{O}$, *and*
- *a unit* $\varepsilon : \mathsf{I} \longrightarrow \mathsf{O}$,

all in $\mathsf{SymSeq}^{\mathfrak{C}}(\mathsf{M})$, *such that the associativity and unity diagrams*

$$
\begin{array}{ccc}
\mathsf{O} \circ \mathsf{O} \circ \mathsf{O} & \xrightarrow{(\mathrm{Id}_{\mathsf{O}}, \mu)} & \mathsf{O} \circ \mathsf{O} \\
{\scriptstyle(\mu, \mathrm{Id}_{\mathsf{O}})}\Big\downarrow & & \Big\downarrow{\scriptstyle\mu} \\
\mathsf{O} \circ \mathsf{O} & \xrightarrow{\hspace{0.8cm}\mu\hspace{0.8cm}} & \mathsf{O}
\end{array}
\qquad
\begin{array}{ccc}
\mathsf{I} \circ \mathsf{O} & \xrightarrow{(\varepsilon, \mathrm{Id}_{\mathsf{O}})} & \mathsf{O} \circ \mathsf{O} & \xleftarrow{(\mathrm{Id}_{\mathsf{O}}, \varepsilon)} & \mathsf{O} \circ \mathsf{I} \\
& {\scriptstyle\cong}\searrow & \Big\downarrow{\scriptstyle\mu} & \swarrow{\scriptstyle\cong} & \\
& & \mathsf{O} & &
\end{array}
\qquad (4.1.11)
$$

are commutative. A morphism of $\mathfrak{C}$-*colored operads is a morphism of the underlying* $\mathfrak{C}$-*colored symmetric sequences that is compatible with the multiplications and the units.*

Notation 4.1.12. If O is a 1-colored operad with color set $\{*\}$, then we write
$$\mathsf{O}(n) = \mathsf{O}(\,{}^{\ \ *}_{*, \ldots, *}\,)$$
for $\big(*, \ldots, *; *\big) \in \Sigma^{\mathrm{op}}_{\{*\}} \times \{*\}$ in which the input profile has length n.

4.2 Operads in Terms of Generating Operations

In Definition 4.1.9 above we defined a colored operad as a monoid with respect to the colored circle product, which is defined in terms of coends (Definition 4.1.3). We can unpack the colored circle product to express a colored operad in terms of a few generating operations from [Yau (2016)] (Section 11.2) and [Yau and Johnson (2015)] (Definition 11.14). In the one-colored topological case, the definition below is due to May [May (1972)].

Definition 4.2.1. A $\mathfrak{C}$-*colored operad* in $(\mathsf{M}, \otimes, \mathbb{1})$ is a triple $(\mathsf{O}, \gamma, \mathbb{1})$ consisting of the following data.

- $\mathsf{O} \in \mathsf{SymSeq}^{\mathfrak{C}}(\mathsf{M})$.
- For $\big(\underline{c} = (c_1, \ldots, c_n); d\big) \in \mathsf{Prof}(\mathfrak{C}) \times \mathfrak{C}$ with $n \geq 1$, $\underline{b}_j \in \mathsf{Prof}(\mathfrak{C})$ for $1 \leq j \leq n$, and $\underline{b} = (\underline{b}_1, \ldots, \underline{b}_n)$ their concatenation, it is equipped with an *operadic composition*

$$\mathsf{O}\big({}^{d}_{\underline{c}}\big) \otimes \bigotimes_{j=1}^{n} \mathsf{O}\big({}^{c_j}_{\underline{b}_j}\big) \xrightarrow{\ \gamma\ } \mathsf{O}\big({}^{d}_{\underline{b}}\big) \in \mathsf{M}. \qquad (4.2.2)$$

- For each $c \in \mathfrak{C}$, it is equipped with a *c-colored unit*

$$\mathbb{1} \xrightarrow{\ \mathbb{1}_c\ } \mathsf{O}\binom{c}{c} \in \mathsf{M}. \tag{4.2.3}$$

This data is required to satisfy the following associativity, unity, and equivariance axioms.

Associativity Suppose that:

- in (4.2.2) $\underline{b}_j = \left(b_1^j, \ldots, b_{k_j}^j \right) \in \mathsf{Prof}(\mathfrak{C})$ with at least one $k_j > 0$;
- $\underline{a}_i^j \in \mathsf{Prof}(\mathfrak{C})$ for each $1 \le j \le n$ and $1 \le i \le k_j$;
- for each $1 \le j \le n$,

$$\underline{a}_j = \begin{cases} \left(\underline{a}_1^j, \ldots, \underline{a}_{k_j}^j \right) & \text{if } k_j > 0, \\ \varnothing & \text{if } k_j = 0 \end{cases}$$

with $\underline{a} = (\underline{a}_1, \ldots, \underline{a}_n)$ their concatenation.

Then the *associativity diagram*

$$\mathsf{O}\binom{d}{\underline{c}} \otimes \left[\bigotimes_{j=1}^n \mathsf{O}\binom{c_j}{\underline{b}_j} \right] \otimes \bigotimes_{j=1}^n \bigotimes_{i=1}^{k_j} \mathsf{O}\binom{b_i^j}{\underline{a}_i^j} \xrightarrow{(\gamma,\mathrm{Id})} \mathsf{O}\binom{d}{\underline{b}} \otimes \bigotimes_{j=1}^n \bigotimes_{i=1}^{k_j} \mathsf{O}\binom{b_i^j}{\underline{a}_i^j} \tag{4.2.4}$$

(with vertical arrow labeled "permute" $\cong$ on the left)

$$\mathsf{O}\binom{d}{\underline{c}} \otimes \bigotimes_{j=1}^n \left[\mathsf{O}\binom{c_j}{\underline{b}_j} \otimes \bigotimes_{i=1}^{k_j} \mathsf{O}\binom{b_i^j}{\underline{a}_i^j} \right]$$

(vertical arrow labeled $(\mathrm{Id}, \otimes_j \gamma)$ on the left; γ on the right)

$$\mathsf{O}\binom{d}{\underline{c}} \otimes \bigotimes_{j=1}^n \mathsf{O}\binom{c_j}{\underline{a}_j} \xrightarrow{\quad \gamma \quad} \mathsf{O}\binom{d}{\underline{a}}$$

in M is commutative.

Unity Suppose $d \in \mathfrak{C}$.

(1) For each $\underline{c} = (c_1, \ldots, c_n) \in \mathsf{Prof}(\mathfrak{C})$ with $n \ge 1$, the *right unity diagram*

$$\mathsf{O}\binom{d}{\underline{c}} \otimes \mathbb{1}^{\otimes n} \xrightarrow{\ \cong\ } \mathsf{O}\binom{d}{\underline{c}} \tag{4.2.5}$$

(vertical arrow labeled $(\mathrm{Id}, \otimes 1_{c_j})$ on the left; $=$ on the right)

$$\mathsf{O}\binom{d}{\underline{c}} \otimes \bigotimes_{j=1}^n \mathsf{O}\binom{c_j}{c_j} \xrightarrow{\ \gamma\ } \mathsf{O}\binom{d}{\underline{c}}$$

in M is commutative.

(2) For each $\underline{b} \in \mathsf{Prof}(\mathfrak{C})$ the *left unity diagram*

$$\mathbb{1} \otimes \mathsf{O}\binom{d}{\underline{b}} \xrightarrow{\ \cong\ } \mathsf{O}\binom{d}{\underline{b}} \tag{4.2.6}$$

(vertical arrow labeled $(1_d, \mathrm{Id})$ on the left; $=$ on the right)

$$\mathsf{O}\binom{d}{d} \otimes \mathsf{O}\binom{d}{\underline{b}} \xrightarrow{\ \gamma\ } \mathsf{O}\binom{d}{\underline{b}}$$

in M is commutative.

Equivariance Suppose that in (4.2.2) $|\underline{b}_j| = k_j \geq 0$.

(1) For each permutation $\sigma \in \Sigma_n$, the *top equivariance diagram*

$$O\binom{d}{\underline{c}} \otimes \bigotimes_{j=1}^{n} O\binom{c_j}{\underline{b}_j} \xrightarrow{(\sigma,\sigma^{-1})} O\binom{d}{\underline{c}\sigma} \otimes \bigotimes_{j=1}^{n} O\binom{c_{\sigma(j)}}{\underline{b}_{\sigma(j)}} \qquad (4.2.7)$$

$$\gamma \downarrow \qquad\qquad\qquad\qquad \downarrow \gamma$$

$$O\binom{d}{\underline{b}_1,\ldots,\underline{b}_n} \xrightarrow{\sigma\langle k_1,\ldots,k_n\rangle} O\binom{d}{\underline{b}_{\sigma(1)},\ldots,\underline{b}_{\sigma(n)}}$$

in M is commutative. The bottom horizontal morphism is the equivariant structure morphism of O corresponding to the block permutation in $\Sigma_{k_1+\cdots+k_n}$ induced by σ that permutes n consecutive blocks of lengths $k_1,\ldots,k_n$. In the top horizontal morphism, σ is the equivariant structure morphism of O corresponding to σ, and σ^{-1} is the left permutation of the n tensor factors.

(2) Given permutations $\tau_j \in \Sigma_{k_j}$ for $1 \leq j \leq n$, the *bottom equivariance diagram*

$$O\binom{d}{\underline{c}} \otimes \bigotimes_{j=1}^{n} O\binom{c_j}{\underline{b}_j} \xrightarrow{(\mathrm{Id},\otimes\tau_j)} O\binom{d}{\underline{c}} \otimes \bigotimes_{j=1}^{n} O\binom{c_j}{\underline{b}_j\tau_j} \qquad (4.2.8)$$

$$\gamma \downarrow \qquad\qquad\qquad\qquad \downarrow \gamma$$

$$O\binom{d}{\underline{b}_1,\ldots,\underline{b}_n} \xrightarrow{\tau_1\oplus\cdots\oplus\tau_n} O\binom{d}{\underline{b}_1\tau_1,\ldots,\underline{b}_n\tau_n}$$

in M is commutative. In the top horizontal morphism, each τ_j is the equivariant structure morphism of O corresponding to $\tau_j \in \Sigma_{k_j}$. The bottom horizontal morphism is the equivariant structure morphism of O corresponding to the block sum $\tau_1 \oplus \cdots \oplus \tau_n \in \Sigma_{k_1+\cdots+k_n}$ induced by the τ_j's.

A morphism of $\mathfrak{C}$-colored operads is a morphism of the underlying $\mathfrak{C}$-colored symmetric sequences that is compatible with the operadic compositions and the colored units in the obvious sense.

Proposition 4.2.9. *The definition of a $\mathfrak{C}$-colored operad in Definition 4.1.9 and in Definition 4.2.1 are equivalent.*

Proof. In Corollary 4.1.10 the domain of the multiplication μ is $O \circ O$, which has entries

$$(O \circ O)\binom{d}{\underline{b}} = \int^{\underline{c}} O\binom{d}{\underline{c}} \otimes O^{\underline{c}}(\underline{b})$$

$$= \int^{\underline{c},\underline{a}_1,\ldots,\underline{a}_m} \Sigma_{\mathfrak{C}}^{\mathrm{op}}(\underline{a};\underline{b}) \cdot O\binom{d}{\underline{c}} \otimes \bigotimes_{j=1}^{m} O\binom{c_j}{\underline{a}_j} \qquad (4.2.10)$$

for $(\underline{b};d) \in \mathsf{Prof}(\mathfrak{C}) \times \mathfrak{C}$, where $\underline{a} = (\underline{a}_1,\ldots,\underline{a}_m)$ is the concatenation. Observe that $\Sigma_{\mathfrak{C}}^{\mathrm{op}}(\underline{a};\underline{b})$ is empty unless $\underline{b} = \underline{a}\sigma$ for some permutation σ, i.e., the concatenation of the $\underline{a}_j$'s is $\underline{b}$ up to a permutation. So the multiplication $\mu : O \circ O \longrightarrow O$ yields the entrywise operadic composition γ (4.2.2). The associativity diagram in (4.1.11) corresponds to the associativity diagram (4.2.4). The top equivariance diagram (4.2.7) corresponds to the $\underline{c}$-variable in the coend (4.2.10) and the fact that μ is a morphism of $\mathfrak{C}$-colored symmetric sequences. Similarly, the bottom equivariance diagram (4.2.8) corresponds to the $\underline{a}_j$ variables in the coend (4.2.10) and the fact that μ is a morphism of $\mathfrak{C}$-colored symmetric sequences.

For each $c \in \mathfrak{C}$, the c-colored unit 1_c in (4.2.3) corresponds to the $(c; c)$-entry of the unit morphism $\varepsilon : I \longrightarrow O$ in Corollary 4.1.10. The unity diagram in (4.1.11) corresponds to the right unity diagram (4.2.5) and the left unity diagram (4.2.6). $\square$

Remark 4.2.11. The reader is cautioned that the definition of a 1-colored operad in [Markl *et. al.* (2002)] (p.41, Definition 1.4) is missing the bottom equivariance axiom (4.2.8). ◇

4.3 Operads in Terms of Partial Compositions

Instead of the operadic composition γ, it is also possible to express a colored operad in terms of binary operations.

Definition 4.3.1. Suppose $\big(\underline{c} = (c_1,\ldots,c_n); d\big) \in \mathsf{Prof}(\mathfrak{C}) \times \mathfrak{C}$ with $n \geq 1$, $\underline{b} \in \mathsf{Prof}(\mathfrak{C})$, and $1 \leq i \leq n$. Define the $\mathfrak{C}$-profile

$$\underline{c} \circ_i \underline{b} = \big(\underbrace{c_1,\ldots,c_{i-1}}_{\varnothing \text{ if } i=1}, \underline{b}, \underbrace{c_{i+1},\ldots,c_n}_{\varnothing \text{ if } i=n} \big).$$

The following is [Yau (2016)] Definition 16.2.1.

Definition 4.3.2. A $\mathfrak{C}$-*colored operad* in $(\mathsf{M}, \otimes, \mathbb{1})$ is a triple $(\mathsf{O}, \circ, \mathbb{1})$ consisting of the following data.

- $\mathsf{O} \in \mathsf{SymSeq}^{\mathfrak{C}}(\mathsf{M})$.
- For $\big(\underline{c} = (c_1,\ldots,c_n); d\big) \in \mathsf{Prof}(\mathfrak{C}) \times \mathfrak{C}$ with $n \geq 1$, $1 \leq i \leq n$, and $\underline{b} \in \mathsf{Prof}(\mathfrak{C})$, it is equipped with a morphism

$$\mathsf{O}\binom{d}{\underline{c}} \otimes \mathsf{O}\binom{c_i}{\underline{b}} \xrightarrow{\;\circ_i\;} \mathsf{O}\binom{d}{\underline{c}\circ_i\underline{b}} \in \mathsf{M} \qquad (4.3.3)$$

  called the $\circ_i$-*composition*.
- For each color $c \in \mathfrak{C}$, it is equipped with a c-colored unit

$$\mathbb{1} \xrightarrow{\;1_c\;} \mathsf{O}\binom{c}{c} \in \mathsf{M}.$$

This data is required to satisfy the following associativity, unity, and equivariance axioms. Suppose $d \in \mathfrak{C}$, $\underline{c} = (c_1,\ldots,c_n) \in \mathsf{Prof}(\mathfrak{C})$, $\underline{b} \in \mathsf{Prof}(\mathfrak{C})$ with length $|\underline{b}| = m$, and $\underline{a} \in \mathsf{Prof}(\mathfrak{C})$ with length $|\underline{a}| = l$.

Associativity There are two associativity axioms.

(1) Suppose $n \geq 2$ and $1 \leq i < j \leq n$. Then the *horizontal associativity diagram* in M

$$\mathsf{O}\binom{d}{\underline{c}} \otimes \mathsf{O}\binom{c_i}{\underline{a}} \otimes \mathsf{O}\binom{c_j}{\underline{b}} \xrightarrow{\ (\circ_i, \mathrm{Id})\ } \mathsf{O}\binom{d}{\underline{c}\circ_i \underline{a}} \otimes \mathsf{O}\binom{c_j}{\underline{b}} \tag{4.3.4}$$

$$\Bigg\downarrow \text{permute} \cong \qquad\qquad\qquad\qquad \Bigg\downarrow \circ_{j-1+l}$$

$$\mathsf{O}\binom{d}{\underline{c}} \otimes \mathsf{O}\binom{c_j}{\underline{b}} \otimes \mathsf{O}\binom{c_i}{\underline{a}}$$

$$\Bigg\downarrow (\circ_j, \mathrm{Id})$$

$$\mathsf{O}\binom{d}{\underline{c}\circ_j\underline{b}} \otimes \mathsf{O}\binom{c_i}{\underline{a}} \xrightarrow{\ \circ_i\ } \mathsf{O}\binom{d}{(\underline{c}\circ_j\underline{b})\circ_i\underline{a}} = \mathsf{O}\binom{d}{(\underline{c}\circ_i\underline{a})\circ_{j-1+l}\underline{b}}$$

is commutative.

(2) Suppose $n, m \geq 1$, $1 \leq i \leq n$, and $1 \leq j \leq m$. Then the *vertical associativity diagram* in M

$$\mathsf{O}\binom{d}{\underline{c}} \otimes \mathsf{O}\binom{c_i}{\underline{b}} \otimes \mathsf{O}\binom{b_j}{\underline{a}} \xrightarrow{\ (\mathrm{Id}, \circ_j)\ } \mathsf{O}\binom{d}{\underline{c}} \otimes \mathsf{O}\binom{c_i}{\underline{b}\circ_j\underline{a}} \tag{4.3.5}$$

$$\Bigg\downarrow (\circ_i, \mathrm{Id}) \qquad\qquad\qquad\qquad\qquad \Bigg\downarrow \circ_i$$

$$\mathsf{O}\binom{d}{\underline{c}\circ_i\underline{b}} \otimes \mathsf{O}\binom{b_j}{\underline{a}} \xrightarrow{\ \circ_{i-1+j}\ } \mathsf{O}\binom{d}{(\underline{c}\circ_i\underline{b})\circ_{i-1+j}\underline{a}} = \mathsf{O}\binom{d}{\underline{c}\circ_i(\underline{b}\circ_j\underline{a})}$$

is commutative.

Unity There are two unity axioms.

(1) The *left unity diagram* in M

$$\mathbb{1} \otimes \mathsf{O}\binom{d}{\underline{c}} \xrightarrow{\ (\mathbb{1}_d, \mathrm{Id})\ } \mathsf{O}\binom{d}{d} \otimes \mathsf{O}\binom{d}{\underline{c}} \tag{4.3.6}$$

$$\searrow{\scriptstyle \cong} \qquad\qquad \Bigg\downarrow \circ_1$$

$$\mathsf{O}\binom{d}{\underline{c}}$$

is commutative.

(2) If $n \geq 1$ and $1 \leq i \leq n$, then the *right unity diagram* in M

$$\mathsf{O}\binom{d}{\underline{c}} \otimes \mathbb{1} \xrightarrow{\ (\mathrm{Id}, \mathbb{1}_{c_i})\ } \mathsf{O}\binom{d}{\underline{c}} \otimes \mathsf{O}\binom{c_i}{c_i} \tag{4.3.7}$$

$$\searrow{\scriptstyle \cong} \qquad\qquad \Bigg\downarrow \circ_i$$

$$\mathsf{O}\binom{d}{\underline{c}}$$

is commutative.

Equivariance Suppose $|\underline{c}| = n \geq 1$, $1 \leq i \leq n$, $\sigma \in \Sigma_n$, and $\tau \in \Sigma_m$. Then the *equivariance diagram* in M

$$\mathsf{O}\binom{d}{\underline{c}} \otimes \mathsf{O}\binom{c_{\sigma(i)}}{\underline{b}} \xrightarrow{\ \circ_{\sigma(i)}\ } \mathsf{O}\binom{d}{\underline{c}\circ_{\sigma(i)}\underline{b}} \tag{4.3.8}$$

$$\Bigg\downarrow (\sigma, \tau) \qquad\qquad\qquad\qquad \Bigg\downarrow \sigma\circ_i\tau$$

$$\mathsf{O}\binom{d}{\underline{c}\sigma} \otimes \mathsf{O}\binom{c_{\sigma(i)}}{\underline{b}\tau} \xrightarrow{\ \circ_i\ } \mathsf{O}\binom{d}{(\underline{c}\sigma)\circ_i(\underline{b}\tau)} = \mathsf{O}\binom{d}{(\underline{c}\circ_{\sigma(i)}\underline{b})(\sigma\circ_i\tau)}$$

is commutative, where

$$\sigma \circ_i \tau = \sigma \langle \underbrace{1, \dots, 1}_{i-1}, m, \underbrace{1, \dots, 1}_{n-i} \rangle \circ \left(\underbrace{id \oplus \cdots \oplus id}_{i-1} \oplus \tau \oplus \underbrace{id \oplus \cdots \oplus id}_{n-i} \right) \in \Sigma_{n+m-1}$$

is the composition of a block sum induced by τ with a block permutation induced by σ that permutes consecutive blocks of the indicated lengths.

A morphism of $\mathfrak{C}$-colored operads is a morphism of the underlying $\mathfrak{C}$-colored symmetric sequences that is compatible with the $\circ_i$-compositions and the colored units in the obvious sense.

Proposition 4.3.9. *The definition of a $\mathfrak{C}$-colored operad in Definition 4.2.1 and in Definition 4.3.2 are equivalent.*

Proof. The proof can be found in [Yau (2016)] Section 16.4. Let us indicate the correspondence of structures. Given a $\mathfrak{C}$-colored operad $(O, \gamma, 1)$ in the sense of Definition 4.2.1, the associated $\circ_i$-composition is the composition

$$O\binom{d}{\underline{c}} \otimes O\binom{c_i}{\underline{b}} \xrightarrow{\hspace{2.5cm} \circ_i \hspace{2.5cm}} O\binom{d}{\underline{c} \circ_i \underline{b}}$$

$$\downarrow \cong \hspace{7cm} \uparrow \gamma$$

$$O\binom{d}{\underline{c}} \otimes 1^{\otimes i-1} \otimes O\binom{c_i}{\underline{b}} \otimes 1^{\otimes n-i} \xrightarrow{\{1_{c_j}\}} O\binom{d}{\underline{c}} \otimes \left[\overset{i-1}{\underset{j=1}{\bigotimes}} O\binom{c_j}{c_j} \right] \otimes O\binom{c_i}{\underline{b}} \otimes \left[\overset{n}{\underset{j=i+1}{\bigotimes}} O\binom{c_j}{c_j} \right]$$

$$(4.3.10)$$

in which the bottom horizontal morphism is the monoidal product of the colored units 1_{c_j} for $1 \le j \ne i \le n$ with the identity morphisms of $O\binom{d}{\underline{c}}$ and $O\binom{c_i}{\underline{b}}$.

Conversely, given a $\mathfrak{C}$-colored operad $(O, \circ, 1)$ in the sense of Definition 4.3.2, the operadic composition γ is recovered as the composition

$$O\binom{d}{\underline{c}} \otimes \overset{n}{\underset{j=1}{\bigotimes}} O\binom{c_j}{\underline{b}_j} \xrightarrow{\hspace{2cm} \gamma \hspace{2cm}} O\binom{d}{\underline{b}} \hspace{2cm} (4.3.11)$$

$$(\circ_1, \mathrm{Id}) \downarrow \hspace{7cm} \uparrow \mathrm{Id}$$

$$O\binom{d}{\underline{c} \circ_1 \underline{b}_1} \otimes \overset{n}{\underset{j=2}{\bigotimes}} O\binom{c_j}{\underline{b}_j} \hspace{2cm} O\binom{d}{((\underline{c} \circ_1 \underline{b}_1) \cdots) \circ_{k_1 + \cdots + k_{n-1} + 1} \underline{b}_n}$$

$$(\circ_{k_1+1}, \mathrm{Id}) \downarrow \hspace{6cm} \uparrow \circ_{k_1 + \cdots + k_{n-1} + 1}$$

$$\cdots \xrightarrow{(\circ_{k_1 + \cdots + k_{n-2} + 1}, \mathrm{Id})} O\binom{d}{((\underline{c} \circ_1 \underline{b}_1) \cdots) \circ_{k_1 + \cdots + k_{n-2} + 1} \underline{b}_{n-1}} \otimes O\binom{c_n}{\underline{b}_n}$$

in which $k_j = |\underline{b}_j|$. $\qquad\qquad\square$

4.4 Operads in Terms of Trees

In Proposition 4.2.9 and Proposition 4.3.9 we observed that there are three equivalent definitions of a colored operad. In this section, we discuss another equivalent description of a colored operad in terms of trees that we will need later to discuss the Boardman-Vogt construction of a colored operad. We will need the following concept about monoidal product [Markl *et. al.* (2002)] (p.64, Definition 1.58).

Definition 4.4.1 (Unordered Monoidal Product). Suppose X is a set with $n \geq 1$ elements.

(1) An *ordering* of X is a bijection

$$\sigma : \{1, \dots, n\} \xrightarrow{\cong} X.$$

The set of all orderings of X is denoted by $\mathsf{Ord}(X)$.
(2) Suppose $A_x \in \mathsf{M}$ is an object for each $x \in X$.

(a) For each ordering σ of X, define the *ordered monoidal product* as

$$\bigotimes_\sigma A_x = A_{\sigma(1)} \otimes \cdots \otimes A_{\sigma(n)} \in \mathsf{M}.$$

For each $\tau \in \Sigma_n$, the symmetry isomorphism in M determines an isomorphism

$$\tau : \bigotimes_\sigma A_x \xrightarrow{\cong} \bigotimes_{\sigma\tau} A_x,$$

which defines a Σ_n-action on the coproduct $\coprod_{\sigma \in \mathsf{Ord}(X)} \bigotimes_\sigma A_x$.
(b) Define the *unordered monoidal product* as the colimit

$$\bigotimes_{x \in X} A_x = \mathrm{colim}\left(\coprod_{\sigma \in \mathsf{Ord}(X)} \bigotimes_\sigma A_x \xrightarrow{\;\;\tau\;\;}_{\tau \in \Sigma_n} \coprod_{\sigma \in \mathsf{Ord}(X)} \bigotimes_\sigma A_x \right). \qquad (4.4.2)$$

Remark 4.4.3. For each ordering σ of X, the natural morphism

$$\bigotimes_\sigma A_x \longrightarrow \bigotimes_{x \in X} A_x$$

from the ordered monoidal product to the unordered monoidal product is an isomorphism. The point of the unordered monoidal product is that we can talk about the iterated monoidal product of the A_x's without first choosing an ordering of the indexing set X.   ◇

As before $\mathfrak{C}$ is a fixed non-empty set. All the trees below are $\mathfrak{C}$-colored trees as in Definition 3.1.15. Recall that $\mathsf{Prof}(\mathfrak{C})$ denotes the set of $\mathfrak{C}$-profiles.

Definition 4.4.4 (Vertex Decorations). Suppose $A \in \mathsf{M}^{\mathsf{Prof}(\mathfrak{C}) \times \mathfrak{C}}$, and T is a tree. Define the *A-decoration of T* as the unordered monoidal product

$$A[T] = \bigotimes_{v \in T} A\big(\mathsf{Prof}(v)\big) = \bigotimes_{v \in T} A\binom{\mathrm{out}(v)}{\mathrm{in}(v)},$$

where $v \in T$ means $v \in \mathsf{Vt}(T)$.

Proposition 4.4.5. *Suppose T is a tree, and H_v is a tree with $\mathsf{Prof}(H_v) = \mathsf{Prof}(v)$ for each vertex v in T. Then for each $A \in \mathsf{M}^{\mathsf{Prof}(\mathfrak{C}) \times \mathfrak{C}}$, there is an isomorphism*

$$A\big[T(H_v)_{v \in T}\big] \cong \bigotimes_{v \in T} A[H_v],$$

where $T(H_v)_{v \in T}$ is the tree substitution in Definition 3.2.6. In particular, in the context of Definition 3.3.1, there is an isomorphism

$$A\big[\mathsf{Graft}(G; \{H_j\})\big] \cong A[G] \otimes A[H_1] \otimes \cdots \otimes A[H_m].$$

Proof. The first isomorphism follows from the decomposition

$$\mathsf{Vt}\big(T(H_v)_{v \in T}\big) = \coprod_{v \in T} \mathsf{Vt}(H_v).$$

The second isomorphism follows from the definition of the grafting as a tree substitution. $\qquad\square$

Notation 4.4.6. In the setting of Definition 4.4.4, we will sometimes abbreviate $A\big(\mathsf{Prof}(v)\big) = A\big(_{\mathsf{in}(v)}^{\mathsf{out}(v)}\big)$ to $A(v)$.

Example 4.4.7. In the context of Example 3.2.13, recall that K is the tree substitution $T(H_u, H_v, H_w)$. There are isomorphisms

$$
\begin{aligned}
A[K] &\cong A[H_u] \otimes A[H_v] \otimes A[H_w] \\
&\cong A(u_1) \otimes A(u_2) \otimes \mathbb{1} \otimes A(w_1) \otimes A(w_2) \\
&\cong A\big(_{a,b,f}^{c}\big) \otimes A\big(_{\varnothing}^{f}\big) \otimes A\big(_{c,g}^{e}\big) \otimes A\big(_{d}^{g}\big)
\end{aligned}
$$

for each $A \in \mathsf{M}^{\mathsf{Prof}(\mathfrak{C}) \times \mathfrak{C}}$. $\qquad\diamond$

Example 4.4.8. In the context of Example 3.3.3, recall that G is the grafting $\mathsf{Graft}(T; H_1, H_2, H_3)$. There are isomorphisms

$$
\begin{aligned}
A[G] &\cong A[T] \otimes A[H_1] \otimes A[H_2] \otimes A[H_3] \\
&\cong A(w) \otimes A(u) \otimes A(v) \otimes A(x_1) \otimes A(x_2) \otimes \mathbb{1} \otimes A(z) \\
&\cong A\big(_{c,d}^{e}\big) \otimes A\big(_{a,b}^{c}\big) \otimes A\big(_{d}^{d}\big) \otimes A\big(_{f,g}^{a}\big) \otimes A\big(_{\varnothing}^{g}\big) \otimes A\big(_{\varnothing}^{d}\big)
\end{aligned}
$$

for each $A \in \mathsf{M}^{\mathsf{Prof}(\mathfrak{C}) \times \mathfrak{C}}$. $\qquad\diamond$

In the next definition of a colored operad, notice (i) the use of the product category $\mathsf{M}^{\mathsf{Prof}(\mathfrak{C}) \times \mathfrak{C}}$ instead of the category $\mathsf{SymSeq}^{\mathfrak{C}}(\mathsf{M})$ of symmetric sequences and (ii) the apparent absence of an equivariance axiom.

Definition 4.4.9. A $\mathfrak{C}$-*colored operad* in M is a pair (O, γ) consisting of

- an object $\mathsf{O} \in \mathsf{M}^{\mathsf{Prof}(\mathfrak{C}) \times \mathfrak{C}}$ and
- an *operadic structure morphism*

$$\mathsf{O}[T] \xrightarrow{\ \gamma_T\ } \mathsf{O}\big(\mathsf{Prof}(T)\big) \in \mathsf{M} \qquad (4.4.10)$$

for each $T \in \underline{\mathsf{Tree}}^{\mathfrak{C}}$

that satisfies the following unity and associativity axioms.

Unity $\gamma_{\mathrm{Cor}_{(\underline{c};d)}}$ is the identity morphism of $\mathsf{O}\binom{d}{\underline{c}}$ for each $(\underline{c};d) \in \mathsf{Prof}(\mathfrak{C}) \times \mathfrak{C}$, where $\mathrm{Cor}_{(\underline{c};d)}$ is the $(\underline{c};d)$-corolla in Example 3.1.21.

Associativity For each tree substitution $T(H_v)_{v \in T}$, the diagram

$$\begin{array}{ccccc}
\mathsf{O}\big[T(H_v)_{v \in T}\big] & \xrightarrow{\cong} & \underset{v \in T}{\otimes} \mathsf{O}[H_v] & \xrightarrow{\underset{v}{\otimes} \gamma_{H_v}} & \underset{v \in T}{\otimes} \mathsf{O}(v) = \mathsf{O}[T] \qquad (4.4.11) \\
\;\;\downarrow{\scriptstyle \gamma_{T(H_v)_{v \in T}}} & & & & \;\;\downarrow{\scriptstyle \gamma_T} \\
\mathsf{O}\big(\mathsf{Prof}(T(H_v)_{v \in T})\big) & \xrightarrow{\hspace{2cm} \mathrm{Id} \hspace{2cm}} & & & \mathsf{O}\big(\mathsf{Prof}(T)\big)
\end{array}$$

is commutative.

A morphism $f : (\mathsf{O}, \gamma^{\mathsf{O}}) \longrightarrow (\mathsf{P}, \gamma^{\mathsf{P}})$ of $\mathfrak{C}$-colored operads is a morphism $f : \mathsf{O} \longrightarrow \mathsf{P} \in \mathsf{M}^{\mathsf{Prof}(\mathfrak{C}) \times \mathfrak{C}}$ such that the diagram

$$\begin{array}{ccc}
\mathsf{O}[T] & \xrightarrow{\underset{v}{\otimes} f} & \mathsf{P}[T] \qquad (4.4.12) \\
\;\;\downarrow{\scriptstyle \gamma_T^{\mathsf{O}}} & & \;\;\downarrow{\scriptstyle \gamma_T^{\mathsf{P}}} \\
\mathsf{O}\big(\mathsf{Prof}(T)\big) & \xrightarrow{\quad f \quad} & \mathsf{P}\big(\mathsf{Prof}(T)\big)
\end{array}$$

is commutative for each $T \in \underline{\mathsf{Tree}}^{\mathfrak{C}}$.

Theorem 4.4.13. *The definitions of a $\mathfrak{C}$-colored operad in Definition 4.2.1 and in Definition 4.4.9 are equivalent.*

Proof. This equivalence is [Yau and Johnson (2015)] Corollary 11.16. Let us describe the correspondence of structures. Suppose (O, γ) is a $\mathfrak{C}$-colored operad in the sense of Definition 4.4.9.

(1) For a pair $(\underline{c}; d) \in \mathsf{Prof}(\mathfrak{C}) \times \mathfrak{C}$ and a permutation $\tau \in \Sigma_{|\underline{c}|}$, the operadic structure morphism

$$\mathsf{O}\binom{d}{\underline{c}} = \mathsf{O}\big[\mathrm{Cor}_{(\underline{c};d)}\tau\big] \xrightarrow{\gamma_{\mathrm{Cor}_{(\underline{c};d)}\tau}} \mathsf{O}\big[\mathsf{Prof}(\mathrm{Cor}_{(\underline{c};d)}\tau)\big] = \mathsf{O}\binom{d}{\underline{c}\tau} ,$$

where $\mathrm{Cor}_{(\underline{c};d)}\tau$ is the permuted corolla in Example 3.1.22, corresponds to the $\mathfrak{C}$-colored symmetric sequence structure in Definition 4.2.1.

(2) For $(\underline{c} = (c_1, \ldots, c_n); d) \in \mathsf{Prof}(\mathfrak{C}) \times \mathfrak{C}$ with $n \geq 1$, $\underline{b}_j \in \mathsf{Prof}(\mathfrak{C})$ for $1 \leq j \leq n$, $\underline{b} = (\underline{b}_1, \ldots, \underline{b}_n)$ their concatenation, and $T = T(\{\underline{b}_j\}; \underline{c}; d)$ the 2-level tree in Example 3.1.23, the operadic structure morphism

$$\mathsf{O}\binom{d}{\underline{c}} \otimes \overset{n}{\underset{j=1}{\otimes}} \mathsf{O}\binom{c_j}{\underline{b}_j} \cong \mathsf{O}[T] \xrightarrow{\gamma_T} \mathsf{O}[\mathsf{Prof}(T)] = \mathsf{O}\binom{d}{\underline{b}}$$

corresponds to the operadic composition γ in (4.2.2).

(3) For each color $c \in \mathfrak{C}$, the operadic structure morphism

$$\mathbb{1} = \mathsf{O}[\uparrow_c] \xrightarrow{\gamma_{\uparrow_c}} \mathsf{O}[\mathsf{Prof}(\uparrow_c)] = \mathsf{O}\binom{c}{c},$$

where $\uparrow_c$ is the c-colored exceptional edge in Example 3.1.18, corresponds to the c-colored unit 1_c in (4.2.3).

The associativity, unity, and equivariance axioms in Definition 4.2.1 are now consequences of the associativity and unity of (O, γ).

Conversely, suppose $(\mathsf{O}, \gamma, 1)$ is a $\mathfrak{C}$-colored operad in the sense of Definition 4.2.1. Reusing the previous paragraph, we first define the operadic structure morphisms

$$\mathsf{O}[G] \xrightarrow{\gamma_G} \mathsf{O}[\mathsf{Prof}(G)] \quad \text{for} \quad G \in \left\{ \mathsf{Cor}_{(\underline{c};d)} \tau, \uparrow_c, T\left(\{\underline{b}_j\}; \underline{c}; d \right) \right\}$$

as the equivariant structure, the c-colored units 1_c, and the operadic composition γ. For a general tree T, a key observation is that it can always be written non-uniquely as an iterated tree substitution involving only permuted corollas, exceptional edges, and 2-level trees. We then use any such tree substitution decomposition of T and the associativity diagram (4.4.11) to define the operadic structure morphism

$$\mathsf{O}[T] \xrightarrow{\gamma_T} \mathsf{O}[\mathsf{Prof}(T)]$$

as an iterated composition of monoidal products of the already defined operadic structure morphisms γ_G. That such a morphism γ_T is well-defined is a consequence of the axioms in Definition 4.2.1. □

Remark 4.4.14. There are two more equivalent descriptions of an operad that we will not need in this book, so we only briefly mention them here.

(1) There is a $\mathsf{Prof}(\mathfrak{C}) \times \mathfrak{C}$-colored operad $\mathsf{Op}^{\mathfrak{C}}$ whose category of algebras is precisely the category of $\mathfrak{C}$-colored operads in M. This is a special case of [Yau and Johnson (2015)] Lemma 14.4. Each entry $\mathsf{Op}^{\mathfrak{C}}\binom{t}{\underline{s}}$, with t and each s_j in $\mathsf{Prof}(\mathfrak{C}) \times \mathfrak{C}$, is a coproduct $\coprod \mathbb{1}$ indexed by pairs (T, σ) with

 - T a $\mathfrak{C}$-colored tree with profile t and
 - σ an ordering of the set $\mathsf{Vt}(T)$ such that $s_j = \mathsf{Prof}(\sigma(j))$.

 Its equivariant structure comes from reordering the set $\mathsf{Vt}(T)$, and its colored units correspond to corollas. Its operadic composition γ is induced by tree substitution with the induced lexicographical ordering on vertices. One checks that $\mathsf{Op}^{\mathfrak{C}}$-algebras are equivalent to $\mathfrak{C}$-colored operads in Definition 4.4.9.

(2) The category of $\mathfrak{C}$-colored operads in M is also canonically isomorphic to the category of M-enriched multicategorical functors from $\mathsf{Op}^{\mathfrak{C}}$ to M. This is a special case of Theorem 14.12 in [Yau and Johnson (2015)], where the reader is referred for the meaning of an enriched multicategorical functor. One checks that such enriched multicategorical functors are also equivalent to $\mathfrak{C}$-colored operads in Definition 4.4.9. ◇

Corollary 4.4.15. *Suppose* O *is a* $\mathfrak{C}$*-colored operad in* M*, and* $(\underline{c}; d) \in \mathsf{Prof}(\mathfrak{C}) \times \mathfrak{C}$*. Then* O *defines a functor*

$$\mathsf{O} : \underline{\mathsf{Tree}}^{\mathfrak{C}}\tbinom{d}{\underline{c}} \longrightarrow \mathsf{M}$$

as follows:

- *Each* $T \in \underline{\mathsf{Tree}}^{\mathfrak{C}}\tbinom{d}{\underline{c}}$ *is sent to* $\mathsf{O}[T]$*.*
- *Each morphism*

$$(H_v)_{v \in T} : T(H_v) \longrightarrow T \in \underline{\mathsf{Tree}}^{\mathfrak{C}}\tbinom{d}{\underline{c}}$$

  is sent to the morphism

$$\mathsf{O}\big[T(H_v)\big] = \bigotimes_{v \in T} \mathsf{O}[H_v] \xrightarrow{\;\overset{\otimes \gamma_{H_v}}{v}\;} \bigotimes_{v \in T} \mathsf{O}(v) = \mathsf{O}[T]$$

  in which γ_{H_v} *is the operadic structure morphism* (4.4.10) *for* H_v*.*

Proof. An identity morphism in $\underline{\mathsf{Tree}}^{\mathfrak{C}}\tbinom{d}{\underline{c}}$ is of the form

$$(\mathrm{Cor}_v)_{v \in T} : T \longrightarrow T.$$

Since γ_{Cor} is the identity morphism for each corolla, the assignment O preserves identity morphisms.

Suppose

$$(H_v)_{v \in T} : K \longrightarrow T \quad \text{and} \quad (G_u)_{u \in K} : E \longrightarrow K$$

are morphisms in $\underline{\mathsf{Tree}}^{\mathfrak{C}}\tbinom{d}{\underline{c}}$ as in Remark 3.2.12. Their composition is

$$\big(H_v(G_u)_{u \in H_v}\big)_{v \in T} : E \longrightarrow T.$$

To see that the assignment O preserves compositions, observe that

$$\begin{aligned}
\mathsf{O}\big(H_v(G_u)_{u \in H_v}\big)_{v \in T} &= \bigotimes_{v \in T} \gamma_{H_v(G_u)_{u \in H_v}} \\
&= \bigotimes_{v \in T}\Big(\gamma_{H_v} \circ \bigotimes_{u \in H_v} \gamma_{G_u}\Big) \\
&= \Big(\bigotimes_{v \in T} \gamma_{H_v}\Big) \circ \Big(\bigotimes_{v \in T} \bigotimes_{u \in H_v} \gamma_{G_u}\Big) \\
&= \mathsf{O}(H_v)_{v \in T} \circ \mathsf{O}(G_u)_{u \in K}.
\end{aligned}$$

The first and the last equalities are the definitions of the assignment O on a morphism. The second equality holds by the associativity axiom (4.4.11) of the $\mathfrak{C}$-colored operad O. $\square$

4.5 Algebras over Operads

In this section, we discuss algebras over colored operads and some relevant examples. Just as monads are important because of their algebras, operads are important mainly because of their algebras.

Notation 4.5.1. For a $\mathfrak{C}$-colored object $X = \{X_c\}_{c \in \mathfrak{C}}$ in M and $\underline{c} = (c_1, \ldots, c_m) \in$ Prof$(\mathfrak{C})$, we will write

$$X_{\underline{c}} = X_{c_1} \otimes \cdots \otimes X_{c_m},$$

which is the initial object $\varnothing$ if $\underline{c}$ is the empty profile.

Lemma 4.5.2. *Suppose* O *is a* $\mathfrak{C}$-*colored operad in* M. *Then it induces a monad whose functor is*

$$\mathsf{O} \circ - : \mathsf{M}^{\mathfrak{C}} \longrightarrow \mathsf{M}^{\mathfrak{C}}$$

and whose multiplication and unit are induced by those of O *as in Corollary 4.1.10*

Proof. Suppose $Y = \{Y_c\}_{c \in \mathfrak{C}}$ is a $\mathfrak{C}$-colored object in M, regarded as a $\mathfrak{C}$-colored symmetric sequence as in (4.1.2). Observe that in (4.1.4) we have

$$Y^{\underline{c}}(\underline{b}) = \begin{cases} Y_{\underline{c}} & \text{if } \underline{b} = \varnothing, \\ \varnothing & \text{if } \underline{b} \neq \varnothing \end{cases}$$

for $\underline{b}, \underline{c} \in$ Prof$(\mathfrak{C})$. Putting this into the definition (4.1.5) of the $\mathfrak{C}$-colored circle product, we obtain

$$(\mathsf{O} \circ Y)\binom{d}{\underline{b}} = \begin{cases} \int^{\underline{c} \in \Sigma_{\mathfrak{C}}} \mathsf{O}\binom{d}{\underline{c}} \otimes Y_{\underline{c}} & \text{if } \underline{b} = \varnothing, \\ \varnothing & \text{if } \underline{b} \neq \varnothing. \end{cases} \qquad (4.5.3)$$

So the restriction of the functor

$$\mathsf{O} \circ - : \mathsf{SymSeq}^{\mathfrak{C}}(\mathsf{M}) \longrightarrow \mathsf{SymSeq}^{\mathfrak{C}}(\mathsf{M})$$

to the full subcategory $\mathsf{M}^{\mathfrak{C}}$ yields a functor $\mathsf{M}^{\mathfrak{C}} \longrightarrow \mathsf{M}^{\mathfrak{C}}$. Corollary 4.1.10 now shows that $\mathsf{O} \circ -$ is a monad in $\mathsf{M}^{\mathfrak{C}}$. □

Definition 4.5.4. Suppose O is a $\mathfrak{C}$-colored operad in M. The category $\mathsf{Alg}_\mathsf{M}(\mathsf{O})$ of O-*algebras* is defined as the category of $(\mathsf{O} \circ -)$-algebras for the monad $\mathsf{O} \circ -$ in $\mathsf{M}^{\mathfrak{C}}$.

We can describe O-algebras more explicitly by unwrapping this definition. The detailed colored operad algebra axioms below are from [Yau (2016)] (Section 13.2) and [Yau and Johnson (2015)] (Corollary 13.37). In the one-colored topological case, the definition below is due to May [May (1972)]. We will use Definition 4.2.1 of a $\mathfrak{C}$-colored operad.

Definition 4.5.5. Suppose $(\mathsf{O}, \gamma, 1)$ is a $\mathfrak{C}$-colored operad in M. An O-*algebra* is a pair (X, λ) consisting of

- a $\mathfrak{C}$-colored object $X = \{X_c\}_{c \in \mathfrak{C}}$ and
- an O-*action structure morphism*

$$\mathsf{O}\binom{d}{\underline{c}} \otimes X_{\underline{c}} \xrightarrow{\ \lambda\ } X_d \in \mathsf{M} \tag{4.5.6}$$

for each $(\underline{c}; d) \in \mathsf{Prof}(\mathfrak{C}) \times \mathfrak{C}$.

It is required that the following associativity, unity, and equivariance axioms hold.

Associativity For $\big(\underline{c} = (c_1, \ldots, c_n); d\big) \in \mathsf{Prof}(\mathfrak{C}) \times \mathfrak{C}$ with $n \geq 1$, $\underline{b}_j \in \mathsf{Prof}(\mathfrak{C})$ for $1 \leq j \leq n$, and $\underline{b} = (\underline{b}_1, \ldots, \underline{b}_n)$ their concatenation, the associativity diagram

$$\mathsf{O}\binom{d}{\underline{c}} \otimes \left[\bigotimes_{j=1}^{n} \mathsf{O}\binom{c_j}{\underline{b}_j}\right] \otimes X_{\underline{b}} \xrightarrow{(\gamma, \mathrm{Id})} \mathsf{O}\binom{d}{\underline{b}} \otimes X_{\underline{b}} \tag{4.5.7}$$

with left vertical arrow labeled permute $\cong$ and

$$\mathsf{O}\binom{d}{\underline{c}} \otimes \bigotimes_{j=1}^{n} \left[\mathsf{O}\binom{c_j}{\underline{b}_j} \otimes X_{\underline{b}_j}\right]$$

left vertical arrow labeled $(\mathrm{Id}, \otimes_j \lambda)$ and

$$\mathsf{O}\binom{d}{\underline{c}} \otimes X_{\underline{c}} \xrightarrow{\ \lambda\ } X_d$$

right vertical arrow labeled λ, in M is commutative.

Unity For each $c \in \mathfrak{C}$, the unity diagram

$$\mathbb{1} \otimes X_c \xrightarrow{\ \cong\ } X_c \tag{4.5.8}$$

with left arrow $(1_c, \mathrm{Id})$, right arrow $=$, and

$$\mathsf{O}\binom{c}{c} \otimes X_c \xrightarrow{\ \lambda\ } X_c$$

in M is commutative.

Equivariance For each $(\underline{c}; d) \in \mathsf{Prof}(\mathfrak{C}) \times \mathfrak{C}$ and each permutation $\sigma \in \Sigma_{|\underline{c}|}$, the equivariance diagram

$$\mathsf{O}\binom{d}{\underline{c}} \otimes X_{\underline{c}} \xrightarrow{(\sigma, \sigma^{-1})} \mathsf{O}\binom{d}{\underline{c}\sigma} \otimes X_{\underline{c}\sigma} \tag{4.5.9}$$

with left arrow λ, right arrow λ, and

$$X_d \xrightarrow{\ =\ } X_d$$

in M is commutative. In the top horizontal morphism, σ^{-1} is the left permutation on the factors in $X_{\underline{c}}$ induced by $\sigma^{-1} \in \Sigma_{|\underline{c}|}$.

A *morphism of* O-*algebras* $f : (X, \lambda) \longrightarrow (Y, \xi)$ is a morphism $f : X \longrightarrow Y$ of $\mathfrak{C}$-colored objects in M such that the diagram

$$\mathsf{O}\binom{d}{\underline{c}} \otimes X_{\underline{c}} \xrightarrow{(\mathrm{Id}, \otimes f)} \mathsf{O}\binom{d}{\underline{c}} \otimes Y_{\underline{c}} \tag{4.5.10}$$

with left arrow λ, right arrow ξ, and

$$X_d \xrightarrow{\ f\ } Y_d$$

in M is commutative for all $(\underline{c}; d) \in \mathsf{Prof}(\mathfrak{C}) \times \mathfrak{C}$.

Proposition 4.5.11. *Suppose* O *is a* $\mathfrak{C}$*-colored operad in* M*. Then the definitions of an* O*-algebra in Definition 4.5.4 and in Definition 4.5.5 are equivalent.*

Proof. This is essentially the same as the proof of Proposition 4.2.9. The key is (4.5.3): For a $\mathfrak{C}$-colored object $X = \{X_c\}_{c \in \mathfrak{C}}$ in M, the $\mathfrak{C}$-colored object $O \circ X$ has entries

$$(O \circ X)_d = \int^{\underline{c} \in \Sigma_\mathfrak{C}} O\binom{d}{\underline{c}} \otimes X_{\underline{c}}$$

for $d \in \mathfrak{C}$. So an $(O \circ -)$-algebra has O-action structure morphisms as in (4.5.6). The equivariance axiom (4.5.9) corresponds to the $\underline{c}$-variable in the coend formula for $(O \circ X)_d$. The associativity axiom (4.5.7) and the unity axiom (4.5.8) correspond to those of an $(O \circ -)$-algebra. $\qquad\square$

The following result is [White and Yau (2018)] Proposition 4.2.1.

Proposition 4.5.12. *Suppose* O *is a* $\mathfrak{C}$*-colored operad in* M*. Then there is a free-forgetful adjunction*

$$\mathsf{M}^{\mathfrak{C}} \underset{U}{\overset{O \circ -}{\rightleftarrows}} \mathsf{Alg}_\mathsf{M}(O)$$

in which the right adjoint U *forgets about the* O*-algebra structure. Moreover, the category* $\mathsf{Alg}_\mathsf{M}(O)$ *is cocomplete (resp., complete), provided* M *is cocomplete (resp., complete).*

Example 4.5.13 (Colored objects as algebras). For the unit $\mathfrak{C}$-colored operad I in (4.1.6), there is an equality

$$\mathsf{Alg}_\mathsf{M}(I) = \mathsf{M}^{\mathfrak{C}},$$

and both functors U and $I \circ -$ are the identity functors.

Example 4.5.14 (Colored endomorphism operads). For each $\mathfrak{C}$-colored object $X = \{X_c\}_{c \in \mathfrak{C}}$ in M, there is a $\mathfrak{C}$-*colored endomorphism operad* $\mathsf{End}(X)$ with entries

$$\mathsf{End}(X)\binom{d}{\underline{c}} = \mathsf{Hom}_\mathsf{M}(X_{\underline{c}}, X_d)$$

for $(\underline{c}; d) \in \mathsf{Prof}(\mathfrak{C}) \times \mathfrak{C}$. Its equivariant structure is induced by permutations of the factors in $X_{\underline{c}}$. Its d-colored unit (4.2.3)

$$\mathbb{1} \longrightarrow \mathsf{Hom}_\mathsf{M}(X_d, X_d)$$

is adjoint to the identity morphism of X_d. Its operadic composition γ (4.2.2) is induced by the $\otimes$-Hom_M-adjunction. Another exercise involving the $\otimes$-Hom_M-adjunction shows that an O-algebra structure (X, θ) is equivalent to a morphism

$$\theta' : O \longrightarrow \mathsf{End}(X)$$

of $\mathfrak{C}$-colored operads. See [Yau (2016)] Sections 13.8 and 13.9 for details. $\qquad\diamond$

Example 4.5.15 (Tree operad). There is a $\mathfrak{C}$-*colored tree operad* $\mathsf{TreeOp}^\mathfrak{C}$ in Set in which each entry $\mathsf{TreeOp}^\mathfrak{C}\binom{d}{\underline{c}}$ is the set of $\mathfrak{C}$-colored trees with profile $(\underline{c}; d)$.

- The c-colored unit is the c-colored exceptional edge $\uparrow_c$ in Example 3.1.18.
- The equivariant structure is given by reordering: If $T \in \mathsf{TreeOp}^{\mathfrak{C}}\binom{d}{c}$ and if $\sigma \in \Sigma_{|c|}$, then $T\sigma \in \mathsf{TreeOp}^{\mathfrak{C}}\binom{d}{c\sigma}$ is the same as T except that its ordering is $\zeta\sigma$, where ζ is the ordering of T.
- The operadic composition γ is given by grafting of trees in Definition 3.3.1.

For each tree $T \in \underline{\mathsf{Tree}}^{\mathfrak{C}}\binom{d}{c}$, the operadic structure morphism

$$\mathsf{TreeOp}^{\mathfrak{C}}[T] = \prod_{v \in T} \mathsf{TreeOp}^{\mathfrak{C}}(v) \xrightarrow{\gamma_T} \mathsf{TreeOp}^{\mathfrak{C}}\binom{d}{c}$$

is given by tree substitution in Definition 3.2.6,

$$\gamma_T \{H_v\}_{v \in T} = T\big(H_v\big)_{v \in T},$$

where each $H_v \in \mathsf{TreeOp}^{\mathfrak{C}}(v)$.   ◇

Example 4.5.16 (Monoids as operads). Suppose (A, μ, ε) is a monoid in M. Then it yields a 1-colored operad A with entries

$$\mathsf{A}(n) = \begin{cases} A & \text{if } n = 1, \\ \varnothing & \text{if } n \neq 1. \end{cases}$$

Its equivariant structure is trivial. The operadic composition γ and the unit are those of the monoid A. In other words, monoids are 1-colored operads concentrated in unary entries.   ◇

Example 4.5.17 (Associative operad). There is a 1-colored operad As in M, called the *associative operad*, with entries

$$\mathsf{As}(n) = \coprod_{\sigma \in \Sigma_n} \mathbb{1}$$

for $n \geq 0$ and unit $\mathbb{1} \longrightarrow \mathsf{As}(1)$ the identity morphism. Its operadic composition γ is induced by the map

$$\Sigma_n \times \Sigma_{k_1} \times \cdots \times \Sigma_{k_n} \longrightarrow \Sigma_{k_1 + \cdots + k_n}$$

that sends $(\sigma; \sigma_1, \ldots, \sigma_n)$ to the composition

$$\sigma(\sigma_1, \ldots, \sigma_n) = \sigma\langle k_1, \ldots, k_n\rangle \circ \big(\sigma_1 \oplus \cdots \oplus \sigma_n\big) \tag{4.5.18}$$

with (i) $\sigma_1 \oplus \cdots \oplus \sigma_n$ the block sum induced by the σ_j and (ii) $\sigma\langle k_1, \ldots, k_n\rangle$ the block permutation induced by σ that permutes n consecutive blocks of lengths $k_1, \ldots, k_n$. Using Proposition 2.6.2(1), one can check that As-algebras are precisely monoids in M.   ◇

Example 4.5.19 (Commutative operad). There is a 1-colored operad Com in M, called the *commutative operad*, with entries

$$\mathsf{Com}(n) = \mathbb{1}$$

for $n \geq 0$, operadic composition induced by the isomorphism $\mathbb{1} \otimes \mathbb{1} \cong \mathbb{1}$, and unit the identity morphism. It follows from Proposition 2.6.2(2) that Com-algebras are precisely commutative monoids in M.   ◇

Example 4.5.20 (Diagrams as operads). Suppose C is a small category with object set $\mathfrak{C}$, and $F : C \longrightarrow M$ is a C-diagram in M. Then F yields an $\mathfrak{C}$-colored operad F in M with entries

$$\mathsf{F}\binom{d}{\underline{c}} = \begin{cases} \coprod_{FC(c,d)} \mathbb{1} & \text{if } \underline{c} = c \in \mathfrak{C}, \\ \varnothing & \text{if } |\underline{c}| \neq 1, \end{cases}$$

for $(\underline{c}; d) \in \mathsf{Prof}(\mathfrak{C}) \times \mathfrak{C}$, where $FC(c,d)$ is the set of morphisms $Ff \in M(Fc, Fd)$ for $f \in C(c,d)$. Since it is concentrated in unary entries, its equivariant structure is trivial. Its colored units come from the identity morphisms in C. Its operadic composition γ arises from the fact that F is a functor. ◇

Example 4.5.21 (Operad for diagrams). Suppose C is a small category with object set $\mathfrak{C}$. There is a $\mathfrak{C}$-colored operad $\mathsf{C}^{\mathsf{diag}}$ in M with entries

$$\mathsf{C}^{\mathsf{diag}}\binom{d}{\underline{c}} = \begin{cases} \coprod_{C(c,d)} \mathbb{1} & \text{if } \underline{c} = c \in \mathfrak{C}, \\ \varnothing & \text{if } |\underline{c}| \neq 1 \end{cases}$$

for $(\underline{c}; d) \in \mathsf{Prof}(\mathfrak{C}) \times \mathfrak{C}$. Its equivariant structure is trivial. Its colored units come from the identity morphisms in C. Its operadic composition γ is induced by the categorical composition in C. One can check that $\mathsf{C}^{\mathsf{diag}}$-algebras are precisely C-diagrams in M. ◇

Example 4.5.22 (Operad for diagrams of monoids). This example is a combination of Examples 4.5.17 and 4.5.21. Suppose C is a small category with object set $\mathfrak{C}$. There is a $\mathfrak{C}$-colored operad $\mathsf{O}_\mathsf{C}^\mathsf{M}$ in M with entries

$$\mathsf{O}_\mathsf{C}^\mathsf{M}\binom{d}{\underline{c}} = \coprod_{\Sigma_n \times \prod_{j=1}^n C(c_j,d)} \mathbb{1} \quad \text{for} \quad \binom{d}{\underline{c}} = \binom{d}{c_1,\ldots,c_n} \in \mathsf{Prof}(\mathfrak{C}) \times \mathfrak{C}.$$

A coproduct summand corresponding to an element $(\sigma, \underline{f}) \in \Sigma_n \times \prod_j C(c_j, d)$ is denoted by $\mathbb{1}_{(\sigma, \underline{f})}$. We will describe the operad structure on $\mathsf{O}_\mathsf{C}^\mathsf{M}$ in terms of the subscripts.

Its equivariant structure sends $\mathbb{1}_{(\sigma,\underline{f})}$ to $\mathbb{1}_{(\sigma\tau,\underline{f}\tau)}$ for $\tau \in \Sigma_{|\underline{c}|}$. Its c-colored unit corresponds to $\mathbb{1}_{(\mathrm{id}_1, \mathrm{Id}_c)}$. Its operadic composition

$$\mathsf{O}_\mathsf{C}^\mathsf{M}\binom{d}{\underline{c}} \otimes \bigotimes_{j=1}^n \mathsf{O}_\mathsf{C}^\mathsf{M}\binom{c_j}{\underline{b}_j} \overset{\gamma}{\longrightarrow} \mathsf{O}_\mathsf{C}^\mathsf{M}\binom{d}{\underline{b}}$$

corresponds to

$$\left((\sigma,\underline{f}); \{(\tau_j,\underline{g}_j)\}_{j=1}^n\right) \longmapsto \left(\sigma(\tau_1,\ldots,\tau_n), (f_1\underline{g}_1,\ldots,f_n\underline{g}_n)\right)$$

where

$$f_j\underline{g}_j = \left(f_j g_{j1},\ldots,f_j g_{jk_j}\right) \in \prod_{i=1}^{k_j} C(b_{ji},d) \quad \text{if} \quad \underline{g}_j = \left(g_{j1},\ldots,g_{jk_j}\right) \in \prod_{i=1}^{k_j} C(b_{ji},c_j)$$

and

$$\sigma(\tau_1, \ldots, \tau_n) = \underbrace{\sigma\langle k_1, \ldots, k_n \rangle}_{\text{block permutation}} \circ \underbrace{(\tau_1 \oplus \cdots \oplus \tau_n)}_{\text{block sum}} \in \Sigma_{k_1 + \cdots + k_n}$$

as in (4.5.18).

There is a canonical isomorphism

$$\mathsf{Alg}_M(O_C^M) \overset{\cong}{\longrightarrow} \mathsf{Mon}(M)^C$$

defined as follows. Each O_C^M-algebra (X, λ) has a restricted structure morphism

for each $(\sigma, \underline{f}) \in \Sigma_n \times \prod_{j=1}^n C(c_j, d)$. For a morphism $f : c \longrightarrow d \in C$, there is a restricted structure morphism

$$X_c \xrightarrow{\lambda_{(\mathrm{id}_1, f)}} X_d \in M.$$

The associativity and unity axioms of (X, λ) imply that this is a C-diagram in M.

For each $c \in C$, the restricted structure morphisms

$$X_c \otimes X_c \xrightarrow{\lambda_{(\mathrm{id}_2, \{\mathrm{Id}_c, \mathrm{Id}_c\})}} X_c \quad \text{and} \quad \mathbb{1} \xrightarrow{\lambda_{(\mathrm{id}_0, *)}} X_c$$

give X_c the structure of a monoid in M, once again by the associativity and unity axioms of (X, λ). One can check that this gives a C-diagram of monoids in M; i.e., the morphisms $\lambda_{(\mathrm{id}_1, f)}$ are compatible with the entrywise monoid structures. In summary, O_C^M is the $\mathfrak{C}$-colored operad whose algebras are C-diagrams of monoids in M. This identification is also given in [Benini *et. al.* (2017)] Theorem 4.26.   ◇

Example 4.5.23 (Operad for diagrams of commutative monoids). This example is a combination of Examples 4.5.19 and 4.5.21 and is a slight modification of Example 4.5.22. Suppose C is a small category with object set $\mathfrak{C}$. There is a $\mathfrak{C}$-colored operad Com^C in M with entries

$$\mathsf{Com}^C\binom{d}{\underline{c}} = \coprod_{\prod_{j=1}^n C(c_j, d)} \mathbb{1} \quad \text{for} \quad \binom{d}{\underline{c}} = \binom{d}{c_1, \ldots, c_n} \in \mathsf{Prof}(\mathfrak{C}) \times \mathfrak{C}.$$

Its operad structure is defined as in Example 4.5.22 by ignoring the first component.

Moreover, with almost the same argument as in Example 4.5.22, one can check that there is a canonical isomorphism

$$\mathsf{Alg}_M(\mathsf{Com}^C) \overset{\cong}{\longrightarrow} \mathsf{Com}(M)^C.$$

To see that the monoid multiplication

$$X_c \otimes X_c \xrightarrow{\mu_c \; = \; \lambda_{\{\mathrm{Id}_c, \mathrm{Id}_c\}}} X_c \in \mathsf{M}$$

is commutative, observe that the pair $\{\mathrm{Id}_c, \mathrm{Id}_c\}$ is fixed by the permutation $(1\,2)$. So the equivariance axiom (4.5.9) implies that μ_c is commutative. In summary, $\mathsf{Com}^{\mathfrak{C}}$ is the $\mathfrak{C}$-colored operad whose algebras are C-diagrams of commutative monoids in M.

◇

Chapter 5

Constructions on Operads

In this chapter, we discuss several important constructions and properties of colored operads. In Section 5.1 to Section 5.3, we discuss the category of algebras over a colored operad under a change of operads and a change of base categories. In Section 5.4 and Section 5.5 we study localizations of colored operads, analogous to localizations of categories, and algebras over localized operads. The material in the last two sections about localizations of operads is new. Localizations of operads are needed later when we discuss the time-slice axiom in prefactorization algebras.

As in the previous chapter, $(\mathsf{M}, \otimes, \mathbb{1})$ is a cocomplete symmetric monoidal closed category with an initial object $\varnothing$.

5.1 Change-of-Operad Adjunctions

In this section, we consider the category of algebras over an operad under an operad morphism. Instead of restricting ourselves to operads with the same color set, we will need to consider morphisms between operads with different color sets. So we first consider operads under a change of colors.

Definition 5.1.1. Suppose $(\mathsf{O}, \gamma^{\mathsf{O}})$ is a $\mathfrak{C}$-colored operad in M in the sense of Definition 4.4.9, and $f : \mathfrak{B} \longrightarrow \mathfrak{C}$ is a map of non-empty sets.

(1) Define the object $f^*\mathsf{O} \in \mathsf{M}^{\mathsf{Prof}(\mathfrak{B}) \times \mathfrak{B}}$ by

$$(f^*\mathsf{O})\binom{d}{\underline{c}} = \mathsf{O}\binom{fd}{f\underline{c}}$$

for $\big(\underline{c} = (c_1, \ldots, c_m); d\big) \in \mathsf{Prof}(\mathfrak{B}) \times \mathfrak{B}$, where $f\underline{c} = (fc_1, \ldots, fc_m)$.

(2) For each $\mathfrak{B}$-colored tree T, define fT as the $\mathfrak{C}$-colored tree obtained from T by applying f to its $\mathfrak{B}$-coloring.

(3) For each $\mathfrak{B}$-colored tree T, define the morphism $\gamma_T^{f^*\mathsf{O}}$ by the commutative dia-

75

gram

$$(f^*O)[T] \xrightarrow{\ \gamma_T^{f^*O}\ } (f^*O)(\mathsf{Prof}(T))$$

$$\mathsf{Id} \Big\downarrow \qquad\qquad\qquad \Big\uparrow \mathsf{Id}$$

$$\bigotimes_{v \in T} O\binom{f\mathrm{out}(v)}{f\mathrm{in}(v)} = O[fT] \xrightarrow{\ \gamma_{fT}^{O}\ } O\big[\mathsf{Prof}(fT)\big].$$

Proposition 5.1.2. *Suppose* (O, γ^O) *is a* $\mathfrak{C}$-*colored operad in* M, *and* $f : \mathfrak{B} \longrightarrow \mathfrak{C}$ *is a map of non-empty sets. Then* (f^*O, γ^{f^*O}) *is a* $\mathfrak{B}$-*colored operad in* M.

Proof. Observe that:

(1) $f\mathsf{Cor}_{(\underline{c};d)} = \mathsf{Cor}_{(f\underline{c};fd)}$ for each $(\underline{c}; d) \in \mathsf{Prof}(\mathfrak{B}) \times \mathfrak{B}$.
(2) $f\big(T(H_v)_{v \in T}\big) = (fT)(fH_u)_{u \in fT}$ for each tree substitution $T(H_v)$ in $\underline{\mathsf{Tree}}^{\mathfrak{B}}$.

Since $\gamma_T^{f^*O} = \gamma_{fT}^O$, the assertion follows from Definition 4.4.9. $\qquad\square$

Definition 5.1.3. *Suppose* O *is a* $\mathfrak{C}$-*colored operad in* M, *and* P *is a* $\mathfrak{D}$-*colored operad in* M. *An operad morphism* $f : O \longrightarrow P$ *is a pair* (f_0, f_1) *consisting of*

- a map $f_0 : \mathfrak{C} \longrightarrow \mathfrak{D}$ and
- a morphism $f_1 : O \longrightarrow f_0^* P$ of $\mathfrak{C}$-colored operads.

The category of all colored operads in M is denoted by $\mathsf{Operad}(\mathsf{M})$. We will sometimes abbreviate both f_0 and f_1 to f.

Unpacking the definition we can express an operad morphism more explicitly as follows.

Proposition 5.1.4. *Suppose* (O, γ^O) *is a* $\mathfrak{C}$-*colored operad, and* (P, γ^P) *is a* $\mathfrak{D}$-*colored operad in* M. *Then an operad morphism* $f : O \longrightarrow P$ *consist of precisely*

- a map $f_0 : \mathfrak{C} \longrightarrow \mathfrak{D}$ and
- a morphism

$$f_1 : O\binom{d}{\underline{c}} \longrightarrow P\binom{f_0 d}{f_0 \underline{c}} \in \mathsf{M}$$

for each $(\underline{c}; d) \in \mathsf{Prof}(\mathfrak{C}) \times \mathfrak{C}$

such that the diagram

$$O[T] \xrightarrow{\ \bigotimes_{v \in T} f_1\ } P[f_0 T]$$

$$\gamma_T^O \Big\downarrow \qquad\qquad\qquad \Big\downarrow \gamma_{f_0 T}^P$$

$$O\big[\mathsf{Prof}(T)\big] \xrightarrow{\ f_1\ } P\big[\mathsf{Prof}(f_0 T)\big]$$

is commutative for each $\mathfrak{C}$-*colored tree* T.

Definition 5.1.5. Suppose $f : O \longrightarrow P$ is an operad morphism with O a $\mathfrak{C}$-colored operad and P a $\mathfrak{D}$-colored operad in M.

(1) For $X \in M^{\mathfrak{D}}$, define the object $f^* X \in M^{\mathfrak{C}}$ by

$$(f^* X)_c = X_{fc} \quad \text{for} \quad c \in \mathfrak{C}.$$

(2) For a P-algebra (X, θ), define the morphism $\theta^{f^* X}$ as the composition in the diagram

$$
\begin{array}{ccc}
O\binom{d}{\underline{c}} \otimes (f^* X)_{\underline{c}} & \xrightarrow{\ \theta^{f^* X}\ } & (f^* X)_d \\[4pt]
{\scriptstyle (f, \mathrm{Id})}\big\downarrow & & \big\uparrow{\scriptstyle \mathrm{Id}} \\[4pt]
P\binom{fd}{f\underline{c}} \otimes X_{f\underline{c}} & \xrightarrow{\quad \theta \quad} & X_{fd}
\end{array}
$$

for each $(\underline{c}; d) \in \mathsf{Prof}(\mathfrak{C}) \times \mathfrak{C}$.

A direct inspection of Definition 4.5.5 yields the following result.

Proposition 5.1.6. *In the context of Definition 5.1.5:*

(1) $\left(f^* X, \theta^{f^* X}\right)$ *is an O-algebra.*
(2) f^* *defines functors*

$$f^* : M^{\mathfrak{D}} \longrightarrow M^{\mathfrak{C}} \quad \text{and} \quad f^* : \mathsf{Alg}_M(P) \longrightarrow \mathsf{Alg}_M(O).$$

Before we discuss the adjunction associated to an operad morphism, let us first consider the following special case on underlying objects.

Lemma 5.1.7. *Suppose* $f : \mathfrak{C} \longrightarrow \mathfrak{D}$ *is a map of non-empty sets. Then there is an adjunction*

$$M^{\mathfrak{C}} \underset{f^*}{\overset{f_!}{\rightleftarrows}} M^{\mathfrak{D}}$$

with left adjoint $f_!$.

Proof. For $Y \in M^{\mathfrak{C}}$, define an object $f_! Y \in M^{\mathfrak{D}}$ by

$$(f_! Y)_d = \coprod_{c \in f^{-1}(d)} Y_c \quad \text{for} \quad d \in \mathfrak{D}.$$

One checks directly that this defines a functor $f_!$ that is a left adjoint of f^*. $\qquad\square$

Theorem 5.1.8. *Suppose* $f : O \longrightarrow P$ *is an operad morphism with O a* $\mathfrak{C}$-colored *operad and P a* $\mathfrak{D}$-colored operad in M. *Then there is an adjunction*

$$\mathsf{Alg}_M(O) \underset{f^*}{\overset{f_!}{\rightleftarrows}} \mathsf{Alg}_M(P) ,$$

called the change-of-operad adjunction, *with left adjoint* $f_!$.

Proof. Consider the solid-arrow diagram

$$
\begin{array}{ccc}
\mathsf{Alg}_\mathsf{M}(\mathsf{O}) & \xleftarrow{\;\;f_!\;\;} & \mathsf{Alg}_\mathsf{M}(\mathsf{P}) \\[2pt]
{\scriptstyle \mathsf{O}\circ-}\big\uparrow\big\downarrow{\scriptstyle U} & \xrightarrow{f^*} & {\scriptstyle \mathsf{P}\circ-}\big\uparrow\big\downarrow{\scriptstyle U} \\[2pt]
\mathsf{M}^{\mathfrak{C}} & \xleftarrow[\;\;f^*\;\;]{\;f_!\;} & \mathsf{M}^{\mathfrak{D}}
\end{array}
$$

with the bottom adjunction from Lemma 5.1.7 and the vertical adjunctions from Proposition 4.5.12. There is an equality

$$U f^* = f^* U : \mathsf{Alg}_\mathsf{M}(\mathsf{P}) \longrightarrow \mathsf{M}^{\mathfrak{C}}.$$

Since the bottom horizontal functor f^* admits a left adjoint $f_!$ and since $\mathsf{Alg}_\mathsf{M}(\mathsf{P})$ is cocomplete, the Adjoint Lifting Theorem [Borceux (1994b)] (Theorem 4.5.6) implies that the top horizontal functor f^* also admits a left adjoint. □

Example 5.1.9 (Free-Forgetful Adjunction). For a $\mathfrak{C}$-colored operad O, the natural morphism $i : \mathsf{I} \longrightarrow \mathsf{O}$, where I is the $\mathfrak{C}$-colored unit operad in (4.1.6), is an operad morphism. In this case, the change-of-operad adjunction $i_! \dashv i^*$ is the free-forgetful adjunction

$$
\mathsf{M}^{\mathfrak{C}} \;\underset{U}{\overset{\mathsf{O}\circ-}{\rightleftarrows}}\; \mathsf{Alg}_\mathsf{M}(\mathsf{O})
$$

in Proposition 4.5.12. ◇

Example 5.1.10. There is an operad morphism $f : \mathsf{As} \longrightarrow \mathsf{Com}$ from the associative operad in Example 4.5.17 to the commutative operad in Example 4.5.19, given entrywise by the morphism

$$\mathsf{As}(n) = \coprod_{\sigma \in \Sigma_n} \mathbb{1} \longrightarrow \mathbb{1} = \mathsf{Com}(n)$$

whose restriction to every copy of $\mathbb{1}$ in $\mathsf{As}(n)$ is the identity morphism. In the change-of-operad adjunction

$$
\mathsf{Mon}(\mathsf{M}) = \mathsf{Alg}_\mathsf{M}(\mathsf{As}) \;\underset{f^*}{\overset{f_!}{\rightleftarrows}}\; \mathsf{Alg}_\mathsf{M}(\mathsf{Com}) = \mathsf{Com}(\mathsf{M})
$$

the right adjoint f^* forgets about the commutativity of a commutative monoid. The left adjoint $f_!$ sends a monoid to the commutative monoid generated by it. For instance, if $\mathsf{M} = \mathsf{Vect}_\mathbb{K}$ and if A is a monoid in M (i.e., a $\mathbb{K}$-algebra), then $f_! A$ is the quotient $\mathbb{K}$-algebra of A by the ideal generated by all the commutators $[a, b] = ab - ba$ with $a, b \in A$. ◇

Example 5.1.11 (Change-of-Monoids). Suppose $f : A \longrightarrow B$ is a morphism of monoids in M. Regarding A and B as 1-colored operads concentrated in unary entries as in Example 4.5.16, we can think of f as an operad morphism. With A

regarded as an operad, A-algebras are precisely A-modules in the sense of Example 2.7.8, and similarly for B-algebras. In the change-of-operad adjunction

$$\mathsf{Alg_M}(A) \underset{f^*}{\overset{f_!}{\rightleftarrows}} \mathsf{Alg_M}(B)$$

the right adjoint f^* is the restriction of the structure morphism to A. If $\mathsf{M} = \mathsf{Vect_K}$, then the left adjoint $f_!$ sends an A-module (X, θ) to the B-module $B \otimes_A X$, where the right A-action on B is induced by f. ◇

Example 5.1.12 (Left Kan Extensions). Suppose $F : \mathsf{C} \longrightarrow \mathsf{D}$ is a functor between small categories with $\mathsf{Ob(C)} = \mathfrak{C}$ and $\mathsf{Ob(D)} = \mathfrak{D}$. Recall from Example 4.5.21 that there is a $\mathfrak{C}$-colored operad $\mathsf{C^{diag}}$ whose algebras are precisely C-diagrams in M. There is an operad morphism

$$F^{\mathrm{diag}} : \mathsf{C^{diag}} \longrightarrow \mathsf{D^{diag}}$$

whose function on color sets $\mathfrak{C} \longrightarrow \mathfrak{D}$ is the object function of the functor F. For a pair of objects $c, d \in \mathfrak{C}$, the morphism

$$\mathsf{C^{diag}}\binom{d}{c} = \coprod_{f \in C(c,d)} \mathbb{1} \longrightarrow \coprod_{g \in D(Fc, Fd)} \mathbb{1} = \mathsf{D^{diag}}\binom{Fd}{Fc}$$

identifies the copy of $\mathbb{1}$ in $\mathsf{C^{diag}}\binom{d}{c}$ corresponding to $f \in C(c,d)$ with the copy of $\mathbb{1}$ in $\mathsf{D^{diag}}\binom{Fd}{Fc}$ corresponding to $Ff \in D(Fc, Fd)$. In the change-of-operad adjunction

$$\mathsf{Fun(C, M)} = \mathsf{Alg_M}(\mathsf{C^{diag}}) \underset{(F^{\mathrm{diag}})^*}{\overset{F^{\mathrm{diag}}_!}{\rightleftarrows}} \mathsf{Alg_M}(\mathsf{D^{diag}}) = \mathsf{Fun(D, M)}$$

the right adjoint $(F^{\mathrm{diag}})^* = \mathsf{Fun}(F, \mathsf{M})$ sends a D-diagram $G : \mathsf{D} \longrightarrow \mathsf{M}$ to the C-diagram $GF : \mathsf{C} \longrightarrow \mathsf{M}$. For a C-diagram $H : \mathsf{C} \longrightarrow \mathsf{M}$, $F^{\mathrm{diag}}_! H$ is the left Kan extension of H along F in Theorem 2.4.11. ◇

5.2 Model Category Structures

5.2.1 *Model Categories*

Before we discuss model category structures on $\mathsf{Alg_M}(\mathsf{O})$ for an operad O, let us first review some basic concepts of model categories, which were originally defined by Quillen [Quillen (1967)]. The reader is referred to the references [Hirschhorn (2003); Hovey (1999); May and Ponto (2012); Schwede and Shipley (2000)] for more details.

The formulation of a model category below is due to [May and Ponto (2012)]. Suppose $f : A \longrightarrow B$ and $g : C \longrightarrow D$ are morphisms in a category M. We write $f \boxempty g$ if for each solid-arrow commutative diagram

$$\begin{array}{ccc} A & \longrightarrow & C \\ f \downarrow & \nearrow & \downarrow g \\ B & \longrightarrow & D \end{array}$$

in M, a dotted arrow exists that makes the entire diagram commutative. For a class $\mathcal{A}$ of morphisms in M, define the classes of morphisms

$$^{\boxtimes}\mathcal{A} = \{f \in M \mid f \boxtimes a \text{ for all } a \in \mathcal{A}\},$$
$$\mathcal{A}^{\boxtimes} = \{g \in M \mid a \boxtimes g \text{ for all } a \in \mathcal{A}\}.$$

A pair $(\mathcal{L}, \mathcal{R})$ of classes of morphisms in M *functorially factors* M if each morphism h in M has a functorial factorization $h = gf$ such that $f \in \mathcal{L}$ and $g \in \mathcal{R}$. A *weak factorization system* in a category M is a pair $(\mathcal{L}, \mathcal{R})$ of classes of morphisms in M such that (i) $(\mathcal{L}, \mathcal{R})$ functorially factors M, (ii) $\mathcal{L} = {}^{\boxtimes}\mathcal{R}$, and (iii) $\mathcal{R} = \mathcal{L}^{\boxtimes}$.

A *model category* is a complete and cocomplete category M equipped with three classes of morphisms $(\mathcal{W}, \mathcal{C}, \mathcal{F})$, called *weak equivalences*, *cofibrations*, and *fibrations*, such that:

- $\mathcal{W}$ has the 2-out-of-3 property. In other words, for any morphisms f and g in M such that the composition gf is defined, if any two of the three morphisms f, g, and gf are in $\mathcal{W}$, then so is the third.
- $(\mathcal{C}, \mathcal{F} \cap \mathcal{W})$ and $(\mathcal{C} \cap \mathcal{W}, \mathcal{F})$ are weak factorization systems.

For a model category $(M, \mathcal{W}, \mathcal{C}, \mathcal{F})$, its *homotopy category* Ho(M) is a $\mathcal{W}$-localization of M as in Definition 2.8.1. For a model category M, its homotopy category always exists.

A model category M is:

(1) *left proper* if weak equivalences are closed under pushouts along cofibrations.
(2) *cofibrantly generated* if (i) it is equipped with two sets $\mathcal{I}$ and $\mathcal{J}$ of morphisms that permit the small object argument [Hirschhorn (2003)] (Definition 10.5.15), (ii) $\mathcal{F} = \mathcal{J}^{\boxtimes}$, and (iii) $\mathcal{F} \cap \mathcal{W} = \mathcal{I}^{\boxtimes}$.
(3) a *monoidal model category* [Schwede and Shipley (2000)] (Definition 3.1) if it is also a symmetric monoidal closed category that satisfies the following *pushout product axiom*:

> Given cofibrations $f : A \longrightarrow B$ and $g : C \longrightarrow D$, the pushout product $f \square g$ in the diagram

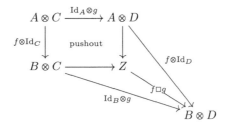

is a cofibration, which is also a weak equivalence if either f or g is also a weak equivalence. Here

$$Z = B \otimes C \coprod_{A \otimes C} A \otimes D$$

is the object of the pushout square.

In [Hovey (1999)] (Definition 4.2.6), a monoidal model category has an extra condition about the monoidal unit, which we do not need in this work.

Example 5.2.1. Here are some basic examples of cofibrantly generated monoidal model categories.

(1) Top is a cofibrantly generated monoidal model category [Hovey (1999)] (Section 2.4) in which a weak equivalence is a weak homotopy equivalence, i.e., a map that induces isomorphisms on all homotopy groups for all choices of base points in the domain. A fibration is a Serre fibration.

(2) The category SSet of simplicial sets is a left proper, cofibrantly generated monoidal model category [Hovey (1999)] (Chapter 3) in which a weak equivalence is a map whose geometric realization is a weak homotopy equivalence. A cofibration is an injection.

(3) For a field $\mathbb{K}$, the category $\mathsf{Chain}_{\mathbb{K}}$ is a left proper, cofibrantly generated monoidal model category [Quillen (1967)] with quasi-isomorphisms as weak equivalences, dimension-wise injections as cofibrations, and dimension-wise surjections as fibrations. The homotopy category of $\mathsf{Chain}_{\mathbb{K}}$ is the derived category of chain complexes of $\mathbb{K}$-vector spaces.

(4) The category Cat of small categories is a left proper, cofibrantly generated monoidal model category [Rezk (2000)], called the *folk model structure*. A weak equivalence is an equivalence of categories, i.e., a functor that is full, faithful, and essentially surjective. A cofibration is a functor that is injective on objects.

◇

Example 5.2.2. For a cofibrantly generated model category M and a small category D, the category M^{D} of D-diagrams in M inherits from M a cofibrantly generated model category structure [Hirschhorn (2003)] (11.6.1) with fibrations and weak equivalences defined entrywise in M. For instance, if $\mathsf{D} = \Sigma_{\mathfrak{C}}^{\mathrm{op}} \times \mathfrak{C}$, then the category

$$\mathsf{SymSeq}^{\mathfrak{C}}(\mathsf{M}) = \mathsf{M}^{\Sigma_{\mathfrak{C}}^{\mathrm{op}} \times \mathfrak{C}}$$

of $\mathfrak{C}$-colored symmetric sequences in M is a model category with weak equivalences and fibrations defined entrywise in M.   ◇

In a model category, an *acyclic (co)fibration* is a morphism that is both a (co)fibration and a weak equivalence. An object Z is *fibrant* if the unique morphism from Z to the terminal object is a fibration. A *fibrant replacement* of an object X is a weak equivalence $X \longrightarrow Z$ such that Z is fibrant. There is a functorial fibrant replacement R given by applying the functorial factorization of the weak factorization system $(\mathcal{C} \cap \mathcal{W}, \mathcal{F})$ to the unique morphism to the terminal object. An object Y is *cofibrant* if the unique morphism from the initial object to Y is a cofibration. A *cofibrant replacement* of an object X is a weak equivalence $Y \longrightarrow X$ such that Y is cofibrant. There is a functorial cofibrant replacement Q given by applying the functorial factorization of the weak factorization system $(\mathcal{C}, \mathcal{F} \cap \mathcal{W})$ to the unique morphism from the initial object.

Suppose $F : \mathsf{M} \rightleftarrows \mathsf{N} : U$ is an adjunction between model categories with left adjoint F. Then (F, U) is called a *Quillen adjunction* if U preserves fibrations and acyclic fibrations. The *total left derived functor*

$$LF : \mathsf{Ho}(\mathsf{M}) \longrightarrow \mathsf{Ho}(\mathsf{N})$$

is defined as the composition

$$\mathsf{Ho}(\mathsf{M}) \xrightarrow{\mathsf{Ho}(Q)} \mathsf{Ho}(\mathsf{M}) \xrightarrow{\mathsf{Ho}(F)} \mathsf{Ho}(\mathsf{N})$$

in which Q is the functorial cofibrant replacement in M. The *total right derived functor*

$$RU : \mathsf{Ho}(\mathsf{N}) \longrightarrow \mathsf{Ho}(\mathsf{M})$$

is defined as the composition

$$\mathsf{Ho}(\mathsf{N}) \xrightarrow{\mathsf{Ho}(R)} \mathsf{Ho}(\mathsf{N}) \xrightarrow{\mathsf{Ho}(U)} \mathsf{Ho}(\mathsf{M})$$

in which R is the functorial fibrant replacement in N. For a Quillen adjunction (F, U), there is a *derived adjunction*

$$\mathsf{Ho}(\mathsf{M}) \underset{RU}{\overset{LF}{\rightleftarrows}} \mathsf{Ho}(\mathsf{N})$$

between the homotopy categories with left adjoint LF.

A Quillen adjunction (F, U) is called a *Quillen equivalence* if for each morphism $f : FX \longrightarrow Y$ with $X \in \mathsf{M}$ cofibrant and $Y \in \mathsf{N}$ fibrant, f is a weak equivalence in N if and only if its adjoint $X \longrightarrow UY$ is a weak equivalence in M. For a Quillen equivalence, the derived adjunction is an adjoint equivalence between the homotopy categories.

Interpretation 5.2.3. The total left derived functor of F is first a cofibrant replacement in the domain of F and then F itself. The total right derived functor of U is first a fibrant replacement in the domain of U and then U itself. For a Quillen equivalence, the two model categories become adjoint equivalent via the derived adjunction after their weak equivalences are inverted. We say that their homotopy theories are equivalent.   ◇

Example 5.2.4. The adjunction

$$\mathsf{SSet} \underset{\mathsf{Sing}}{\overset{|-|}{\rightleftarrows}} \mathsf{Top}$$

involving the geometric realization functor and the singular simplicial set functor is a Quillen equivalence.   ◇

5.2.2 Model Structure on Algebra Categories

For more in-depth discussion of model category structure on the category of algebras over a colored operad, the reader may consult [Batanin and Berger (2017); Berger and Moerdijk (2007); Fresse (2009); White and Yau (2018)].

Definition 5.2.5. Suppose M is a monoidal model category, and O is a $\mathfrak{C}$-colored operad in M.

(1) O is *admissible* if $\mathsf{Alg_M}(O)$ admits a model category structure in which a morphism $f = \{f_c\}_{c \in \mathfrak{C}} \in M^{\mathfrak{C}}$ is:

- a weak equivalence if and only if f_c is a weak equivalence in M for each $c \in \mathfrak{C}$;
- a fibration if and only if f_c is a fibration in M for each $c \in \mathfrak{C}$;
- a cofibration if and only if $f \boxempty g$ for all morphisms $g \in \mathsf{Alg_M}(O)$ that are both weak equivalences and fibrations.

(2) O is *well-pointed* if the c-colored unit $1_c : \mathbb{1} \longrightarrow O\binom{c}{c}$ is a cofibration for each $c \in \mathsf{C}$.

(3) Suppose M is also cofibrantly generated. The operad O is called Σ-*cofibrant* if its underlying $\mathfrak{C}$-colored symmetric sequence is a cofibrant object in $\mathsf{SymSeq}^{\mathfrak{C}}(\mathsf{M})$.

(4) A morphism $f : O \longrightarrow P$ of $\mathfrak{C}$-colored operads in M is a *weak equivalence* if each entry of f is a weak equivalence in M.

Example 5.2.6. In the model categories SSet, $\mathsf{Chain}_{\mathbb{K}}$ where $\mathbb{K}$ has characteristic 0, Cat, and Top, every colored operad is admissible. The proofs for SSet and $\mathsf{Chain}_{\mathbb{K}}$ are in [White and Yau (2018)] (Section 8). The method of proof for the SSet case also works for Cat. For Top and many other model categories, the admissibility of all colored operads is proved in [Batanin and Berger (2017); Berger and Moerdijk (2007)]. Furthermore, in SSet, $\mathsf{Chain}_{\mathbb{K}}$, and Cat, every colored operad is well-pointed. In $\mathsf{Chain}_{\mathbb{K}}$ every colored operad is Σ-cofibrant, which is a consequence of Maschke's Theorem. ◇

The following comparison result is [Berger and Moerdijk (2007)] Theorem 4.1 in the general colored case and [Berger and Moerdijk (2003)] Theorem 4.4 in the one-colored case.

Theorem 5.2.7. *Suppose* M *is a monoidal model category, and* $f : O \longrightarrow P$ *is a morphism between admissible* $\mathfrak{C}$-*colored operads in* M.

(1) *The change-of-operad adjunction* $f_! \dashv f^*$ *in Theorem 5.1.8 is a Quillen adjunction.*

(2) *Suppose in addition that:*

(a) M *is left proper and cofibrantly generated with* $\mathbb{1}$ *cofibrant.*

(b) f *is a weak equivalence between well-pointed and* Σ-*cofibrant* $\mathfrak{C}$-*colored operads.*

Then the change-of-operad adjunction is a Quillen equivalence.

Example 5.2.8. If $f : \mathsf{O} \longrightarrow \mathsf{P}$ is a weak equivalence of $\mathfrak{C}$-colored operads in $\mathsf{Chain}_{\mathbb{K}}$, where $\mathbb{K}$ has characteristic 0, then the change-of-operad adjunction $f_! \dashv f^*$ is a Quillen equivalence. ◇

Remark 5.2.9. In [White and Yau (2019)] there are general results extending the homotopical change-of-operad adjunction in Theorem 5.2.7 to situations where O and P are colored operads in different monoidal model categories. We will not need those results in this book, so we refer the interested reader to [White and Yau (2019)]. ◇

5.3   Changing the Base Categories

We will later need to consider operads transferred from one category to another category. The following result is a special case of [Yau and Johnson (2015)] Theorem 12.11 and Corollary 12.13.

Theorem 5.3.1. *Suppose $F : \mathsf{M} \longrightarrow \mathsf{N}$ is a symmetric monoidal functor between symmetric monoidal closed categories.*

(1) F prolongs to a functor
$$F_* : \mathsf{Operad}^{\mathfrak{C}}(\mathsf{M}) \longrightarrow \mathsf{Operad}^{\mathfrak{C}}(\mathsf{N})$$
for every non-empty set $\mathfrak{C}$.

(2) Suppose F admits a right adjoint G that is also a symmetric monoidal functor. Then the prolonged functors
$$\mathsf{Operad}^{\mathfrak{C}}(\mathsf{M}) \underset{G_*}{\overset{F_*}{\rightleftarrows}} \mathsf{Operad}^{\mathfrak{C}}(\mathsf{N})$$
form an adjunction.

Proof. Let us describe the operad structure of $F_*\mathsf{O}$ for a $\mathfrak{C}$-colored operad (O, γ) in M. For a pair $(\underline{c}; d) \in \mathsf{Prof}(\mathfrak{C}) \times \mathfrak{C}$, we define
$$(F_*\mathsf{O})\binom{d}{\underline{c}} = F\mathsf{O}\binom{d}{\underline{c}} \in \mathsf{N}.$$
For each $\mathfrak{C}$-colored tree $T \in \underline{\mathsf{Tree}}^{\mathfrak{C}}\binom{d}{\underline{c}}$, we define the operadic structure morphism $\gamma_T^{F_*\mathsf{O}}$ as the composition in the diagram

$$
\begin{array}{ccc}
(F_*\mathsf{O})[T] & \xrightarrow{\gamma_T^{F_*\mathsf{O}}} & (F_*\mathsf{O})\binom{d}{\underline{c}} \\
{\scriptstyle \mathrm{Id}} \downarrow & & \uparrow {\scriptstyle F\gamma_T} \\
\underset{v \in T}{\otimes} F\mathsf{O}(v) & \xrightarrow{F_2} & F\big(\mathsf{O}[T]\big)
\end{array}
$$

in which $\mathsf{O}(v) = \mathsf{O}\binom{\mathrm{out}(v)}{\mathrm{in}(v)}$. The bottom horizontal morphism is an iteration of the monoidal structure of F. One can now check that $(F_*\mathsf{O}, \gamma^{F_*\mathsf{O}})$ is a $\mathfrak{C}$-colored operad in N in the sense of Definition 4.4.9. ∎

Notation 5.3.2. In the setting of Theorem 5.3.1, for a $\mathfrak{C}$-colored operad O in M, we will often write the image $F_*\mathsf{O}$ as O^N.

Example 5.3.3 (Set operads to enriched operads). There is an adjunction of symmetric monoidal functors

$$\mathsf{Set} \underset{\mathsf{M}(\mathbb{1},-)}{\overset{\coprod_{(-)}\mathbb{1}}{\rightleftarrows}} \mathsf{M}$$

in which the left adjoint is strong symmetric monoidal. For a $\mathfrak{C}$-colored operad O in Set, its image in M has entries

$$\mathsf{O}^\mathsf{M}\binom{d}{\underline{c}} = \coprod_{\mathsf{O}\binom{d}{\underline{c}}} \mathbb{1}$$

for $(\underline{c}; d) \in \mathsf{Prof}(\mathfrak{C}) \times \mathfrak{C}$. ◇

Example 5.3.4 (Change-of-rings). For a map $f : A \longrightarrow B$ of associative and commutative rings, there is an adjunction of symmetric monoidal functors

$$\mathsf{Chain}_A \underset{\mathsf{Res}}{\overset{f_!}{\rightleftarrows}} \mathsf{Chain}_B$$

with the left adjoint $f_! = - \otimes_A B$ and the right adjoint induced by restriction of scalars along f. For a $\mathfrak{C}$-colored operad O in Chain_A, its image O^B in Chain_B has entries

$$\mathsf{O}^B\binom{d}{\underline{c}} = \mathsf{O}\binom{d}{\underline{c}} \otimes_A B$$

for $(\underline{c}; d) \in \mathsf{Prof}(\mathfrak{C}) \times \mathfrak{C}$. ◇

Example 5.3.5. Both the geometric realization functor

$$|-| : \mathsf{SSet} \longrightarrow \mathsf{Top}$$

and its right adjoint, the singular simplicial set functor, are symmetric monoidal [Hovey (1999)] (Proposition 4.2.17). ◇

5.4 Localizations of Operads

In this section, we define the operad analogue of localizations of categories in Section 2.8. For a category C and a set S of morphisms in C, recall that the S-localization $\mathsf{C}[S^{-1}]$ is the category obtained from C by adjoining formal inverses f^{-1} for $f \in S$. For an operad O in Set and a set S of unary elements, we will show in this section that there is an analogous S-localization $\mathsf{O}[S^{-1}]$ in which formal inverses s^{-1} for $s \in S$ are added. The importance of this construction is what it does on algebras. In Section 5.5 we will show that $\mathsf{O}[S^{-1}]$-algebras are precisely the O-algebras in which the structure morphisms corresponding to elements in S are isomorphisms.

Motivation 5.4.1. We will need localized operads later when we discuss the time-slice axiom in prefactorization algebras from an operad viewpoint. The time-slice axiom is an invertibility condition that says that certain structure morphisms are isomorphisms. Applied to the colored operad for prefactorization algebras, we will see that algebras over the localized operad have the desired invertible structure morphisms.   ◇

As before $\mathfrak{C}$ is an arbitrary but fixed non-empty set. We will use Definition 4.2.1 of a $\mathfrak{C}$-colored operad below.

Definition 5.4.2. Suppose $(\mathsf{O}, \gamma, 1)$ is a $\mathfrak{C}$-colored operad in Set.

(1) Elements in $\mathsf{O}\binom{d}{c}$ for $c, d \in \mathfrak{C}$ are called *unary elements*.
(2) A unary element $x \in \mathsf{O}\binom{d}{c}$ is said to be *invertible* if there exists a unary element $y \in \mathsf{O}\binom{c}{d}$, called an *inverse of* x, such that

$$\gamma(y; x) = 1_c \quad \text{for} \quad \mathsf{O}\binom{c}{d} \times \mathsf{O}\binom{d}{c} \xrightarrow{\ \gamma\ } \mathsf{O}\binom{c}{c} \, ,$$

$$\gamma(x; y) = 1_d \quad \text{for} \quad \mathsf{O}\binom{d}{c} \times \mathsf{O}\binom{c}{d} \xrightarrow{\ \gamma\ } \mathsf{O}\binom{d}{d} \, .$$

Since an inverse of a unary element x is unique if it exists, we will write it as x^{-1}.

The next definition is the operad version of a localization of a category in Definition 2.8.1.

Definition 5.4.3. Suppose O is a $\mathfrak{C}$-colored operad in Set, and S is a set of unary elements in O. An *S-localization of* O, if it exists, is a pair $(\mathsf{O}[S^{-1}], \ell)$ consisting of

- a $\mathfrak{C}$-colored operad $\mathsf{O}[S^{-1}]$ in Set and
- a morphism of $\mathfrak{C}$-colored operads $\ell : \mathsf{O} \longrightarrow \mathsf{O}[S^{-1}]$

that satisfies the following two properties.

(1) $\ell(s)$ is invertible for each $s \in S$.
(2) $(\mathsf{O}[S^{-1}], \ell)$ is initial with respect to the previous property: If $f : \mathsf{O} \longrightarrow \mathsf{P}$ is an operad morphism such that $f(s)$ is invertible for each $s \in S$, then there exists a unique operad morphism

$$f' : \mathsf{O}[S^{-1}] \longrightarrow \mathsf{P} \quad \text{such that} \quad f = f'\ell.$$

$$\begin{array}{ccc}
\mathsf{O} & \xrightarrow{\ \ell\ } & \mathsf{O}[S^{-1}] \\
{\scriptstyle \forall f}\big\downarrow & \nearrow_{\scriptstyle \exists! \, f'} & \\
\mathsf{P} & &
\end{array} \qquad\qquad (5.4.4)$$

$$f(S) \text{ invertible}$$

In this setting, ℓ is called the *S-localization morphism*.

Remark 5.4.5. In Definition 5.4.3:

(1) By the universal property (5.4.4), an S-localization of O, if it exists, is unique up to a unique isomorphism.
(2) We may assume that S is closed under operadic composition. Indeed, if x and y are invertible elements, then $z = \gamma(y;x)$, if it is defined, is also invertible with inverse $\gamma(x^{-1};y^{-1})$. So if S_* denotes the closure of S under operadic composition, then the properties defining $\mathsf{O}[S^{-1}]$ and $\mathsf{O}[S_*^{-1}]$ are equivalent. ◇

The next observation is the operad version of Theorem 2.8.3. Its proof is an adaptation of the proof of Theorem 2.8.3 by replacing linear graphs with trees.

Theorem 5.4.6. *Suppose O is a $\mathfrak{C}$-colored operad in* Set, *and S is a set of unary elements in O. Then an S-localization of O exists.*

Proof. Without loss of generality, we may assume that S is closed under the operadic composition γ in O. Choose a set S^{-1} that is disjoint from O and consists of symbols x^{-1} for $x \in S$. For $c, d \in \mathfrak{C}$, the subset of S^{-1} consisting of x^{-1} with $x \in S \cap \mathsf{O}(\begin{smallmatrix}c\\d\end{smallmatrix})$ is denoted by $S^{-1}(\begin{smallmatrix}c\\d\end{smallmatrix})$. We define an S-localization O' of O as follows.

For $(\underline{c};d) \in \mathsf{Prof}(\mathfrak{C}) \times \mathfrak{C}$, $\mathsf{O}'(\begin{smallmatrix}d\\\underline{c}\end{smallmatrix})$ is the set of equivalence classes of pairs (T, ϕ) in which:

- $T \in \underline{\mathsf{Tree}}^{\mathfrak{C}}(\begin{smallmatrix}d\\\underline{c}\end{smallmatrix})$.
- $\phi : \mathsf{Vt}(T) \longrightarrow \mathsf{O} \sqcup S^{-1}$ is a function that satisfies the following conditions:
    - $\phi(v) \in \mathsf{O}(\begin{smallmatrix}d\\\underline{c}\end{smallmatrix})$ if $\mathsf{Prof}(v) = (\begin{smallmatrix}d\\\underline{c}\end{smallmatrix})$ with $|\underline{c}| \neq 1$.
    - $\phi(v) \in \mathsf{O}(\begin{smallmatrix}d\\c\end{smallmatrix}) \sqcup S^{-1}(\begin{smallmatrix}d\\c\end{smallmatrix})$ if $\mathsf{Prof}(v) = (\begin{smallmatrix}d\\c\end{smallmatrix})$ with $c, d \in \mathfrak{C}$.
    - If u and v are adjacent vertices in T, then one of $\phi(u)$ and $\phi(v)$ is in O with the other in S^{-1}.

Intuitively, the function ϕ decorates the vertices in T by elements in $\mathsf{O} \sqcup S^{-1}$ with the correct profiles such that adjacent vertices cannot be both decorated by O or both by S^{-1}.

The equivalence relation $\sim$ on such pairs (T, ϕ) is generated by the following four types of identifications.

(1) Suppose u and v are adjacent unary vertices in T such that

$$\phi(u) \in S \cap \mathsf{O}(\begin{smallmatrix}c\\d\end{smallmatrix}) \quad \text{and} \quad \phi(v) = \phi(u)^{-1} \in S^{-1}(\begin{smallmatrix}d\\c\end{smallmatrix}).$$

Then we identify

$$(T, \phi) \sim (T', \phi')$$

in which:

- Without the $\mathfrak{C}$-coloring, $T' = T(\uparrow_u, \uparrow_v)$ with the exceptional edge $\uparrow_u$ (resp., $\uparrow_v$) substituted into u (resp., v). The $\mathfrak{C}$-coloring of T' is inherited from that of T.
- ϕ' is the restriction of ϕ to $\mathsf{Vt}(T') = \mathsf{Vt}(T) \smallsetminus \{u, v\}$.

(2) Suppose $v \in \mathsf{Vt}(T)$ with $\mathsf{Prof}(v) = \binom{c}{c}$ for some $c \in \mathfrak{C}$ and that
$$\phi(v) = 1_c \in \mathsf{O}\binom{c}{c}.$$

(a) If v has no adjacent vertices, then $T = \mathsf{Cor}_{(c;c)}$ with unique vertex v. In this case, we identify
$$(T, \phi) \sim (\uparrow_c, \varnothing)$$
in which $\varnothing$ is the trivial function with domain $\mathsf{Vt}(\uparrow_c) = \varnothing$.

(b) If v has only one adjacent vertex u, then one of the two flags in v is a leg in T, while the other flag in v is part of an internal edge adjacent to u. In this case, we identify
$$(T, \phi) \sim \big(T(\uparrow_v), \phi'\big)$$
with
- $\uparrow_v = \uparrow_c$ substituted into v;
- ϕ' the restriction of ϕ to $\mathsf{Vt}\big(T(\uparrow_v)\big) = \mathsf{Vt}(T) \smallsetminus \{v\}$.

(c) If v has two adjacent vertices u and w, then one flag in v is part of an internal edge $e = \{e_\pm\}$ adjacent to u, and the other flag in v is part of an internal edge $f = \{f_\pm\}$ adjacent to w. Moreover, both u and w are unary vertices such that
$$\phi(u) = x^{-1} \quad \text{and} \quad \phi(w) = y^{-1} \in S^{-1}$$
for some $x, y \in S$. Switching the names u and w if necessary, we may assume that $v = \{e_+, f_-\}$, so the relevant part of (T, ϕ) looks like:

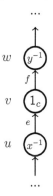

Suppose T' is the tree obtained from T by (i) removing the four flags $\{e_\pm, f_\pm\}$ and (ii) redefining $\{\mathsf{out}(w), \mathsf{in}(u)\}$ as a single unary vertex t with
$$\mathsf{in}(t) = \mathsf{in}(u) \quad \text{and} \quad \mathsf{out}(t) = \mathsf{out}(w).$$
We identify $(T, \phi) \sim (T', \phi')$ in which
$$\phi' : \mathsf{Vt}(T') = \{t\} \sqcup \mathsf{Vt}(T) \smallsetminus \{u, v, w\} \longrightarrow \mathsf{O} \sqcup S^{-1}$$
is the restriction of ϕ away from t and
$$\phi'(t) = \gamma(x; y)^{-1} \in S^{-1}.$$

(3) Suppose $v \in \mathsf{Vt}(T)$ with $\mathsf{Prof}(v) = \binom{c}{c}$ for some $c \in \mathfrak{C}$ and that
$$\phi(v) = 1_c^{-1} \in S^{-1}\binom{c}{c}.$$
This can only happen if $1_c \in S$.

(a) If v has no adjacent vertices, then $T = \mathsf{Cor}_{(c;c)}$ with unique vertex v, and we identify $(T, \phi) \sim (\uparrow_c, \varnothing)$ as in Case (2)(a) above.

(b) If v has only one adjacent vertex u, then $(T, \phi) \sim (T(\uparrow_v), \phi')$ as in Case (2)(b) above.

(c) If v has two adjacent vertices u and w, then
$$\phi(u) = x \quad \text{and} \quad \phi(w) = y \in \mathsf{O}.$$
Proceeding as in Case (2)(c) above, the relevant part of (T, ϕ) now looks like:

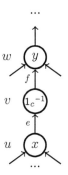

Suppose T' is the tree obtained from T by (i) removing the four flags $\{e_\pm, f_\pm\}$ and (ii) redefining a single vertex
$$t = \big\{w \smallsetminus \{f_+\}, u \smallsetminus \{e_-\}\big\}$$
with
$$\mathsf{out}(t) = \mathsf{out}(w) \quad \text{and} \quad \mathsf{in}(t) = \mathsf{in}(w) \circ_i \mathsf{in}(u).$$
Here we assume f_+ is the ith input of w, and $\circ_i$ was defined in Definition 4.3.1. We identify $(T, \phi) \sim (T', \phi')$ in which
$$\phi' : \mathsf{Vt}(T') = \{t\} \sqcup \mathsf{Vt}(T) \smallsetminus \{u, v, w\} \longrightarrow \mathsf{O} \sqcup S^{-1}$$
is the restriction of ϕ away from t and
$$\phi'(t) = y \circ_i x \in \mathsf{O},$$
which was defined in (4.3.10)

(4) Suppose v is a vertex in T with $\phi(v) \in \mathsf{O}\binom{b}{a}$ and $\sigma \in \Sigma_{|\underline{a}|}$. Suppose T^σ is the tree that is the same as T except that its ordering at v is the composition $\zeta_v \sigma$, where ζ_v is the ordering at v in T. Then we identify
$$(T, \phi) \sim (T^\sigma, \phi^\sigma)$$
in which
$$\phi^\sigma(u) = \begin{cases} \phi(u) & \text{if } u \neq v, \\ \phi(v)\sigma & \text{if } u = v. \end{cases}$$

The equivalence class of (T, ϕ) is denoted by $[(T, \phi)]$.

We will use Definition 4.3.2 of a $\mathfrak{C}$-colored operad in the rest of this proof. Next we define the $\mathfrak{C}$-colored operad structure on O'. For $c \in \mathfrak{C}$ the c-colored unit is the equivalence class of $(\uparrow_c, \varnothing)$. The equivariant structure is induced by reordering the inputs

$$(T, \phi)\sigma = (T\sigma, \phi),$$

where $T\sigma$ is the same as T except that its ordering is $\zeta\sigma$ with ζ the ordering of T. This equivariant structure is well-defined in the sense that it respects the equivalence relation $\sim$ that defines O'.

For the $\circ_i$-composition in (4.3.3), suppose that (T, ϕ) represents an equivalence class in $\mathsf{O}'\binom{d}{\underline{c}}$ with $|\underline{c}| = n \geq 1$ and that (T', ϕ') represents an equivalence class in $\mathsf{O}'\binom{c_i}{\underline{b}}$. First define

$$T \circ_i T' = \mathsf{Graft}\big(T; \uparrow_{c_1}, \ldots, \uparrow_{c_{i-1}}, T', \uparrow_{c_{i+1}}, \ldots, \uparrow_{c_n}\big) \in \underline{\mathsf{Tree}}^{\mathfrak{C}}\big(\begin{smallmatrix}d\\\underline{c}\circ_i\underline{b}\end{smallmatrix}\big), \qquad (5.4.7)$$

which is a grafting as in Definition 3.3.1.

- If $(T', \phi') = (\uparrow_{c_i}, \varnothing)$, then $T \circ_i \uparrow_{c_i} = T$, and we define

$$[(T, \phi)] \circ_i [(\uparrow_{c_i}, \varnothing)] = [(T, \phi)].$$

- If $(T, \phi) = (\uparrow_d, \varnothing)$, then we similarly define

$$[(\uparrow_d, \varnothing)] \circ_i [(T', \phi')] = [(T', \phi')].$$

- If neither T nor T' is an exceptional edge, then $T \circ_i T'$ has exactly one internal edge $e = \{e_{\pm}\}$ that is neither an internal edge in T nor in T'. If e is oriented from the vertex u to the vertex v, then $u \in \mathsf{Vt}(T')$ and $v \in \mathsf{Vt}(T)$. So a portion of $T \circ_i T'$ looks like:

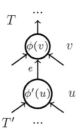

  - If one of $\phi'(u)$ and $\phi(v)$ is in S^{-1} with the other in O, then we define

$$[(T, \phi)] \circ_i [(T', \phi')] = \big[(T \circ_i T', \phi \circ_i \phi')\big] \qquad (5.4.8)$$

  in which $\phi \circ_i \phi'$ is induced by ϕ and ϕ' via the decomposition

$$\mathsf{Vt}\big(T \circ_i T'\big) = \mathsf{Vt}(T) \sqcup \mathsf{Vt}(T').$$

– If both $\phi'(u)$ and $\phi(v)$ are in O, we first define $G \in \underline{\mathsf{Tree}}^{\mathfrak{C}}\binom{d}{\underline{c} \circ_i \underline{b}}$ as the tree obtained from $T \circ_i T'$ by (i) removing the two flags in e and (ii) redefining a vertex

$$t = \{v \smallsetminus \{e_+\}, u \smallsetminus \{e_-\}\}$$

with

$$\mathsf{out}(t) = \mathsf{out}(v) \quad \text{and} \quad \mathsf{in}(t) = \mathsf{in}(v) \circ_j \mathsf{in}(u).$$

Here we assume e_+ is the jth input of v. We define

$$[(T, \phi)] \circ_i [(T', \phi')] = [(G, \varphi)] \tag{5.4.9}$$

in which

$$\mathsf{Vt}(G) = \{t\} \sqcup \big[\mathsf{Vt}(T) \smallsetminus \{v\}\big] \sqcup \big[\mathsf{Vt}(T') \smallsetminus \{u\}\big] \xrightarrow{\ \varphi\ } \mathsf{O} \sqcup S^{-1}$$

is the restrictions of ϕ and ϕ' away from t and

$$\varphi(t) = \phi(v) \circ_j \phi'(u) \in \mathsf{O}.$$

– If both $\phi'(u) = x^{-1}$ and $\phi(v) = y^{-1}$ are in S^{-1}, then both u and v are unary vertices. Using the same tree G as in the previous case, we define

$$[(T, \phi)] \circ_i [(T', \phi')] = [(G, \pi)]$$

in which π is the restrictions of ϕ and ϕ' away from t and

$$\pi(t) = \gamma(x; y)^{-1} \in S^{-1}.$$

The $\mathfrak{C}$-colored operad axioms of O and the unity and associativity of tree substitution (as in Corollary 3.2.8) imply that the $\circ_i$-composition above is indeed well-defined, i.e., independent of the choices of the representatives (T, ϕ) and (T', ϕ') in their equivalence classes. Furthermore, O' satisfies the axioms in Definition 4.3.2 because of the existence of the $\mathfrak{C}$-colored tree operad in Example 4.5.15, so O' is a $\mathfrak{C}$-colored operad in Set.

Now we define a morphism

$$\ell : \mathsf{O} \longrightarrow \mathsf{O}' \in \mathsf{M}^{\mathsf{Prof}(\mathfrak{C}) \times \mathfrak{C}}$$

by setting

$$\ell(x) = \big[(\mathrm{Cor}_x, \phi_x)\big] \tag{5.4.10}$$

for $x \in \mathsf{O}\binom{d}{\underline{c}}$ in which:

- Cor_x is the corolla $\mathrm{Cor}_{(\underline{c};d)}$ in Example 3.1.21.
- ϕ_x sends the unique vertex in Cor_x to x.

This defines a morphism $\ell : O \longrightarrow O'$ of $\mathfrak{C}$-colored operads. Indeed, ℓ respects the c-colored units and the equivariant structures by the identifications of type (2)(a) and type (4) above. It respects the $\circ_i$-composition by the definition (5.4.9).

By the identification of type (1) above, for each $s \in S \cap O\binom{c}{d}$, its image

$$\ell(s) = \left[\left(\mathrm{Lin}_{(d,c)}, \phi_s\right)\right]$$

is an invertible unary element with inverse

$$\left[\left(\mathrm{Lin}_{(c,d)}, \phi_{s^{-1}}\right)\right]$$

in which $\phi_{s^{-1}}$ sends the unique vertex to s^{-1}. Recall from Example 3.1.19 that $\mathrm{Lin}_{(c,d)} = \mathrm{Cor}_{(c;d)}$ is the linear graph with one vertex and profile $(c;d)$.

Finally, to prove the universal property (5.4.4), suppose $f : O \longrightarrow P$ is an operad morphism with P a $\mathfrak{D}$-colored operad in Set such that $f(s)$ is invertible for each $s \in S$. The requirement that the diagram (5.4.4) be commutative forces us to make the following definition of $f' : O' \longrightarrow P$:

- We define $f' = f : \mathfrak{C} \longrightarrow \mathfrak{D}$ on colors.
- For each $[(T, \phi)] \in O'$, we define

$$f'[(T, \phi)] = \gamma^P_{fT}\left(f'\phi(v)\right)_{v \in T}$$

  in which:

  (1) fT is the $\mathfrak{D}$-colored tree obtained from T by applying $f : \mathfrak{C} \longrightarrow \mathfrak{D}$ to its $\mathfrak{C}$-coloring.
  (2) For each $v \in T$,

$$f'\phi(v) = \begin{cases} f\phi(v) & \text{if } \phi(v) \in O, \\ (fx)^{-1} & \text{if } \phi(v) = x^{-1} \in S^{-1} \text{ for some } x \in S. \end{cases}$$

The operad axioms of P imply that (i) f' is entrywise well-defined, i.e., independent of the choice of a representative (T, ϕ) in its equivalence class, and that (ii) it is an operad morphism. The diagram (5.4.4) is commutative by construction. As we mentioned above, the uniqueness of f' is guaranteed by the commutativity of the diagram (5.4.4). Therefore, we have shown that O' is an S-localization of O. $\square$

5.5   Algebras over Localized Operads

In this section, we consider algebras over a localization of a colored operad. For a $\mathfrak{C}$-colored operad O in Set, recall that O^M is the image of O in M via the change-of-category functor

$$(-)^M : \mathsf{Operad}^{\mathfrak{C}}(\mathsf{Set}) \longrightarrow \mathsf{Operad}^{\mathfrak{C}}(\mathsf{M})$$

induced by the strong symmetric monoidal functor $\coprod_{(-)} \mathbb{1} : \mathsf{Set} \longrightarrow \mathsf{M}$. This is an instance of Theorem 5.3.1. First we consider what it means to be an algebra over a Set-operad in M.

Theorem 5.5.1. *Suppose* $(O, \gamma, 1)$ *is a* $\mathfrak{C}$*-colored operad in* Set. *Then an* O^{M}*-algebra is precisely a pair* (X, θ) *consisting of*

- *a* $\mathfrak{C}$*-colored object* $X \in \mathsf{M}^{\mathfrak{C}}$ *and*
- *a morphism*

$$X_{\underline{c}} \xrightarrow{\ \theta_p\ } X_d \in \mathsf{M}$$

for each $p \in O\binom{d}{\underline{c}}$ *with* $(\underline{c}; d) \in \mathsf{Prof}(\mathfrak{C}) \times \mathfrak{C}$

that satisfies the following axioms.

Associativity *For* $\left(\underline{c} = (c_1, \dots, c_n); d\right) \in \mathsf{Prof}(\mathfrak{C}) \times \mathfrak{C}$ *with* $n \geq 1$, $\underline{b}_j \in \mathsf{Prof}(\mathfrak{C})$ *for* $1 \leq j \leq n$, $\underline{b} = (\underline{b}_1, \dots, \underline{b}_n)$, $p \in O\binom{d}{\underline{c}}$, *and* $q_j \in O\binom{c_j}{\underline{b}_j}$, *the associativity diagram*

$$(5.5.2)$$

$$
\begin{array}{ccc}
X_{\underline{b}} & \xrightarrow{\ \theta_{\gamma(p; q_1, \dots, q_n)}\ } & X_d \\
\Big\| & & \Big\uparrow{\scriptstyle \theta_p} \\
X_{\underline{b}_1} \otimes \cdots \otimes X_{\underline{b}_n} & \xrightarrow[\ \underset{j}{\otimes} \theta_{q_j}\ ]{} & X_{c_1} \otimes \cdots \otimes X_{c_n}
\end{array}
$$

in M *is commutative.*

Unity *For each* $c \in \mathfrak{C}$, $\theta_{1_c} = \mathrm{Id}_{X_c}$.

Equivariance *For each* $(\underline{c}; d) \in \mathsf{Prof}(\mathfrak{C}) \times \mathfrak{C}$ *and each permutation* $\sigma \in \Sigma_{|\underline{c}|}$, *the equivariance diagram*

$$(5.5.3)$$

$$
\begin{array}{ccc}
X_{\underline{c}} & \xrightarrow[\cong]{\ \sigma^{-1}\ } & X_{\underline{c}\sigma} \\
{\scriptstyle \theta_p}\Big\downarrow & & \Big\downarrow{\scriptstyle \theta_{p\sigma}} \\
X_d & =\!\!=\!\!= & X_d
\end{array}
$$

in M *is commutative.*

A morphism of O*-algebras* $f : (X, \theta) \longrightarrow (Y, \xi)$ *is a morphism* $f : X \longrightarrow Y$ *of* $\mathfrak{C}$*-colored objects in* M *such that the diagram*

$$(5.5.4)$$

$$
\begin{array}{ccc}
X_{\underline{c}} & \xrightarrow{\ \otimes f\ } & Y_{\underline{c}} \\
{\scriptstyle \theta_p}\Big\downarrow & & \Big\downarrow{\scriptstyle \xi_p} \\
X_d & \xrightarrow[\ f\ ]{} & Y_d
\end{array}
$$

in M *is commutative for all* $(\underline{c}; d) \in \mathsf{Prof}(\mathfrak{C}) \times \mathfrak{C}$ *and* $p \in O\binom{d}{\underline{c}}$.

Proof. Each entry of O is a coproduct

$$O^{\mathsf{M}}\binom{d}{\underline{c}} = \coprod_{O\binom{d}{\underline{c}}} \mathbb{1}.$$

If (X, θ) is an O^{M}-algebra in the sense of Definition 4.5.5, then for each $p \in \mathsf{O}\binom{d}{\underline{c}}$ it has an induced O-action structure morphism

$$
\begin{array}{ccc}
X_{\underline{c}} & \xrightarrow{\quad \theta_p \quad} & X_d \\
\cong \downarrow & & \uparrow \theta \\
\mathbb{1} \otimes X_{\underline{c}} \xrightarrow[\text{inclusion}]{p} \underset{\mathsf{O}\binom{d}{\underline{c}}}{\coprod} \left(\mathbb{1} \otimes X_{\underline{c}}\right) & \xrightarrow{\quad \cong \quad} & \mathsf{O}^{\mathsf{M}}\binom{d}{\underline{c}} \otimes X_{\underline{c}}
\end{array}
$$

in which the bottom left horizontal morphism is the coproduct summand inclusion corresponding to p. In terms of these structure morphisms θ_p, the axioms stated above are simply those in Definition 4.5.5. The converse also holds because a morphism

$$
\mathsf{O}^{\mathsf{M}}\binom{d}{\underline{c}} \otimes X_{\underline{c}} \longrightarrow X_d
$$

is unique determined by the morphisms θ_p as p runs through $\mathsf{O}\binom{d}{\underline{c}}$. $\square$

Recall the change-of-operad adjunction in Theorem 5.1.8. In the next result, we consider the change-of-operad adjunction induced by ℓ^{M}, which is the image of a localization morphism ℓ as in Definition 5.4.3 under the change-of-category functor $(-)^{\mathsf{M}}$.

Theorem 5.5.5. *Suppose* O *is a* $\mathfrak{C}$-*colored operad in* Set, *and* S *is a set of unary elements in* O. *Consider the change-of-operad adjunction*

$$
\mathsf{Alg}_{\mathsf{M}}(\mathsf{O}^{\mathsf{M}}) \underset{(\ell^{\mathsf{M}})^*}{\overset{\ell^{\mathsf{M}}_!}{\rightleftarrows}} \mathsf{Alg}_{\mathsf{M}}\left(\mathsf{O}[S^{-1}]^{\mathsf{M}}\right)
$$

induced by the image in M *of the* S-*localization morphism* $\ell : \mathsf{O} \longrightarrow \mathsf{O}[S^{-1}]$.

(1) The right adjoint $(\ell^{\mathsf{M}})^*$ *is full and faithful.*
(2) The counit of the adjunction

$$
\epsilon : \ell^{\mathsf{M}}_!(\ell^{\mathsf{M}})^* \xrightarrow{\;\cong\;} \mathrm{Id}_{\mathsf{Alg}_{\mathsf{M}}\left(\mathsf{O}[S^{-1}]^{\mathsf{M}}\right)}
$$

is a natural isomorphism.

Proof. By Theorem 5.5.1 an O^{M}-algebra morphism is a morphism of the underlying colored objects that respects the structure morphisms θ_p, in the sense of (5.5.4), as p runs through all of O, and similarly for an $\mathsf{O}[S^{-1}]^{\mathsf{M}}$-algebra morphism. It follows that the right adjoint $(\ell^{\mathsf{M}})^*$ is faithful.

To see that $(\ell^{\mathsf{M}})^*$ is full, it is enough to prove the following statement.

Given $\mathsf{O}[S^{-1}]^{\mathsf{M}}$-algebras (X, θ^X) and (Y, θ^Y) and a morphism $f : X \longrightarrow Y$ of the underlying colored objects that respects the structure morphisms $\theta_{\ell(p)}$ for $p \in \mathsf{O}$, then f is a morphism of $\mathsf{O}[S^{-1}]^{\mathsf{M}}$-algebras.

We prove this statement by the following series of reductions.

(1) By the equivariance axiom (5.5.3) of $O[S^{-1}]^M$-algebras, if f respects θ_q for some $q \in O[S^{-1}]$, then it also respects $\theta_{q\sigma}$ for all permutations σ for which $q\sigma$ is defined.

(2) By Theorem 3.3.4, the definition (5.4.8) of $\circ_i$ in $O[S^{-1}]$, the associativity (5.5.2) of $O[S^{-1}]^M$-algebras, and the previous step, if f respects $\theta_{[(\mathrm{Cor}_x, \phi_x)]}$ for all $x \in O \sqcup S^{-1}$, then f is a morphism of $O[S^{-1}]^M$-algebras. Here Cor_x is the corolla with the same profile as x, and ϕ_x sends the unique vertex in Cor_x to x. The reader is reminded that the operadic composition γ in any colored operad can be written in terms of the various $\circ_i$-compositions as in (4.3.11).

(3) By the definition of ℓ in (5.4.10), we are assuming that f respects the structure morphism $\theta_{[(\mathrm{Cor}_x, \phi_x)]}$ for all $x \in O$. The associativity and unity of an $O[S^{-1}]^M$-algebra imply that, for each $s \in S$, the structure morphism

$$\theta_{\ell(s)} = \theta_{[(\mathrm{Cor}_s, \phi_s)]}$$

is an isomorphism with inverse $\theta_{[(\mathrm{Cor}_{s^{-1}}, \phi_{s^{-1}})]}$. Since f respects $\theta_{\ell(s)}$ for all $s \in S$, it also respects $\theta_{[(\mathrm{Cor}_{s^{-1}}, \phi_{s^{-1}})]}$.

This finishes the proof of the first assertion. The second assertion about the counit is a consequence of the first assertion and [Mac Lane (1998)] IV.3 Theorem 1. □

Theorem 5.5.6. *In the setting of Theorem 5.5.5, an O^M-algebra (X, θ) is in the image of the right adjoint $(\ell^M)^*$ if and only if the structure morphisms θ_s are isomorphisms for all $s \in S$.*

Proof. We already noted in the previous proof that, for an $O[S^{-1}]^M$-algebra, the structure morphism $\theta_{\ell(s)}$ is an isomorphism for each $s \in S$. So for an O^M-algebra in the image of $(\ell^M)^*$, θ_s must be an isomorphism for each $s \in S$.

For the converse, observe that by Theorem 3.3.4 and the axioms in Theorem 5.5.1, the structure morphisms

$$\{\theta_q : q \in O[S^{-1}]\}$$

for an $O[S^{-1}]^M$-algebra are uniquely determined by the subset

$$\{\theta_{\ell(x)} : x \in O\}.$$

So for an O^M-algebra (X, θ) in which the structure morphisms θ_s are isomorphisms for all $s \in S$, we can first define

$$\theta_{[(\mathrm{Cor}_x, \phi_x)]} = \theta_x \quad \text{for} \quad x \in O.$$

Then we use the associativity axiom (5.5.2) and the equivariance axiom (5.5.3) to define a general θ_q for $q \in O[S^{-1}]$. The assumptions on (X, θ) ensure that these structure morphisms θ_q are well-defined and that they satisfy the axioms in Theorem 5.5.1 for an $O[S^{-1}]^M$-algebra. □

Remark 5.5.7. By Theorem 5.5.5 and Theorem 5.5.6, we may regard the category $\mathsf{Alg_M}(\mathsf{O}[S^{-1}]^\mathsf{M})$ of $\mathsf{O}[S^{-1}]^\mathsf{M}$-algebras as the full subcategory of $\mathsf{Alg_M}(\mathsf{O}^\mathsf{M})$ consisting of the O^M-algebras in which the structure morphisms θ_s are isomorphisms for all $s \in S$.   ◇

Chapter 6

Boardman-Vogt Construction of Operads

In this chapter we define the Boardman-Vogt construction of a colored operad in a symmetric monoidal category as an entrywise coend and study its naturality properties.

6.1 Overview

The Boardman-Vogt construction was originally defined for a colored topological operad (without using the term operad) in [Boardman and Vogt (1972)] and also in [Vogt (2003)]. Our one-step formulation of the Boardman-Vogt construction in terms of a coend will be important when we apply it to the colored operads for algebraic quantum field theories and prefactorization algebras. The very explicit nature of our coend definition will allow us to elucidate the structures in homotopy algebraic quantum field theories and homotopy prefactorization algebras.

As we will explain later, the Boardman-Vogt construction WO is a resolution of the original colored operad O. Its algebras are algebras over O up to coherent higher homotopies. When O is a colored operad for algebraic quantum field theories or prefactorization algebras, WO-algebras are homotopy algebraic quantum field theories or homotopy prefactorization algebras. For instance, suppose O is the colored operad C^{diag} for C-diagrams. If X is an O-algebra and if the composition $f \circ g$ is defined in O, then $X_f \circ X_g$ is equal to $X_{f \circ g}$ by the associativity axiom of O-algebras. If Y is a homotopy coherent C-diagram, i.e., a WO-algebra, then both $Y_f \circ Y_g$ and $Y_{f \circ g}$ are defined, but they are not equal in general. Instead, there is another WO-algebra structure morphism of Y that is a homotopy from $Y_{f \circ g}$ to $Y_f \circ Y_g$. There are other WO-algebra structure morphisms that relate these homotopies, and so forth. We will discuss homotopy coherent diagrams in Section 7.3.

The Boardman-Vogt construction of a colored operad is defined in Section 6.2 and Section 6.3. An augmentation of the Boardman-Vogt construction over the original colored operad is defined in Section 6.4. The augmentation induces a change-of-operad adjunction, which allows us to go back and forth between algebras of the original colored operad and of the Boardman-Vogt construction. In

Section 6.5 we construct an entrywise section of the augmentation and use it to show that, in familiar cases, the augmentation is a weak equivalence. In particular, over $\mathsf{Chain}_\mathbb{K}$ the change-of-operad adjunction induced by the augmentation of a colored operad is always a Quillen equivalence. Let us emphasize that, in order to define the Boardman-Vogt construction and to understand the structure of its algebras, a model structure on the base category and that the augmentation is a weak equivalence are not necessary.

In Section 6.6 we discuss a natural filtration of the Boardman-Vogt construction. This filtration is not needed for applications to homotopy algebraic quantum field theories and homotopy prefactorization algebras, so the reader may skip this section safely. One main point of this filtration is to show that, for one-colored operads, our one-step coend definition of the Boardman-Vogt construction is isomorphic to the sequential colimit definition by Berger and Moerdijk. In [Berger and Moerdijk (2006)] the Boardman-Vogt construction of a one-colored operad was defined as the sequential colimit of an inductively defined sequence of morphisms, each being the pushout of some square involving the previous inductive step. For the Boardman-Vogt construction of more general objects, including dioperad, properads, wheeled operads, and wheeled properads, the reader is referred to [Yau and Johnson (2017)].

Our coend definition of the Boardman-Vogt construction uses the language of trees from Chapter 3. As before $(\mathsf{M}, \otimes, \mathbb{1})$ is a cocomplete symmetric monoidal closed category with an initial object $\varnothing$, and $\mathfrak{C}$ is an arbitrary non-empty set whose elements are called colors.

6.2   Commutative Segments

To define the Boardman-Vogt construction of a colored operad, we will equip the internal edges in trees with a suitable length using the following concept from [Berger and Moerdijk (2006)] (Definition 4.1). Recall the concept of a monoid in Section 2.6.

Definition 6.2.1. A *segment* in M is a tuple $(J, \mu, 0, 1, \epsilon)$ in which:

- $(J, \mu, 0)$ is a monoid in M.
- $1 : \mathbb{1} \longrightarrow J$ is an absorbing element.
- $\epsilon : J \longrightarrow \mathbb{1}$ is a counit.

A *commutative segment* is a segment whose multiplication μ is commutative.

Remark 6.2.2. More explicitly, in a (commutative) segment, J is a (commutative) monoid with multiplication $\mu : J \otimes J \longrightarrow J$ and unit $0 : \mathbb{1} \longrightarrow J$. To say that

$1 : \mathbb{1} \longrightarrow J$ is an absorbing element means that the diagram

$$
\begin{array}{ccccc}
\mathbb{1} \otimes J & \xrightarrow{\ (1,\mathrm{Id})\ } & J \otimes J & \xleftarrow{\ (\mathrm{Id},1)\ } & J \otimes \mathbb{1} \\
{\scriptstyle (\mathrm{Id},\epsilon)}\downarrow & & \downarrow{\scriptstyle \mu} & & \downarrow{\scriptstyle (\epsilon,\mathrm{Id})} \\
\mathbb{1} \otimes \mathbb{1} \xrightarrow{\ \cong\ } \mathbb{1} & \xrightarrow{\ 1\ } & J & \xleftarrow{\ 1\ } & \mathbb{1} \xleftarrow{\ \cong\ } \mathbb{1} \otimes \mathbb{1}
\end{array}
$$

is commutative. The counit ϵ makes the diagrams

$$
\begin{array}{ccc}
J \otimes J & \xrightarrow{\ (\epsilon,\epsilon)\ } & \mathbb{1} \otimes \mathbb{1} \\
{\scriptstyle \mu}\downarrow & & \downarrow{\scriptstyle \cong} \\
J & \xrightarrow{\ \epsilon\ } & \mathbb{1}
\end{array}
\qquad
\begin{array}{ccc}
\mathbb{1} & \xrightarrow{\ 0\ } & J \\
{\scriptstyle 1}\downarrow\ \ {\scriptstyle \mathrm{Id}}\searrow & & \downarrow{\scriptstyle \epsilon} \\
J & \xrightarrow{\ \epsilon\ } & \mathbb{1}
\end{array}
$$

commutative. A commutative segment provides a concept of homotopy from the 0-end $0 : \mathbb{1} \longrightarrow J$ to the 1-end $1 : \mathbb{1} \longrightarrow J$. ◇

Example 6.2.3. There is always a *trivial commutative segment* $\mathbb{1}$ with $0, 1, \epsilon = \mathrm{Id}_{\mathbb{1}}$ and $\mu : \mathbb{1} \otimes \mathbb{1} \cong \mathbb{1}$ the canonical isomorphism. ◇

Example 6.2.4. Here are some examples of non-trivial commutative segments.

(1) In Top the unit interval $[0,1]$ equipped with the multiplication
$$\mu(a,b) = \max\{a,b\}$$
is a commutative segment.

(2) In Cat the category
$$J = \left\{\, 0 \xleftrightarrow{\ \cong\ } 1 \,\right\}$$
with two objects $\{0,1\}$ and a unique isomorphism from 0 to 1 is a commutative segment with the multiplication induced by the maximum operation.

(3) In SSet the simplicial interval, that is, the representable simplicial set $\Delta^1 = \Delta(-,[1])$, is a commutative segment with the multiplication induced by the maximum operation.

(4) In Chain$_{\mathbb{K}}$ with $\mathbb{K}$ a field of characteristic 0, the normalized chain complex $J = N\Delta^1$ of Δ^1 is a commutative segment whose structure is uniquely determined by that on the simplicial interval Δ^1 and the monoidal structure of the normalized chain functor [Weibel (1997)] (8.3.6 page 265). More explicitly, J is a 2-stage chain complex

$$
\cdots \longrightarrow 0 \longrightarrow \mathbb{K} \xrightarrow{\ (+,-)\ } \mathbb{K} \oplus \mathbb{K} \longrightarrow 0 \longrightarrow \cdots
$$

with $\mathbb{K}$ in degree 1 and $\mathbb{K} \oplus \mathbb{K}$ in degree 0. The morphisms $0, 1 : \mathbb{K} \longrightarrow J$ correspond to the two copies of $\mathbb{K}$ in degree 0 in J, and the counit $\epsilon : J \longrightarrow \mathbb{K}$ is the identity morphism on each copy of $\mathbb{K}$ in degree 0. The mapping cylinder of a chain complex C (see, e.g., [Weibel (1997)] Exercise 1.5.3) is $J \otimes C$. Two chain maps $f, g : C \longrightarrow D$ are chain homotopic if and only if there is an extension $J \otimes C \longrightarrow D$ whose restrictions to C via 0 and 1 are f and g, respectively. We leave it to the reader to write down explicit formulas for the multiplication μ on J.

For the categories Top, SSet, Chain$_\mathbb{K}$, and Cat, unless otherwise specified, we will always use these commutative segments.   ◇

We will use the language of trees from Chapter 3. In particular, recall that for a tree T, $|T|$ denotes the set of internal edges in T. Also recall the exceptional edges in Example 3.1.18 and the substitution category $\underline{\mathsf{Tree}}^{\mathfrak{C}}$ in Definition 3.2.11.

Definition 6.2.5. Suppose $(J, \mu, 0, 1, \epsilon)$ is a commutative segment in M. For each $(\underline{c}; d) \in \mathsf{Prof}(\mathfrak{C}) \times \mathfrak{C}$, define a functor

$$\mathsf{J} : \underline{\mathsf{Tree}}^{\mathfrak{C}}\left(\tbinom{d}{\underline{c}}\right)^{\mathsf{op}} \longrightarrow \mathsf{M}$$

by the unordered monoidal product

$$\mathsf{J}[T] = \bigotimes_{e \in |T|} J = J^{\otimes |T|}$$

for $T \in \underline{\mathsf{Tree}}^{\mathfrak{C}}\left(\tbinom{d}{\underline{c}}\right)$. For each morphism

$$(H_v)_{v \in T} : T(H_v) \longrightarrow T \in \underline{\mathsf{Tree}}^{\mathfrak{C}}\left(\tbinom{d}{\underline{c}}\right),$$

the morphism

$$\mathsf{J}[T] \longrightarrow \mathsf{J}[T(H_v)] \in \mathsf{M}$$

is induced by:

- $0 : \mathbb{1} \longrightarrow J$ for each internal edge in each H_v, which must become an internal edge in $T(H_v)$;
- the multiplication $\mu : J \otimes J \longrightarrow J$ if H_v is an exceptional edge and if v is adjacent to two other vertices in T;
- the counit $\epsilon : J \longrightarrow \mathbb{1}$ if H_v is an exceptional edge and if v is adjacent to only one other vertex in T;
- the identity morphism of $\mathbb{1}$ if H_v is an exceptional edge and if v is not adjacent to any other vertices in T (i.e., T is a linear graph $\mathsf{Lin}_{(d,d)}$).

Remark 6.2.6. The absorbing element $1 : \mathbb{1} \longrightarrow J$ is not needed to define the functor $\mathsf{J} : \underline{\mathsf{Tree}}^{\mathfrak{C}}\left(\tbinom{d}{\underline{c}}\right)^{\mathsf{op}} \longrightarrow \mathsf{M}$.   ◇

Pick a commutative segment $(J, \mu, 0, 1, \epsilon)$ in M. The following observation will be needed to define the operad structure on the Boardman-Vogt construction. We will use the morphism $1 : \mathbb{1} \longrightarrow J$ of the commutative segment.

Lemma 6.2.7. *Suppose $T(H_v)_{v \in T}$ is a tree substitution of $\mathfrak{C}$-colored trees, and S is the set of internal edges in $T(H_v)_{v \in T}$ that are not in any of the H_v. Then there is a morphism*

$$\bigotimes_{v \in T} \mathsf{J}[H_v] \xrightarrow{\ \pi\ } \mathsf{J}\big[T(H_v)_{v \in T}\big]$$

of the form $\left(\bigotimes_S 1\right) \otimes \mathrm{Id}_{\bigotimes_{\sqcup_{v \in T} |H_v|} J}$ up to isomorphism.

Proof. Each internal edge in each H_v becomes a unique internal edge in the tree substitution $T(H_v)_{v \in T}$, and there is a decomposition

$$|T(H_v)| = S \sqcup \coprod_{v \in T} |H_v|.$$

The morphism π is the composition

$$
\begin{array}{ccc}
\displaystyle\bigotimes_{v \in T} J[H_v] & \xrightarrow{\quad \pi \quad} & J\big[T(H_v)_{v \in T}\big] \\[2mm]
\cong \Big\downarrow & & \Big\| \\[2mm]
\displaystyle\Big(\bigotimes_S \mathbb{1}\Big) \otimes \Big(\bigotimes_{\coprod_{v \in T} |H_v|} J\Big) & \xrightarrow{(\otimes_S 1, \mathrm{Id})} & \displaystyle\bigotimes_{|T(H_v)|} J
\end{array}
$$

in which $1 : \mathbb{1} \longrightarrow J$ is part of the commutative segment. $\square$

Interpretation 6.2.8. Intuitively, the morphism π in Lemma 6.2.7 assigns length 1 to each new internal edge, i.e., those in $T(H_v)_{v \in T}$ that are not in any of the H_v. $\diamond$

Example 6.2.9. Consider the morphism

$$(H_u, H_v, H_w) : K \longrightarrow T \in \underline{\mathsf{Tree}}^{\mathfrak{C}}$$

in Example 3.2.13. Counting the number of internal edges, we have

$$
\begin{aligned}
J[T] &\cong J_c \otimes J_d, & J[K] &\cong J_c \otimes J_f \otimes J_g, \\
J[H_u] &= J_f, & J[H_v] &= \mathbb{1}, \quad \text{and} \quad J[H_w] = J_g,
\end{aligned}
$$

in which we use J_c to denote a copy of J corresponding to a c-colored internal edge. The morphism $J[T] \longrightarrow J[K]$ is the composition in the following diagram.

$$
\begin{array}{ccc}
J[T] \cong J_c \otimes J_d & \longrightarrow & J_c \otimes J_f \otimes J_g \cong J[K] \\[2mm]
\cong \Big\downarrow & & \Big\uparrow \cong \\[2mm]
J_c \otimes \mathbb{1} \otimes \mathbb{1} \otimes J_d & \xrightarrow{(\mathrm{Id},0,0,\epsilon)} & J_c \otimes J_f \otimes J_g \otimes \mathbb{1}
\end{array}
$$

Each of H_u and H_w has one internal edge. This accounts for the morphisms $0 : \mathbb{1} \longrightarrow J_f$ and $0 : \mathbb{1} \longrightarrow J_g$. The tree H_v is the exceptional edge $\uparrow_d$, and v is adjacent to only one vertex in T. This accounts for the counit $\epsilon : J_d \longrightarrow \mathbb{1}$.

Since $K = T(H_u, H_v, H_w)$, the morphism π in Lemma 6.2.7 is

$$J[H_u] \otimes J[H_v] \otimes J[H_w] \cong \mathbb{1} \otimes J_f \otimes J_g \xrightarrow{(1,\mathrm{Id})} J_c \otimes J_f \otimes J_g \cong J[K]$$

with $1 : \mathbb{1} \longrightarrow J_c$ corresponding to the c-colored internal edge in K. $\diamond$

Example 6.2.10. Suppose L is the linear graph

in Example 3.1.19 with three vertices $\{v_1, v_2, v_3\}$, two internal edges, and each flag having color c. Suppose $H_{v_i} = \uparrow_c$ for each i, so there is a morphism

$$L(H_{v_i})_{i=1}^3 = \uparrow_c \xrightarrow{(H_{v_i})} L$$

in $\underline{\mathsf{Tree}}^{\mathfrak{C}}\binom{c}{c}$. The morphism $\mathsf{J}[L] \longrightarrow \mathsf{J}[L(H_{v_i})_{i=1}^3]$ is given by either composition in the commutative diagram

$$
\begin{array}{ccc}
\mathsf{J}[L] \cong J \otimes J & \xrightarrow{(\epsilon, \epsilon)} & \mathbb{1} \otimes \mathbb{1} \\
{\scriptstyle \mu} \downarrow & & \downarrow {\scriptstyle \cong} \\
J & \xrightarrow{\quad \epsilon \quad} & \mathbb{1} = \mathsf{J}[L(H_{v_i})].
\end{array}
$$

The composition $\epsilon \mu$ corresponds to the factorization

$$L(H_{v_i})_{i=1}^3 = \Big(\big(L(H_{v_2}) \big)(H_{v_1}) \Big)(H_{v_3}),$$

while the other composition corresponds to the factorization

$$L(H_{v_i})_{i=1}^3 = \Big(\big(L(H_{v_1}) \big)(H_{v_3}) \Big)(H_{v_2}).$$

The morphism π in Lemma 6.2.7 is the isomorphism

$$\mathsf{J}[H_{v_1}] \otimes \mathsf{J}[H_{v_2}] \otimes \mathsf{J}[H_{v_3}] = \mathbb{1} \otimes \mathbb{1} \otimes \mathbb{1} \xrightarrow{\;\cong\;} \mathbb{1} = \mathsf{J}\big[L(H_{v_i})_{i=1}^3\big].$$

$\diamond$

6.3   Coend Definition of the BV Construction

In this section, we define the Boardman-Vogt construction of a colored operad in a symmetric monoidal category as an entrywise coend. Pick a commutative segment $(J, \mu, 0, 1, \epsilon)$ in M. Recall the concept of a coend in Definition 2.3.7.

Definition 6.3.1. Suppose O is a $\mathfrak{C}$-colored operad in M. For each $(\underline{c}; d) \in \mathsf{Prof}(\mathfrak{C}) \times \mathfrak{C}$, define an object

$$\mathsf{WO}\binom{d}{\underline{c}} = \int^{T \in \underline{\mathsf{Tree}}^{\mathfrak{C}}\binom{d}{\underline{c}}} \mathsf{J}[T] \otimes \mathsf{O}[T] \in \mathsf{M} \tag{6.3.2}$$

with $\mathsf{J} : \underline{\mathsf{Tree}}^{\mathfrak{C}}\binom{d}{\underline{c}}^{\mathsf{op}} \longrightarrow \mathsf{M}$ the functor in Definition 6.2.5 and $\mathsf{O} : \underline{\mathsf{Tree}}^{\mathfrak{C}}\binom{d}{\underline{c}} \longrightarrow \mathsf{M}$ the functor in Corollary 4.4.15.

- We call $\mathsf{WO} \in \mathsf{M}^{\mathsf{Prof}(\mathfrak{C}) \times \mathfrak{C}}$ the *Boardman-Vogt construction*, or *BV construction*, of O.
- For $T \in \underline{\mathsf{Tree}}^{\mathfrak{C}}\binom{d}{\underline{c}}$, we write

$$\omega_T : \mathsf{J}[T] \otimes \mathsf{O}[T] \longrightarrow \mathsf{WO}\binom{d}{\underline{c}}$$

   for the natural morphism.

Interpretation 6.3.3. Intuitively, each entry of the Boardman-Vogt construction WO is made up of decorated trees $J[T] \otimes O[T]$, with each internal edge decorated by the commutative segment J and each vertex decorated by the entry of O with the same profile. When O is a colored topological operad, one can check that the above definition of WO agrees with the original one in [Boardman and Vogt (1972); Vogt (2003)] in terms of a quotient. This is proved in [Yau and Johnson (2017)] Example 3.4.7. ◇

Example 6.3.4. In the setting of Example 3.2.13, we have:

$$J[H_u] \otimes O[H_u] = J_f \otimes O(\genfrac{}{}{0pt}{}{c}{a,b,f}) \otimes O(\genfrac{}{}{0pt}{}{f}{\varnothing}),$$

$$J[H_v] \otimes O[H_v] = \mathbb{1} \otimes \mathbb{1},$$

$$J[H_w] \otimes O[H_w] = J_g \otimes O(\genfrac{}{}{0pt}{}{e}{c,g}) \otimes O(\genfrac{}{}{0pt}{}{g}{d}),$$

$$J[T] \otimes O[T] \cong J_c \otimes J_d \otimes O(\genfrac{}{}{0pt}{}{e}{c,d}) \otimes O(\genfrac{}{}{0pt}{}{c}{a,b}) \otimes O(\genfrac{}{}{0pt}{}{d}{d}),$$

$$J[K] \otimes O[K] \cong J_c \otimes J_f \otimes J_g \otimes O(\genfrac{}{}{0pt}{}{e}{c,g}) \otimes O(\genfrac{}{}{0pt}{}{c}{a,b,f}) \otimes O(\genfrac{}{}{0pt}{}{f}{\varnothing}) \otimes O(\genfrac{}{}{0pt}{}{g}{d}).$$

◇

Example 6.3.5. If $\underline{c} = (c_0, \ldots, c_n)$ with $n \geq 1$ and if $\text{Lin}_{\underline{c}}$ is the linear graph in Example 3.1.19, then we have

$$J[\text{Lin}_{\underline{c}}] \otimes O[\text{Lin}_{\underline{c}}] \cong \left(\bigotimes_{j=1}^{n-1} J_{c_j} \right) \otimes \left(\bigotimes_{i=1}^{n} O(\genfrac{}{}{0pt}{}{c_i}{c_{i-1}}) \right).$$

◇

For a $\mathfrak{C}$-colored operad O and a vertex v in a $\mathfrak{C}$-colored tree, recall our notation $O(v) = O\big(\genfrac{}{}{0pt}{}{\text{out}(v)}{\text{in}(v)}\big)$ and $\underline{\text{Tree}}^{\mathfrak{C}}(v) = \underline{\text{Tree}}^{\mathfrak{C}}\big(\genfrac{}{}{0pt}{}{\text{out}(v)}{\text{in}(v)}\big)$.

Lemma 6.3.6. *Suppose* O *is a* $\mathfrak{C}$-*colored operad in* M, *and* T *is a* $\mathfrak{C}$-*colored tree. Then there is a canonical isomorphism*

$$WO[T] \cong \int^{\{H_v\} \in \prod_{v \in T} \underline{\text{Tree}}^{\mathfrak{C}}(v)} \left(\bigotimes_{v \in T} J[H_v] \right) \otimes \left(\bigotimes_{v \in T} O[H_v] \right).$$

Proof. By definition there are canonical isomorphisms

$$WO[T] = \bigotimes_{v \in T} WO(v)$$

$$= \bigotimes_{v \in T} \left(\int^{H_v \in \underline{\text{Tree}}^{\mathfrak{C}}(v)} J[H_v] \otimes O[H_v] \right)$$

$$\cong \int^{\{H_v\} \in \prod_{v \in T} \underline{\text{Tree}}^{\mathfrak{C}}(v)} \bigotimes_{v \in T} \left(J[H_v] \otimes O[H_v] \right)$$

$$\cong \int^{\{H_v\} \in \prod_{v \in T} \underline{\text{Tree}}^{\mathfrak{C}}(v)} \left(\bigotimes_{v \in T} J[H_v] \right) \otimes \left(\bigotimes_{v \in T} O[H_v] \right).$$

The first isomorphism uses the naturality of coends. The second isomorphism uses the symmetry in M. □

Using Lemma 6.3.6, next we define the operad structure on the Boardman-Vogt construction. We will use Definition 4.4.9 of a $\mathfrak{C}$-colored operad.

Definition 6.3.7. Suppose O is a $\mathfrak{C}$-colored operad in M. For each pair $(\underline{c}; d) \in$ $\mathsf{Prof}(\mathfrak{C}) \times \mathfrak{C}$ and each $\mathfrak{C}$-colored tree T with profile $(\underline{c}; d)$, define the morphism

$$\gamma_T : \mathsf{WO}[T] \longrightarrow \mathsf{WO}\binom{d}{\underline{c}}$$

by insisting that the diagram

$$\left(\bigotimes_{v \in T} \mathsf{J}[H_v] \right) \otimes \left(\bigotimes_{v \in T} \mathsf{O}[H_v] \right) \xrightarrow{(\pi, \cong)} \mathsf{J}\big[T(H_v)_{v \in T}\big] \otimes \mathsf{O}\big[T(H_v)_{v \in T}\big] \qquad (6.3.8)$$

$$\big\{\omega_{H_v}\big\}_{v \in T} \downarrow \qquad\qquad\qquad\qquad \downarrow \omega_{T(H_v)_{v \in T}}$$

$$\mathsf{WO}[T] \xrightarrow{\qquad\qquad \gamma_T \qquad\qquad} \mathsf{WO}\binom{d}{\underline{c}}$$

be commutative for each $\{H_v\} \in \prod_{v \in T} \underline{\mathsf{Tree}}^{\mathfrak{C}}(v)$. In the top horizontal morphism, π is the morphism in Lemma 6.2.7, and the isomorphism is from Proposition 4.4.5. The left vertical natural morphism is from Lemma 6.3.6.

Lemma 6.3.9. *The morphism γ_T in (6.3.8) is well-defined.*

Proof. For each vertex v in T, suppose

$$(D_{vu})_{u \in H_v} : H_v(D_{vu})_{u \in H_v} \longrightarrow H_v$$

is a morphism in $\underline{\mathsf{Tree}}^{\mathfrak{C}}(v)$. In the following diagram, we will abbreviate H_v to H and D_{vu} to D, with v and u running through $\mathsf{Vt}(T)$ and $\mathsf{Vt}(H_v)$, respectively. By Lemma 6.3.6 it suffices to show that the outermost diagram in

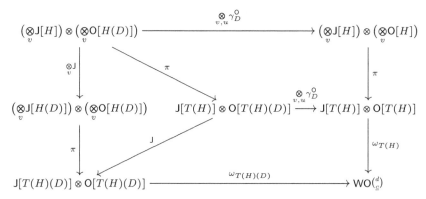

is commutative, in which identity morphisms and isomorphisms are omitted. The morphism

$$\gamma^{\mathsf{O}}_{D_{vu}} : \mathsf{O}[D_{vu}] \longrightarrow \mathsf{O}(u)$$

is the operadic structure morphism of O for D_{vu} in (4.4.10). The top left vertical morphism

$$\mathsf{J} : \mathsf{J}[H_v] \longrightarrow \mathsf{J}[H_v(D_{vu})]$$

is the image under the functor J in Definition 6.2.5 of the morphism $(D_{vu})_{u \in H_v}$.
Similarly, the slanted morphism

$$\mathsf{J} : \mathsf{J}[T(H_v)] \longrightarrow \mathsf{J}[T(H_v)(D_{vu})]$$

is the image under the functor J of the morphism

$$(D_{vu})_{v,u} : T(H_v)(D_{vu}) \longrightarrow T(H_v)$$

in $\underline{\mathsf{Tree}}^{\mathfrak{C}}\binom{d}{\underline{c}}$. The lower right trapezoid is commutative by the coend definition of $\mathsf{WO}\binom{d}{\underline{c}}$. The left triangle and the top trapezoid are commutative by inspection. $\square$

Interpretation 6.3.10. Intuitively, the morphism γ_T in (6.3.8) is given by sub-stituting decorated trees $\{\mathsf{J}[H_v] \otimes \mathsf{O}[H_v]\}_{v \in T}$ into T, with new internal edges (i.e., those in $T(H_v)_{v \in T}$ that are not in any of the H_v) given length 1. $\diamond$

Theorem 6.3.11. *Suppose O is a $\mathfrak{C}$-colored operad in M. With the operadic struc-ture morphisms γ_T in Definition 6.3.7, WO is a $\mathfrak{C}$-colored operad.*

Proof. For a corolla $\mathsf{Cor}_{(\underline{c};d)}$ with unique vertex v, since

$$\mathsf{Cor}_{(\underline{c};d)}(H_v) = H_v,$$

the top horizontal morphism in (6.3.8) is the identity morphism. So $\gamma_{\mathsf{Cor}_{(\underline{c};d)}}$ is the identity morphism.

To prove the associativity axiom (4.4.11), suppose $T(H_v)_{v \in T}$ is a tree substitu-tion with $T \in \underline{\mathsf{Tree}}^{\mathfrak{C}}\binom{d}{\underline{c}}$. We want to show that the diagram

$$
\begin{array}{ccc}
\underset{v \in T}{\otimes} \mathsf{WO}[H_v] & \xrightarrow{\otimes \gamma_{H_v}} & \underset{v \in T}{\otimes} \mathsf{WO}(v) = \mathsf{WO}[T] \\
\cong \big\downarrow & & \big\downarrow \gamma_T \\
\mathsf{WO}[T(H_v)_{v \in T}] & \xrightarrow{\gamma_{T(H_v)}} & \mathsf{WO}\binom{d}{\underline{c}}
\end{array}
\tag{6.3.12}
$$

is commutative. By Lemma 6.3.6 and the naturality of coends, there are canonical isomorphisms

$$
\underset{v \in T}{\bigotimes} \mathsf{WO}[H_v] \cong \underset{v \in T}{\bigotimes} \int^{\{D_{vu}\} \in \prod_{u \in H_v} \underline{\mathsf{Tree}}^{\mathfrak{C}}(u)} \left(\underset{u \in H_v}{\bigotimes} \mathsf{J}[D_{vu}] \right) \otimes \left(\underset{u \in H_v}{\bigotimes} \mathsf{O}[D_{vu}] \right)
$$

$$
\cong \int^{\{D_{vu}\} \in \prod_{v \in T, u \in H_v} \underline{\mathsf{Tree}}^{\mathfrak{C}}(u)} \left(\underset{v,u}{\bigotimes} \mathsf{J}[D_{vu}] \right) \otimes \left(\underset{v,u}{\bigotimes} \mathsf{O}[D_{vu}] \right).
$$

In the following diagram, as in the proof of Lemma 6.3.9, we will abbreviate H_v to H and D_{vu} to D, with v and u running through $\mathsf{Vt}(T)$ and $\mathsf{Vt}(H_v)$, respectively, and omit identity morphisms and isomorphisms. To prove the commutativity of the diagram (6.3.12), it is enough to show that the outermost diagram in

$$
\left(\bigotimes_{v,u} \mathsf{J}[D]\right) \otimes \left(\bigotimes_{v,u} \mathsf{O}[D]\right) \xrightarrow{\;\overset{\otimes \pi}{v}\;} \left(\bigotimes_{v,u} \mathsf{J}[H(D)]\right) \otimes \left(\bigotimes_{v,u} \mathsf{O}[H(D)]\right)
$$

with vertical maps π on the left and π on the right leading to

$$
\mathsf{J}[T(H)(D)] \otimes \mathsf{O}[T(H)(D)]
$$

$$
\mathsf{J}[T(H)(D)] \otimes \mathsf{O}[T(H)(D)] \xrightarrow{\;\omega_{T(H)(D)}\;} \mathsf{WO}\binom{d}{\underline{c}}
$$

with the Id triangle and $\omega_{T(H)(D)}$ on the right.

is commutative. The upper trapezoid is commutative by inspection, and the triangle is commutative by definition.   □

Remark 6.3.13. By Theorem 6.3.11 the Boardman-Vogt construction can be iterated. In other words, given a $\mathfrak{C}$-colored operad O, since WO is a $\mathfrak{C}$-colored operad, it has a Boardman-Vogt construction $\mathsf{W}(\mathsf{WO})$, which is a $\mathfrak{C}$-colored operad. Then $\mathsf{W}(\mathsf{WO})$ has a Boardman-Vogt construction $\mathsf{W}(\mathsf{W}(\mathsf{WO}))$, which is again a $\mathfrak{C}$-colored operad, and so forth. However, we do not know of any applications of these iterated Boardman-Vogt constructions.   ◇

We can also express the operad structure on the Boardman-Vogt construction WO in terms of the generating operations in Definition 4.2.1. Using Theorem 4.4.13 and its proof on WO and Definition 6.3.7, we infer the following result.

Proposition 6.3.14. *Suppose* O *is a* $\mathfrak{C}$-*colored operad in* M.

(1) For each $c \in \mathfrak{C}$, *the* c-*colored unit of* WO *is the composition*

$$
\mathbb{1} = \mathsf{WO}[\uparrow_c] \xrightarrow{\;\cong\;} \mathbb{1} \otimes \mathbb{1} = \mathsf{J}[\uparrow_c] \otimes \mathsf{O}[\uparrow_c] \xrightarrow{\;\omega_{\uparrow c}\;} \mathsf{WO}\binom{c}{c}
$$

in which $\uparrow_c$ *is the* c-*colored exceptional edge in Example 3.1.18.*

(2) For each pair $(\underline{c}; d) \in \mathsf{Prof}(\mathfrak{C}) \times \mathfrak{C}$ *and permutation* $\sigma \in \Sigma_{|\underline{c}|}$, *the equivariant structure of* WO *is uniquely determined by the commutative diagrams*

$$
\begin{array}{ccc}
\mathsf{J}[T] \otimes \mathsf{O}[T] & \xrightarrow{\;\mathrm{Id}\;} & \mathsf{J}[T\sigma] \otimes \mathsf{O}[T\sigma] \\[4pt]
{\scriptstyle \omega_T}\big\downarrow & & \big\downarrow{\scriptstyle \omega_{T\sigma}} \\[4pt]
\mathsf{WO}\binom{d}{\underline{c}} & \xrightarrow{\;\sigma\;} & \mathsf{WO}\binom{d}{\underline{c}\sigma}
\end{array}
$$

for $T \in \underline{\mathsf{Tree}}^{\mathfrak{C}}\binom{d}{\underline{c}}$, *where* $T\sigma \in \underline{\mathsf{Tree}}^{\mathfrak{C}}\binom{d}{\underline{c}\sigma}$ *is the same as* T *except that its ordering is* $\zeta_T\sigma$ *with* ζ_T *the ordering of* T.

(3) For $\left(\underline{c} = (c_1, \ldots, c_n); d\right) \in \mathsf{Prof}(\mathfrak{C}) \times \mathfrak{C}$ *with* $n \geq 1$, $\underline{b}_j \in \mathsf{Prof}(\mathfrak{C})$ *for* $1 \leq j \leq n$, *and* $\underline{b} = (\underline{b}_1, \ldots, \underline{b}_n)$, *the operadic composition* γ *of* WO *is uniquely determined by*

the commutative diagrams

$$J[T] \otimes O[T] \otimes \bigotimes_{j=1}^{n} \left(J[T_j] \otimes O[T_j] \right) \xrightarrow[\cong]{\text{permute}} \left(J[T] \otimes \bigotimes_{j=1}^{n} J[T_j] \right) \otimes \left(O[T] \otimes \bigotimes_{j=1}^{n} O[T_j] \right)$$

with left vertical map $\left(\omega_T, \otimes_j \omega_{T_j} \right)$, right side $(\pi, \cong)$ to $J[G] \otimes O[G]$, then ω_G,

$$WO\binom{d}{c} \otimes \bigotimes_{j=1}^{n} WO\binom{c_j}{b_j} \xrightarrow{\quad \gamma \quad} WO\binom{d}{b}$$

for $T \in \underline{\text{Tree}}^{\mathfrak{C}}\binom{d}{c}$, $T_j \in \underline{\text{Tree}}^{\mathfrak{C}}\binom{c_j}{b_j}$ *for* $1 \le j \le n$, *and*

$$G = \text{Graft}(T; T_1, \ldots, T_n) \in \underline{\text{Tree}}^{\mathfrak{C}}\binom{d}{b}$$

the grafting (3.3.1). *Here* $\pi = \otimes_S 1$ *is the morphism in Lemma 6.2.7 for the grafting* G.

Remark 6.3.15. In the last part of Proposition 6.3.14, the morphism π is $\otimes_S 1$, in which $1 : \mathbb{1} \longrightarrow J$ is part of the commutative segment J. The set S is defined as the set of internal edges in the grafting G that are neither in T nor in any of the T_j. For example, if neither T nor any of the T_j is an exceptional edge, then S has exactly n elements, one for each input of T. $\diamond$

6.4 Augmentation

In this section, we observe that the Boardman-Vogt construction is augmented over the identity functor on the category of colored operads. The augmentation induces a change-of-operad adjunction between the category of algebras over the Boardman-Vogt construction and the category of algebras over the original colored operad. Furthermore, this adjunction is natural with respect to operad morphisms. In the next section, we will see that in $\mathsf{Chain}_{\mathbb{K}}$ the change-of-operad adjunction induced by the augmentation is always a Quillen equivalence.

Recall the concept of an operad morphism in Definition 5.1.3. First we define what the Boardman-Vogt construction does to an operad morphism.

Lemma 6.4.1. *Suppose* $f : (O, \gamma^O) \longrightarrow (P, \gamma^P)$ *is an operad morphism with* O *a* $\mathfrak{C}$-*colored operad and* P *a* $\mathfrak{D}$-*colored operad. Then there is an induced operad morphism*

$$Wf : WO \longrightarrow WP$$

that is $f : \mathfrak{C} \longrightarrow \mathfrak{D}$ *on colors and is entrywise defined by the commutative diagram*

$$J[T] \otimes O[T] \xrightarrow{\left(\text{Id}, \otimes_{v \in T} f \right)} J[fT] \otimes P[fT]$$

with left vertical ω_T, right vertical ω_{fT}, bottom $WO\binom{d}{c} \xrightarrow{Wf} WP\binom{fd}{fc}$.

*for $(\underline{c}; d) \in \mathsf{Prof}(\mathfrak{C}) \times \mathfrak{C}$ and $T \in \underline{\mathsf{Tree}}^{\mathfrak{C}}(\substack{d \\ \underline{c}})$, where $fT \in \underline{\mathsf{Tree}}^{\mathfrak{D}}(\substack{fd \\ f\underline{c}})$ is obtained from T
by applying f to its $\mathfrak{C}$-coloring.*

Proof. To see that the morphism $\mathsf{W}f$ is entrywise well-defined, it is enough to show
that the outermost diagram in

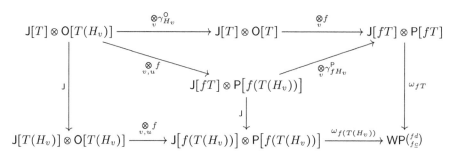

is commutative for each $T \in \underline{\mathsf{Tree}}^{\mathfrak{C}}(\substack{d \\ \underline{c}})$ and $\{H_v\} \in \prod_{v \in T} \underline{\mathsf{Tree}}^{\mathfrak{C}}(v)$, in which v and u
run through $\mathsf{Vt}(T)$ and $\mathsf{Vt}(H_v)$, respectively. The top triangle is commutative by
Proposition 5.1.4 because f is an operad morphism. The left trapezoid is commutative by inspection. The right trapezoid is commutative by the coend definition of
$\mathsf{WP}(\substack{fd \\ f\underline{c}})$ because there is a morphism

$$(fH_v) : f(T(H_v)) = (fT)(fH_v) \longrightarrow fT$$

in $\underline{\mathsf{Tree}}^{\mathfrak{D}}$.

To see that $\mathsf{W}f$ respects the operadic structure morphisms in Definition 6.3.7,
by Lemma 6.3.6, it is enough to show that the outermost diagram in

$$
\begin{array}{ccc}
\left(\underset{v}{\otimes}\mathsf{J}[H_v]\right) \otimes \left(\underset{v}{\otimes}\mathsf{O}[H_v]\right) & \xrightarrow{\underset{v,u}{\otimes} f} & \left(\underset{v}{\otimes}\mathsf{J}[fH_v]\right) \otimes \left(\underset{v}{\otimes}\mathsf{P}[fH_v]\right) \\
\Big\downarrow{\scriptstyle\pi} & & \Big\downarrow{\scriptstyle\pi} \\
\mathsf{J}[T(H_v)] \otimes \mathsf{O}[T(H_v)] & & \mathsf{J}[f(T(H_v))] \otimes \mathsf{P}[f(T(H_v))] \\
\underset{v,u}{\otimes} f\Big\downarrow & \xrightarrow{\;\;\mathrm{Id}\;\;} & \Big\downarrow{\scriptstyle\omega_{f(T(H_v))}} \\
\mathsf{J}[f(T(H_v))] \otimes \mathsf{P}[f(T(H_v))] & \xrightarrow{\;\omega_{f(T(H_v))}\;} & \mathsf{WP}(\substack{fd \\ f\underline{c}})
\end{array}
$$

is commutative. Both sub-diagrams are commutative by definition.   □

Interpretation 6.4.2. Intuitively, the morphism $\mathsf{W}f$ sends each decorated tree
$\mathsf{J}[T] \otimes \mathsf{O}[T]$ to the decorated tree $\mathsf{J}[fT] \otimes \mathsf{P}[fT]$ by applying f at each vertex. The
internal edges in T and in fT are canonically identified, so $\mathsf{J}[T]$ and $\mathsf{J}[fT]$ are the
same.   ◇

Recall from Definition 5.1.3 that $\mathsf{Operad}(\mathsf{M})$ denotes the category of all colored
operads in M.

Proposition 6.4.3. *The Boardman-Vogt construction defines a functor*

$$W : \mathsf{Operad}(M) \longrightarrow \mathsf{Operad}(M)$$

that preserves color sets.

Proof. The assignment on objects is defined by Theorem 6.3.11, and the assignment on morphisms is defined by Lemma 6.4.1. The Boardman-Vogt construction of a $\mathfrak{C}$-colored operad is a $\mathfrak{C}$-colored operad. The Boardman-Vogt construction preserves identity morphisms and composition of operad morphisms by the definition in Lemma 6.4.1. □

Next we define an augmentation of the Boardman-Vogt construction, which will allow us to compare the Boardman-Vogt construction with the original colored operad.

Theorem 6.4.4. *There is a natural transformation*

$$\eta : W \longrightarrow \mathrm{Id}_{\mathsf{Operad}(M)}$$

such that, for each $\mathfrak{C}$-colored operad (O, γ^O) in M, the operad morphism

$$\eta : WO \longrightarrow O$$

fixes colors and is defined entrywise by the commutative diagrams

$$
\begin{array}{ccccc}
J[T] \otimes O[T] & \xrightarrow{\underset{|T|}{(\otimes \epsilon, \mathrm{Id})}} & \mathbb{1}[T] \otimes O[T] & \xrightarrow{\cong} & O[T] \\
{\scriptstyle \omega_T} \downarrow & & & & \downarrow {\scriptstyle \gamma_T^O} \\
WO\binom{d}{\underline{c}} & & \xrightarrow{\quad\quad \eta \quad\quad} & & O\binom{d}{\underline{c}}
\end{array}
$$

for $(\underline{c}; d) \in \mathsf{Prof}(\mathfrak{C}) \times \mathfrak{C}$ and $T \in \underline{\mathsf{Tree}}^{\mathfrak{C}}\binom{d}{\underline{c}}$.

Proof. To see that the morphism η is entrywise well-defined, suppose

$$(H_v)_{v \in T} : T(H_v) \longrightarrow T$$

is a morphism in $\underline{\mathsf{Tree}}^{\mathfrak{C}}\binom{d}{\underline{c}}$. It is enough to show that the outermost diagram in

$$
\begin{array}{ccccc}
J[T] \otimes O[T(H_v)] & \xrightarrow{\otimes \gamma^O_{H_v}} & J[T] \otimes O[T] & \xrightarrow{\underset{|T|}{\otimes \epsilon}} & \mathbb{1}[T] \otimes O[T] \\
{\scriptstyle J} \downarrow & {\scriptstyle \cong \, \otimes \epsilon} \searrow & & & \downarrow {\scriptstyle \cong} \\
J[T(H_v)] \otimes O[T(H_v)] & & O[T(H_v)] & \xrightarrow{\otimes \gamma^O_{H_v}} & O[T] \\
{\scriptstyle \underset{|T(H_v)|}{\otimes} \epsilon} \downarrow & & \| & & \downarrow {\scriptstyle \gamma_T^O} \\
\mathbb{1}[T(H_v)] \otimes O[T(H_v)] & \xrightarrow{\cong} & O[T(H_v)] & \xrightarrow{\gamma^O_{T(H_v)}} & O\binom{d}{\underline{c}}
\end{array}
$$

is commutative. The top trapezoid is commutativity by definition. The left trapezoid is commutative by the fact that ϵ is the counit of the commutative segment J.

The lower right square is commutative by the associativity (4.4.11) of the operadic structure morphism γ^{O}.

To see that η is a morphism of $\mathfrak{C}$-colored operads, we must show that the diagram

$$
\begin{array}{ccc}
\mathsf{WO}[T] & \xrightarrow{\underset{v}{\otimes}\eta} & \mathsf{O}[T] \\
\gamma_T \downarrow & & \downarrow \gamma_T^{\mathsf{O}} \\
\mathsf{WO}\binom{d}{\underline{c}} & \xrightarrow{\eta} & \mathsf{O}\binom{d}{\underline{c}}
\end{array}
$$

is commutative for each $T \in \underline{\mathsf{Tree}}^{\mathfrak{C}}\binom{d}{\underline{c}}$. By Lemma 6.3.6, it is enough to show that the outermost diagram in

$$
\begin{array}{ccccc}
\left(\underset{v\in T}{\bigotimes}\mathsf{J}[H_v]\right)\otimes\left(\underset{v\in T}{\bigotimes}\mathsf{O}[H_v]\right) & \xrightarrow{\underset{v}{\otimes}(\cong\,\otimes\,\epsilon)} & \underset{v\in T}{\bigotimes}\mathsf{O}[H_v] & \xrightarrow{\underset{v}{\otimes}\gamma_{H_v}^{\mathsf{O}}} & \underset{v\in T}{\bigotimes}\mathsf{O}(v) \\
\pi\downarrow & & & & \| \\
\mathsf{J}[T(H_v)]\otimes\mathsf{O}[T(H_v)] & & \cong\downarrow & & \mathsf{O}[T] \\
\otimes\,\epsilon\downarrow & & & & \downarrow\gamma_T^{\mathsf{O}} \\
\mathbb{1}[T(H_v)]\otimes\mathsf{O}[T(H_v)] & \xrightarrow{\cong} & \mathsf{O}[T(H_v)] & \xrightarrow{\gamma_{T(H_v)}^{\mathsf{O}}} & \mathsf{O}\binom{d}{\underline{c}}
\end{array}
$$

is commutative. The left sub-diagram is commutative by the fact that ϵ is the counit of J. The right sub-diagram is commutative by the associativity of γ^{O}.

Finally, suppose $f : \mathsf{O} \longrightarrow \mathsf{P}$ is an operad morphism with $(\mathsf{O},\gamma^{\mathsf{O}})$ a $\mathfrak{C}$-colored operad and $(\mathsf{P},\gamma^{\mathsf{P}})$ a $\mathfrak{D}$-colored operad as in Lemma 6.4.1. To show that the diagram

$$
\begin{array}{ccc}
\mathsf{WO} & \xrightarrow{\mathsf{W}f} & \mathsf{WP} \\
\eta^{\mathsf{O}}\downarrow & & \downarrow\eta^{\mathsf{P}} \\
\mathsf{O} & \xrightarrow{f} & \mathsf{P}
\end{array}
\qquad (6.4.5)
$$

is commutative, it suffices to prove it in a typical $(\underline{c};d)$-entry. By the coend definition of $\mathsf{WO}\binom{d}{\underline{c}}$, it is enough to show that the outermost diagram in

$$
\begin{array}{ccc}
\mathsf{J}[T]\otimes\mathsf{O}[T] & \xrightarrow{\underset{v}{\otimes}f} & \mathsf{J}[fT]\otimes\mathsf{P}[fT] \\
\cong_{\mathsf{O}}\otimes\,\epsilon\downarrow & & \downarrow\cong_{\mathsf{O}}\otimes\,\epsilon \\
\mathsf{O}[T] & \xrightarrow{\underset{v}{\otimes}f} & \mathsf{P}[fT] \\
\gamma_T^{\mathsf{O}}\downarrow & & \downarrow\gamma_{fT}^{\mathsf{P}} \\
\mathsf{O}\binom{d}{\underline{c}} & \xrightarrow{f} & \mathsf{P}\binom{fd}{f\underline{c}}
\end{array}
$$

is commutative for each $T \in \underline{\mathsf{Tree}}^{\mathfrak{C}}\binom{d}{\underline{c}}$. The top square is commutative by naturality. The bottom square is commutative by Proposition 5.1.4. $\qquad\square$

Definition 6.4.6. For each $\mathfrak{C}$-colored operad O in M, we call the morphism $\eta : \mathsf{WO} \longrightarrow \mathsf{O}$ of $\mathfrak{C}$-colored operads the *augmentation* of the Boardman-Vogt construction.

The following change-of-operad adjunction is a special case of Theorem 5.1.8.

Corollary 6.4.7. *For each $\mathfrak{C}$-colored operad O in M, the augmentation $\eta :$ $\mathsf{WO} \longrightarrow \mathsf{O}$ induces an adjunction*

$$\mathsf{Alg_M(WO)} \xrightleftharpoons[\eta^*]{\eta_!} \mathsf{Alg_M(O)}$$

with left adjoint $\eta_!$.

Interpretation 6.4.8. This change-of-operad adjunction says that each O-algebra pulls back to a WO-algebra via the augmentation $\eta : \mathsf{WO} \longrightarrow \mathsf{O}$. Conversely, the left adjoint $\eta_!$ rectifies each WO-algebra to an O-algebra. Looking ahead, when O is a colored operad for algebraic quantum field theories or prefactorization algebras, the change of operad adjunction will allow us to go back and forth between algebraic quantum field theories (resp., prefactorization algebras) and homotopy algebraic quantum field theories (resp., homotopy prefactorization algebras). ◇

When applied to the commutative diagram (6.4.5) above, the change-of-operad adjunction in Theorem 5.1.8 yields the following result.

Corollary 6.4.9. *Suppose $f : \mathsf{O} \longrightarrow \mathsf{P}$ is an operad morphism in M. Then there is a diagram of change-of-operad adjunctions*

$$
\begin{array}{ccc}
\mathsf{Alg_M(WO)} & \xrightleftharpoons[(\mathsf{W}f)^*]{(\mathsf{W}f)_!} & \mathsf{Alg_M(WP)} \\[2mm]
\eta_!^{\mathsf{O}} \big\uparrow\big\downarrow (\eta^{\mathsf{O}})^* & & \eta_!^{\mathsf{P}} \big\uparrow\big\downarrow (\eta^{\mathsf{P}})^* \\[2mm]
\mathsf{Alg_M(O)} & \xrightleftharpoons[f^*]{f_!} & \mathsf{Alg_M(P)}
\end{array}
$$

in which

$$f_! \circ \eta_!^{\mathsf{O}} = \eta_!^{\mathsf{P}} \circ (\mathsf{W}f)_! \quad and \quad (\eta^{\mathsf{O}})^* \circ f^* = (\mathsf{W}f)^* \circ (\eta^{\mathsf{P}})^*.$$

Remark 6.4.10. The equality

$$f_! \circ \eta_!^{\mathsf{O}} = \eta_!^{\mathsf{P}} \circ (\mathsf{W}f)_!$$

says that the left adjoint diagram is commutative. Similarly, the equality

$$(\eta^{\mathsf{O}})^* \circ f^* = (\mathsf{W}f)^* \circ (\eta^{\mathsf{P}})^*$$

says that the right adjoint diagram is commutative. ◇

6.5   Homotopy Morita Equivalence

In this section, we construct an entrywise section of the augmentation, called the standard section, that preserves some of the operad structure, but is not an operad morphism in general. Using the standard section, we will observe that in familiar model categories such as Top, SSet, and $\mathsf{Chain}_{\mathbb{K}}$, the augmentation is a weak equivalence from the Boardman-Vogt construction to the original colored operad. Moreover, in $\mathsf{Chain}_{\mathbb{K}}$ the augmentation is always a homotopy Morita equivalence; i.e., the change-of-operad adjunction induced by the augmentation is a Quillen equivalence.

Definition 6.5.1. Suppose O is a $\mathfrak{C}$-colored operad in M. The *standard section* is the morphism
$$\xi : \mathsf{O} \longrightarrow \mathsf{WO} \in \mathsf{M}^{\mathsf{Prof}(\mathfrak{C}) \times \mathfrak{C}}$$
defined entrywise as the composition
$$\mathsf{O}\binom{d}{\underline{c}} \xrightarrow{\ \cong\ } \mathsf{J}\big[\mathrm{Cor}_{(\underline{c};d)}\big] \otimes \mathsf{O}\big[\mathrm{Cor}_{(\underline{c};d)}\big] \xrightarrow{\ \omega_{\mathrm{Cor}_{(\underline{c};d)}}\ } \mathsf{WO}\binom{d}{\underline{c}} \in \mathsf{M}$$
for $(\underline{c};d) \in \mathsf{Prof}(\mathfrak{C}) \times \mathfrak{C}$, where $\mathrm{Cor}_{(\underline{c};d)}$ is the $(\underline{c};d)$-corolla in Example 3.1.21.

First we observe that the standard section is an entrywise right inverse of the augmentation.

Proposition 6.5.2. *Suppose* O *is a* $\mathfrak{C}$-*colored operad in* M. *Then the diagram*

$$
\begin{array}{ccc}
\mathsf{O} & \xrightarrow{\ \xi\ } & \mathsf{WO} \\
 & {\scriptstyle \mathrm{Id}}\searrow & \Big\downarrow{\scriptstyle \eta} \\
 & & \mathsf{O}
\end{array}
$$

in $\mathsf{M}^{\mathsf{Prof}(\mathfrak{C}) \times \mathfrak{C}}$ *is commutative.*

Proof. For each pair $(\underline{c};d) \in \mathsf{Prof}(\mathfrak{C}) \times \mathfrak{C}$, the $(\underline{c};d)$-entry of the composition $\eta \circ \xi$ is the top-right composition in the commutative diagram

$$
\begin{array}{ccc}
\mathsf{O}\binom{d}{\underline{c}} \xrightarrow{\ \cong\ } \mathsf{J}\big[\mathrm{Cor}_{(\underline{c};d)}\big] \otimes \mathsf{O}\big[\mathrm{Cor}_{(\underline{c};d)}\big] & \xrightarrow{\ \omega_{\mathrm{Cor}_{(\underline{c};d)}}\ } & \mathsf{WO}\binom{d}{\underline{c}} \\
{\scriptstyle \mathrm{Id}}\searrow \quad \Big\downarrow{\scriptstyle \cong} & & \Big\downarrow{\scriptstyle \eta} \\
\mathsf{O}\big[\mathrm{Cor}_{(\underline{c};d)}\big] & \xrightarrow{\ \gamma^{\mathsf{O}}_{\mathrm{Cor}_{(\underline{c};d)}}\ } & \mathsf{O}\binom{d}{\underline{c}}.
\end{array}
$$

We finish the proof by noting that $\gamma^{\mathsf{O}}_{\mathrm{Cor}_{(\underline{c};d)}}$ is the identity morphism on $\mathsf{O}\binom{d}{\underline{c}}$ by the unity axiom in Definition 4.4.9. $\square$

One might hope that the standard section is a morphism of colored operads, but we will see that this is not the case in general. However, the standard section does preserve some of the operad structure.

Proposition 6.5.3. *Suppose* O *is a* $\mathfrak{C}$*-colored operad in* M. *Then the standard section*

$$\xi : \mathsf{O} \longrightarrow \mathsf{WO} \in \mathsf{M}^{\mathsf{Prof}(\mathfrak{C}) \times \mathfrak{C}}$$

in Definition 6.5.1 preserves the equivariant structure and the colored units.

Proof. As we explained in the proof of Theorem 4.4.13, the equivariant structure comes from the operadic structure morphisms $\gamma_{\mathsf{Cor}_{(\underline{c};d)}\tau}$, where $\mathsf{Cor}_{(\underline{c};d)}\tau$ is the permuted corolla in Example 3.1.22. Similarly, the colored units are the operadic structure morphisms $\gamma_{\uparrow_c}$ for the exceptional edges in Example 3.1.18.

To show that the standard section preserves these operadic structure morphisms, consider more generally a $\mathfrak{C}$-colored tree T with profile $(\underline{c}; d)$. The standard section preserves the operadic structure morphism for T if and only if the outermost diagram in

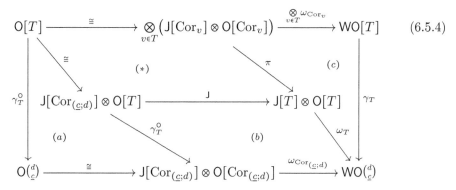

$$(6.5.4)$$

is commutative, in which Cor_v is the corolla with the same profile as v. The sub-diagram (a) is commutative by definition. The sub-diagram (b) is commutative by the coend definition of $\mathsf{WO}\binom{d}{\underline{c}}$ because

$$(T) : T = \mathsf{Cor}_{(\underline{c};d)}(T) \longrightarrow \mathsf{Cor}_{(\underline{c};d)}$$

is a morphism in $\underline{\mathsf{Tree}}^{\mathfrak{C}}\binom{d}{\underline{c}}$. The sub-diagram (c) is commutative by the definition (6.3.8) of γ_T in WO.

In the sub-diagram (∗), the morphism π is defined in Lemma 6.2.7 and is isomorphic to $\otimes_{|T|} 1$ with $1 : \mathbb{1} \longrightarrow J$ a part of the commutative segment J and $|T|$ the set of internal edges in T. The morphism J is isomorphic to $\otimes_{|T|} 0$, where $0 : \mathbb{1} \longrightarrow J$ is also a part of the commutative segment. If T is either a permuted corolla or an exceptional edge, then the set $|T|$ is empty. In this case, both π and J are the identity morphism of $\mathbb{1}$, so (∗) is also commutative. $\square$

Remark 6.5.5. One can see from the diagram (6.5.4) that the standard section does not preserve the operadic structure morphism γ_T in general. Indeed, in the sub-diagram (∗), the morphisms $\pi = \otimes_{|T|} 1$ and $\mathsf{J} = \otimes_{|T|} 0$ are different for most T. Intuitively, the morphism π assigns length 1 to every internal edge in T, while J assigns length 0 to every internal edge in T. $\diamond$

In the rest of this section, we will compare the categories of algebras over a colored operad and over its Boardman-Vogt construction. Recall from Definition 5.2.5 the concept of a weak equivalence between $\mathfrak{C}$-colored operads. Next we observe that in familiar cases, the augmentation is a weak equivalence.

Proposition 6.5.6. *Suppose* M *is* Top, SSet, Chain$_\mathbb{K}$, *or* Cat *with the model category structure in Example 5.2.1 and with the commutative segment in Example 6.2.4, and* O *is a* $\mathfrak{C}$-*colored operad in* M. *Then the augmentation* $\eta : \mathsf{WO} \longrightarrow \mathsf{O}$ *is a weak equivalence.*

Proof. Let us consider the case M = Top with $J = [0,1]$; the other cases are proved similarly. By Proposition 6.5.2, we already know that

$$\eta \circ \xi = \mathrm{Id}_\mathsf{O} \in \mathsf{M}^{\mathrm{Prof}(\mathfrak{C}) \times \mathfrak{C}}.$$

It remains to show that

$$\xi \circ \eta : \mathsf{WO} \longrightarrow \mathsf{WO}$$

is homotopic to the identity morphism. For each $p \in [0,1]$, define H_p by the commutative diagrams

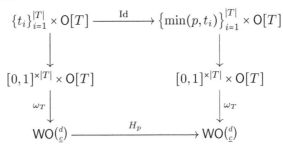

for $T \in \underline{\mathsf{Tree}}^{\mathfrak{C}}\binom{d}{\underline{c}}$ and $t_i \in [0,1]$ for $1 \leq i \leq |T|$. In other words, replace every internal edge length by its minimum with p. Then H_1 is the identity morphism, and $H_0 = \xi \circ \eta$ by the coend definition of $\mathsf{WO}\binom{d}{\underline{c}}$. So $\{H_p\}_{p \in [0,1]}$ defines a homotopy from $\xi \circ \eta$ to the identity morphism. $\square$

Remark 6.5.7. A statement and a proof similar to Proposition 6.5.6 for Top were first given in [Boardman and Vogt (1972); Vogt (2003)]. $\diamond$

In abstract algebra, two unital associative rings are said to be *Morita equivalent* if their categories of left modules are equivalent. Using the category of algebras, a similar concept of Morita equivalence also makes sense for colored operads. Moreover, in the presence of a model category structure in the base category, it makes sense to consider a homotopy version of a Morita equivalence.

Definition 6.5.8. Suppose $f : \mathsf{O} \longrightarrow \mathsf{P}$ is an operad morphism in a monoidal model category M with O and P admissible. Then we say that f is a *homotopy Morita equivalence* if the change-of-operad adjunction

$$\mathsf{Alg}_\mathsf{M}(\mathsf{O}) \underset{f^*}{\overset{f_!}{\rightleftarrows}} \mathsf{Alg}_\mathsf{M}(\mathsf{P})$$

is a Quillen equivalence.

Remark 6.5.9. Suppose given an operad morphism $f : O \longrightarrow P$ between admissible colored operads, such as the augmentation $\eta : \mathsf{WO} \longrightarrow O$ of a colored operad O in $\mathsf{M} = \mathsf{Top}$, SSet, $\mathsf{Chain}_{\mathbb{K}}$, or Cat. Then the change-of-operad adjunction $f_! \dashv f^*$ is already a Quillen adjunction. Indeed, fibrations and acyclic fibrations in the algebra categories are defined entrywise in M, so they are preserved by the right adjoint f^*. Therefore, the concept of a homotopy Morita equivalence is well-defined. $\diamond$

Combining Proposition 6.5.6 and Example 5.2.8, we obtain the following result that says that the augmentation of each colored operad over $\mathsf{Chain}_{\mathbb{K}}$ is a homotopy Morita equivalence.

Corollary 6.5.10. *Suppose O is a $\mathfrak{C}$-colored operad in $\mathsf{M} = \mathsf{Chain}_{\mathbb{K}}$, where $\mathbb{K}$ is a field of characteristic zero. Then the augmentation $\eta : \mathsf{WO} \longrightarrow O$ is a homotopy Morita equivalence. In other words, the change-of-operad adjunction*

$$\mathsf{Alg}_{\mathsf{M}}(\mathsf{WO}) \underset{\eta^*}{\overset{\eta_!}{\rightleftarrows}} \mathsf{Alg}_{\mathsf{M}}(O)$$

induced by the augmentation $\eta : \mathsf{WO} \longrightarrow O$ is a Quillen equivalence.

Since Quillen equivalences have the 2-out-of-3 property, combining Corollary 6.4.9 and Corollary 6.5.10, we infer that the Boardman-Vogt construction preserves homotopy Morita equivalences over $\mathsf{Chain}_{\mathbb{K}}$.

Corollary 6.5.11. *Suppose $f : O \longrightarrow P$ is a homotopy Morita equivalence in $\mathsf{M} = \mathsf{Chain}_{\mathbb{K}}$, where $\mathbb{K}$ is a field of characteristic zero. Then $\mathsf{W}f : \mathsf{WO} \longrightarrow \mathsf{WP}$ is also a homotopy Morita equivalence.*

Remark 6.5.12. Although Corollary 6.5.10 and Corollary 6.5.11 are only stated for $\mathsf{Chain}_{\mathbb{K}}$, this is sufficient for most applications to (homotopy) algebraic quantum field theories and (homotopy) prefactorization algebras, which are often considered over $\mathsf{Chain}_{\mathbb{K}}$. $\diamond$

6.6   Filtration

In this section, we discuss a natural filtration of the Boardman-Vogt construction. None of this is needed for applications to algebraic quantum field theories and prefactorization algebras. The rest of this book is independent of this section, so the reader may skip this section safely.

In the coend definition of the Boardman-Vogt construction in Definition 6.3.1, we used the substitution category $\underline{\mathsf{Tree}}^{\mathfrak{C}}\binom{d}{\underline{c}}$ in Definition 3.2.11. To obtain a natural filtration of the Boardman-Vogt construction, we will use smaller substitution categories.

Definition 6.6.1. For each pair $(\underline{c}; d) \in \mathsf{Prof}(\mathfrak{C}) \times \mathfrak{C}$ and each $n \geq 0$, define the nth *substitution category* $\underline{\mathsf{Tree}}_n^{\mathfrak{C}}\binom{d}{\underline{c}}$ as the full subcategory of the substitution category

$\underline{\mathsf{Tree}}^{\mathfrak{C}}\binom{d}{\underline{c}}$ consisting of $\mathfrak{C}$-colored trees with profile $\binom{d}{\underline{c}}$ and with at most n internal edges.

Example 6.6.2. If $\underline{c} \neq d$, then $\underline{\mathsf{Tree}}_0^{\mathfrak{C}}\binom{d}{\underline{c}}$ contains only permuted corollas with profile $\binom{d}{\underline{c}}$. If $\underline{c} = d$, then $\underline{\mathsf{Tree}}_0^{\mathfrak{C}}\binom{d}{d}$ contains only the linear graph $\mathrm{Lin}_{(d,d)}$ and the d-colored exceptional edge $\uparrow_d$. ◇

Definition 6.6.3. Suppose O is a $\mathfrak{C}$-colored operad in M, and $n \geq 0$. Define the object $\mathsf{W}_n\mathsf{O} \in \mathsf{M}^{\mathsf{Prof}(\mathfrak{C}) \times \mathfrak{C}}$ entrywise as the coend

$$\mathsf{W}_n\mathsf{O}\binom{d}{\underline{c}} = \int^{T \in \underline{\mathsf{Tree}}_n^{\mathfrak{C}}\binom{d}{\underline{c}}} \mathsf{J}[T] \otimes \mathsf{O}[T]$$

for $(\underline{c}; d) \in \mathsf{Prof}(\mathfrak{C}) \times \mathfrak{C}$. Here

$$\mathsf{J} : \underline{\mathsf{Tree}}_n^{\mathfrak{C}}\binom{d}{\underline{c}}^{\mathsf{op}} \longrightarrow \mathsf{M} \quad \text{and} \quad \mathsf{O} : \underline{\mathsf{Tree}}_n^{\mathfrak{C}}\binom{d}{\underline{c}} \longrightarrow \mathsf{M}$$

are the restrictions of the functors in Definition 6.2.5 and Corollary 4.4.15, respectively.

- We call $\mathsf{W}_n\mathsf{O} \in \mathsf{M}^{\mathsf{Prof}(\mathfrak{C}) \times \mathfrak{C}}$ the *nth filtration of the Boardman-Vogt construction* of O.
- For $T \in \underline{\mathsf{Tree}}_n^{\mathfrak{C}}\binom{d}{\underline{c}}$, we write

$$\omega_T : \mathsf{J}[T] \otimes \mathsf{O}[T] \longrightarrow \mathsf{W}_n\mathsf{O}\binom{d}{\underline{c}}$$

  for the natural morphism.

Proposition 6.6.4. *Suppose O is a $\mathfrak{C}$-colored operad in M. Then there is a natural diagram*

$$\mathsf{O} \cong \mathsf{W}_0\mathsf{O} \xrightarrow{\iota_1} \mathsf{W}_1\mathsf{O} \xrightarrow{\iota_2} \mathsf{W}_2\mathsf{O} \xrightarrow{\iota_3} \cdots \longrightarrow \operatorname*{colim}_{n \geq 1} \mathsf{W}_n\mathsf{O} \cong \mathsf{WO}$$

in $\mathsf{M}^{\mathsf{Prof}(\mathfrak{C}) \times \mathfrak{C}}$, in which ι_n is defined entrywise by the subcategory inclusion

$$\underline{\mathsf{Tree}}_{n-1}^{\mathfrak{C}}\binom{d}{\underline{c}} \subset \underline{\mathsf{Tree}}_n^{\mathfrak{C}}\binom{d}{\underline{c}}.$$

Proof. The morphism $\mathsf{O} \longrightarrow \mathsf{W}_0\mathsf{O}$ in $\mathsf{M}^{\mathsf{Prof}(\mathfrak{C}) \times \mathfrak{C}}$ defined entrywise as the composition

$$\mathsf{O}\binom{d}{\underline{c}} \xrightarrow{\cong} \mathsf{J}\big[\mathrm{Cor}_{(\underline{c};d)}\big] \otimes \mathsf{O}\big[\mathrm{Cor}_{(\underline{c};d)}\big] \xrightarrow{\omega_{\mathrm{Cor}_{(\underline{c};d)}}} \mathsf{W}_0\mathsf{O}\binom{d}{\underline{c}}$$

and the morphism $\mathsf{W}_0\mathsf{O} \longrightarrow \mathsf{O}$ in $\mathsf{M}^{\mathsf{Prof}(\mathfrak{C}) \times \mathfrak{C}}$ defined entrywise by the commutative diagrams

$$\begin{array}{ccc} \mathsf{J}[T] \otimes \mathsf{O}[T] & \xrightarrow{\cong} & \mathsf{O}[T] \\ {\scriptstyle \omega_T}\downarrow & & \downarrow{\scriptstyle \gamma_T^{\mathsf{O}}} \\ \mathsf{W}_0\mathsf{O}\binom{d}{\underline{c}} & \longrightarrow & \mathsf{O} \end{array}$$

for $T \in \underline{\mathsf{Tree}}_0^{\mathfrak{C}}\binom{d}{\underline{c}}$ are mutual inverses by the coend definition of $\mathsf{W}_0\mathsf{O}\binom{d}{\underline{c}}$. The last isomorphism follows from the isomorphism

$$\operatorname*{colim}_{n \geq 1} \underline{\mathsf{Tree}}_n^{\mathfrak{C}}\binom{d}{\underline{c}} \xrightarrow{\cong} \underline{\mathsf{Tree}}^{\mathfrak{C}}\binom{d}{\underline{c}}$$

of categories for each $(\underline{c}; d) \in \mathsf{Prof}(\mathfrak{C}) \times \mathfrak{C}$. $\square$

To understand the above filtration better, we will decompose each morphism ι_n further as a pushout. To define such a pushout, we will need the following definitions. Recall the exceptional edges in Example 3.1.18 and the permuted corollas in Example 3.1.22. The intuitive meaning of the concepts in the following definition is explained in Remark 6.6.6.

Definition 6.6.5. Suppose O is a $\mathfrak{C}$-colored operad in M, and $n \geq 1$.

(1) For each $(\underline{c}; d) \in \mathsf{Prof}(\mathfrak{C}) \times \mathfrak{C}$, define $\underline{\mathsf{Tree}}^{\mathfrak{C}}_{=n}\binom{d}{\underline{c}}$ as the subcategory of $\underline{\mathsf{Tree}}^{\mathfrak{C}}_{n}\binom{d}{\underline{c}}$ consisting of $\mathfrak{C}$-colored trees with profile $\binom{d}{\underline{c}}$ and with exactly n internal edges, in which a morphism $(H_v)_{v \in T} : K \longrightarrow T$ must have H_v a permuted corolla for each $v \in \mathsf{Vt}(T)$.

(2) Define $\mathsf{W}_{=n}\mathsf{O} \in \mathsf{M}^{\mathsf{Prof}(\mathfrak{C}) \times \mathfrak{C}}$ entrywise as the coend

$$\mathsf{W}_{=n}\mathsf{O}\binom{d}{\underline{c}} = \int^{T \in \underline{\mathsf{Tree}}^{\mathfrak{C}}_{=n}\binom{d}{\underline{c}}} J[T] \otimes \mathsf{O}[T]$$

for $(\underline{c}; d) \in \mathsf{Prof}(\mathfrak{C}) \times \mathfrak{C}$, in which

$$J : \underline{\mathsf{Tree}}^{\mathfrak{C}}_{n}\binom{d}{\underline{c}}^{\mathsf{op}} \longrightarrow \mathsf{M} \quad \text{and} \quad \mathsf{O} : \underline{\mathsf{Tree}}^{\mathfrak{C}}_{n}\binom{d}{\underline{c}} \longrightarrow \mathsf{M}$$

are the restrictions of the functors in Definition 6.2.5 and Corollary 4.4.15, respectively.

(3) In a $\mathfrak{C}$-colored tree T, a *tunnel* is a vertex v with $|\mathsf{in}(v)| = 1$ such that the input and the output have the same color. For a tunnel v whose input has color c, we will write $\uparrow_v$ for the exceptional edge $\uparrow_c$. The set of all tunnels in T is denoted by $\mathsf{Tun}(T)$. Define the object $\mathsf{O}^-[T]$ and the morphism

$$\mathsf{O}^-[T] = \operatorname*{colim}_{\varnothing \neq S \subseteq \mathsf{Tun}(T)} \mathsf{O}\big[T(\uparrow_v)_{v \in S}\big] \xrightarrow{\ \alpha_T\ } \mathsf{O}[T] \in \mathsf{M}$$

in which the colimit is indexed by the category of non-empty subsets of $\mathsf{Tun}(T)$, where a morphism $S \longrightarrow S'$ is a subset inclusion $S' \subseteq S$. The morphisms that define the colimit and the morphism α_T are induced by the functor O in Corollary 4.4.15.

(4) For a $\mathfrak{C}$-colored tree $T \in \underline{\mathsf{Tree}}^{\mathfrak{C}}_{n}\binom{d}{\underline{c}}$, define the *decomposition category* $\mathsf{D}(T)$ in which an object is a morphism

$$(H_v)_{v \in K} : T = K(H_v)_{v \in K} \longrightarrow K \quad \text{in} \quad \underline{\mathsf{Tree}}^{\mathfrak{C}}_{n}\binom{d}{\underline{c}}$$

such that $\mathsf{Vt}(H_v) \neq \varnothing$ for all $v \in \mathsf{Vt}(K)$ and that at least one H_u has $|H_u| \geq 1$. A morphism

$$(G_u)_{u \in K'} : \big(T \xrightarrow{(H'_u)_{u \in K'}} K' \big) \longrightarrow \big(T \xrightarrow{(H_v)_{v \in K}} K \big) \quad \text{in} \quad \mathsf{D}(T)$$

is a morphism

$$K = K'(G_u)_{u \in K'} \xrightarrow{(G_u)_{u \in K'}} K' \quad \text{in} \quad \underline{\mathsf{Tree}}^{\mathfrak{C}}_{n}\binom{d}{\underline{c}}$$

such that $\mathsf{Vt}(G_u) \neq \varnothing$ for all $u \in K'$ and that the diagram

$$
\begin{array}{ccc}
T = K(H_v)_{v \in K} & \xrightarrow{(H_v)_{v \in K}} & K = K'(G_u)_{u \in K'} \\
\text{Id} \downarrow & & \downarrow (G_u)_{u \in K'} \\
T = K'(H'_u)_{u \in K'} & \xrightarrow{(H'_u)_{u \in K'}} & K'
\end{array}
$$

in $\underline{\mathsf{Tree}}_n^{\mathfrak{C}}\binom{d}{\underline{c}}$ is commutative. Identity morphisms and composition are induced by those in $\underline{\mathsf{Tree}}_n^{\mathfrak{C}}\binom{d}{\underline{c}}$.

(5) For a $\mathfrak{C}$-colored tree $T \in \underline{\mathsf{Tree}}_n^{\mathfrak{C}}\binom{d}{\underline{c}}$, define the object $\mathsf{J}^-[T]$ and the morphism

$$
\mathsf{J}^-[T] = \operatorname*{colim}_{(H_v)_{v \in K} \in \mathsf{D}(T)} \mathsf{J}[K] \xrightarrow{\ \beta_T\ } \mathsf{J}[T] \in \mathsf{M}
$$

induced by the functor J in Definition 6.2.5.

(6) For $T \in \underline{\mathsf{Tree}}_n^{\mathfrak{C}}\binom{d}{\underline{c}}$, define the morphism δ_T as the pushout product $\beta_T \,\square\, \alpha_T$ in the diagram

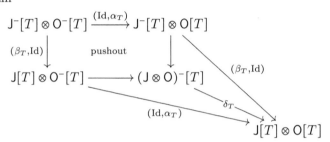

in M in which $(\mathsf{J} \otimes \mathsf{O})^-[T]$ is defined as the pushout of the square.

(7) Define the object $\mathsf{W}'_{n-1}\mathsf{O}\binom{d}{\underline{c}}$ and the morphism ρ

$$
\mathsf{W}'_{n-1}\mathsf{O}\binom{d}{\underline{c}} = \operatorname*{colim}_{T \in \underline{\mathsf{Tree}}_{=n}^{\mathfrak{C}}\binom{d}{\underline{c}}} (\mathsf{J} \otimes \mathsf{O})^-[T] \xrightarrow{\ \rho\ } \mathsf{W}_{n-1}\mathsf{O}\binom{d}{\underline{c}} \in \mathsf{M}
$$

in which the colimit is defined using the equivariant structure of O. The morphism ρ is defined by the commutative diagrams

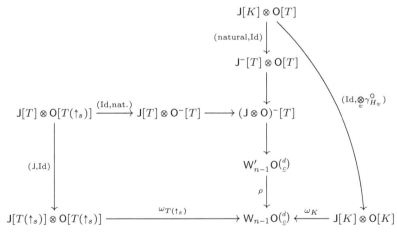

for

- $T \in \underline{\mathsf{Tree}}^{\mathfrak{C}}_{=n}\binom{d}{c}$,
- $\varnothing \neq S \subseteq \mathsf{Tun}(T)$ with $s \in S$, and
- objects $(H_v)_{v \in K} : T \longrightarrow K$ in $\mathsf{D}(T)$.

(8) Define the morphism δ by the commutative diagrams

$$
\begin{array}{ccc}
(\mathsf{J} \otimes \mathsf{O})^-[T] & \xrightarrow{\ \delta_T\ } & \mathsf{J}[T] \otimes \mathsf{O}[T] \\[2pt]
\text{natural} \Big\downarrow & & \Big\downarrow{\omega_T} \\[2pt]
\mathsf{W}'_{n-1}\mathsf{O}\binom{d}{c} & \xrightarrow{\ \ \delta\ \ } & \mathsf{W}_{=n}\mathsf{O}\binom{d}{c}
\end{array}
$$

for $T \in \underline{\mathsf{Tree}}^{\mathfrak{C}}_{=n}\binom{d}{c}$.

Interpretation 6.6.6. Let us explain the intuitive meaning of the concepts in the previous definition.

- In the category $\underline{\mathsf{Tree}}^{\mathfrak{C}}_{=n}\binom{d}{c}$, a morphism is only allowed to change the ordering at each vertex. In particular, there is no effect on the set of internal edges.
- $\mathsf{W}_{=n}\mathsf{O}\binom{d}{c}$ is the coend of the decorated trees $\mathsf{J}[T] \otimes \mathsf{O}[T]$ over the category $\underline{\mathsf{Tree}}^{\mathfrak{C}}_{=n}\binom{d}{c}$. The J variable is unaffected by the morphisms in $\underline{\mathsf{Tree}}^{\mathfrak{C}}_{=n}\binom{d}{c}$ because they do not change the set of internal edges.
- $\mathsf{O}^-[T]$ is the sub-object of the vertex-decorated tree $\mathsf{O}[T] = \bigotimes_{v \in T} \mathsf{O}(v)$ in which at least one tunnel is decorated by the corresponding colored unit of O.
- $\mathsf{J}^-[T]$ is the sub-object of the internal edge-decorated tree $\mathsf{J}[T] = \mathsf{J}^{\otimes|T|}$ in which at least one internal edge is assigned length $0 : \mathbb{1} \longrightarrow \mathsf{J}$.
- The pushout $(\mathsf{J} \otimes \mathsf{O})^-[T]$ is the sub-object of the decorated tree $\mathsf{J}[T] \otimes \mathsf{O}[T]$ such that
  - at least one tunnel is decorated by the corresponding colored unit of O,
  - or at least one internal edge is assigned length 0,
  - or both.
- $\mathsf{W}'_{n-1}\mathsf{O}\binom{d}{c}$ is the colimit of these sub-objects over the category $\underline{\mathsf{Tree}}^{\mathfrak{C}}_{=n}\binom{d}{c}$. The morphism

$$
\delta : \mathsf{W}'_{n-1}\mathsf{O}\binom{d}{c} \longrightarrow \mathsf{W}_{=n}\mathsf{O}\binom{d}{c}
$$

  is the sum of the sub-object inclusions over the category $\underline{\mathsf{Tree}}^{\mathfrak{C}}_{=n}\binom{d}{c}$.
- The morphism

$$
\rho : \mathsf{W}'_{n-1}\mathsf{O}\binom{d}{c} \longrightarrow \mathsf{W}_{n-1}\mathsf{O}\binom{d}{c}
$$

  reduces the number of internal edges in each decorated tree in its domain using
  - the functor J for a morphism $T(\uparrow_s)_{s \in S} \longrightarrow T$;
  - the functor O for an object $(H_v)_{v \in K} : T \longrightarrow K$ in $\mathsf{D}(T)$. ◇

The main categorical property of the filtration in Proposition 6.6.4 is the following observation.

Theorem 6.6.7. *Suppose* O *is a* $\mathfrak{C}$-*colored operad in* M, $n \geq 1$, *and* $(\underline{c}; d) \in \mathsf{Prof}(\mathfrak{C}) \times$ $\mathfrak{C}$. *Then there is a pushout*

$$
\begin{array}{ccc}
\mathsf{W}'_{n-1}\mathsf{O}(^d_{\underline{c}}) & \xrightarrow{\ \delta\ } & \mathsf{W}_{=n}\mathsf{O}(^d_{\underline{c}}) \\
\downarrow{\scriptstyle\rho} & & \downarrow \\
\mathsf{W}_{n-1}\mathsf{O}(^d_{\underline{c}}) & \xrightarrow{\ \iota_n\ } & \mathsf{W}_n\mathsf{O}(^d_{\underline{c}})
\end{array}
$$

in M, *in which the right vertical morphism is induced by the subcategory inclusion* $\underline{\mathsf{Tree}}^{\mathfrak{C}}_{=n}(^d_{\underline{c}}) \subset \underline{\mathsf{Tree}}^{\mathfrak{C}}_n(^d_{\underline{c}})$.

Proof. The commutativity of the square follows from the definition of $\mathsf{W}_n\mathsf{O}(^d_{\underline{c}})$ as a coend over the category $\underline{\mathsf{Tree}}^{\mathfrak{C}}_n(^d_{\underline{c}})$. To see that it has the universal property of a pushout, first note that a $\mathfrak{C}$-colored tree $T \in \underline{\mathsf{Tree}}^{\mathfrak{C}}_n(^d_{\underline{c}})$ is either in $\underline{\mathsf{Tree}}^{\mathfrak{C}}_{=n}(^d_{\underline{c}})$ or in $\underline{\mathsf{Tree}}^{\mathfrak{C}}_{n-1}(^d_{\underline{c}})$, but not both. Suppose given a commutative solid-arrow diagram

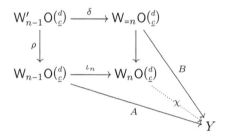

in M for some object Y and morphisms A and B. Then the only possible extension χ must be defined by (i) the commutative diagram

$$
\begin{array}{ccc}
\mathsf{J}[T] \otimes \mathsf{O}[T] & \xrightarrow{\ \omega_T\ } & \mathsf{W}_{=n}\mathsf{O}(^d_{\underline{c}}) \\
\downarrow{\scriptstyle\omega_T} & & \downarrow{\scriptstyle B} \\
\mathsf{W}_n\mathsf{O}(^d_{\underline{c}}) & \xrightarrow{\ \chi\ } & Y
\end{array}
$$

if $T \in \underline{\mathsf{Tree}}^{\mathfrak{C}}_{=n}(^d_{\underline{c}})$, and (ii) the commutative diagram

$$
\begin{array}{ccc}
\mathsf{J}[T] \otimes \mathsf{O}[T] & \xrightarrow{\ \omega_T\ } & \mathsf{W}_{n-1}\mathsf{O}(^d_{\underline{c}}) \\
\downarrow{\scriptstyle\omega_T} & & \downarrow{\scriptstyle A} \\
\mathsf{W}_n\mathsf{O}(^d_{\underline{c}}) & \xrightarrow{\ \chi\ } & Y
\end{array}
$$

if $T \in \underline{\mathsf{Tree}}^{\mathfrak{C}}_{n-1}(^d_{\underline{c}})$. Using the coend definition of $\mathsf{W}_n\mathsf{O}(^d_{\underline{c}})$, one checks that this candidate χ is indeed a morphism $\mathsf{W}_n\mathsf{O}(^d_{\underline{c}}) \longrightarrow Y$ that uniquely extends both A and B. $\qquad\square$

Remark 6.6.8. Proposition 6.6.4 and Theorem 6.6.7 together imply that, in the one-colored case, our coend definition of the Boardman-Vogt construction is isomorphic to the one given by Berger and Moerdijk [Berger and Moerdijk (2006)].

The main difference is that in [Berger and Moerdijk (2006)] WO is entrywise defined as a sequential colimit as in the filtration in Proposition 6.6.4, in which the morphisms ι_n are inductively defined using a pushout similar to the one in Theorem 6.6.7. In contrast, our coend definition of WO, which does not appear in [Berger and Moerdijk (2006)], describes the Boardman-Vogt construction in one step. The coend definition is crucial for our understanding of WO-algebras, as in the Coherence Theorem 7.2.1. It will also be important for our study of homotopy algebraic quantum field theories and homotopy prefactorization algebras. ◇

Remark 6.6.9. In nice enough situations (e.g., when M = Chain$_{\mathbb{K}}$ with $\mathbb{K}$ a field of characteristic zero), one can use the pushouts in Theorem 6.6.7 to prove that each entry of each ι_n is an acyclic cofibration, so the same is true for the standard section $\xi : \mathsf{O} \longrightarrow \mathsf{WO}$. Furthermore, the augmentation $\eta : \mathsf{WO} \longrightarrow \mathsf{O}$ is a cofibrant replacement of O in the model category of $\mathfrak{C}$-colored operads in M. In the one-colored case, these properties are proved in [Berger and Moerdijk (2006)] Sections 4 and 5. In the general colored case and for even more general objects than colored operads, these properties are proved in [Yau and Johnson (2017)] Chapters 3-7. We refer the interested reader to these sources for more details. For applications to (homotopy) algebraic quantum field theories and (homotopy) prefactorization algebras, we will not need to use these homotopical properties. ◇

Chapter 7

Algebras over the Boardman-Vogt Construction

This chapter is about the structure of algebras over the Boardman-Vogt construction of a colored operad and some key examples. The categorical setting is the same as before, so $(\mathsf{M}, \otimes, \mathbb{1})$ is a cocomplete symmetric monoidal closed category with an initial object $\varnothing$. Unless otherwise specified, $\mathfrak{C}$ is an arbitrary non-empty set.

7.1 Overview

Since we intend to apply the Boardman-Vogt construction to colored operads for algebraic quantum field theories and prefactorization algebras, it is crucial that we be able to describe explicitly the structure of algebras over the Boardman-Vogt construction.

In Section 7.2 we prove a coherence theorem for algebras over the Boardman-Vogt construction, which describes such an algebra explicitly in terms of certain structure morphisms and four axioms. Our coend definition of the Boardman-Vogt construction plays an important role here. It allows us to phrase the structure morphisms and axioms explicitly and non-inductively in terms of trees and tree substitution. We will use this coherence theorem many times in the rest of this book. The remaining sections of this chapter contain key examples that will be relevant in the discussion of homotopy algebraic quantum field theories and homotopy prefactorization algebras.

In Section 7.3 we explain the structure in homotopy coherent diagrams, which are algebras over the Boardman-Vogt construction of the C-diagram operad. A homotopy coherent C-diagram in M is a relaxed version of a C-diagram in M in which functoriality is replaced by specified homotopies that are also structure morphisms. An algebraic quantum field theory and a prefactorization algebra each has an underlying C-diagram in M. Therefore, a homotopy algebraic quantum field theory and a homotopy prefactorization algebra each has an underlying homotopy coherent C-diagram.

In Section 7.4 we discuss homotopy inverses in homotopy coherent C-diagrams. In a C-diagram X in M, if f is an isomorphism in C, then the structure morphism $X(f)$ is also invertible. In a homotopy coherent C-diagram, this invertibility is expressed homotopically with specified homotopies that are also structure morphisms. This will be important when we discuss a homotopy version of the time-slice axiom in homotopy algebraic quantum field theories and homotopy prefactorization algebras.

In Section 7.5 we discuss A_∞-algebras, which are algebras over the Boardman-Vogt construction of the associative operad As. They are monoids up to coherent higher homotopies. An algebraic quantum field theory is, in particular, a diagram of monoids. Therefore, in a homotopy algebraic quantum field theory, each entry is an A_∞-algebra.

In Section 7.6 we discuss E_∞-algebras, which are algebras over the Boardman-Vogt construction of the commutative operad Com. They are commutative monoids up to coherent higher homotopies. Commutative monoids appear in some entries of prefactorization algebras. In a homotopy prefactorization algebra, certain entries are E_∞-algebras.

In Section 7.7 we discuss homotopy coherent diagrams of A_∞-algebras. Every algebraic quantum field theory has an underlying C-diagram of monoids. So every homotopy algebraic quantum field theory has an underlying homotopy coherent C-diagram of A_∞-algebras. Roughly speaking, for a C-diagram of monoids, there are two directions in which homotopy can happen, namely the diagram direction and the monoid direction. A homotopy coherent C-diagram of A_∞-algebras combines both of these directions. In particular, it has an underlying homotopy coherent C-diagram in M as well as an underlying objectwise A_∞-algebra structure.

In Section 7.8 we discuss homotopy coherent diagrams of E_∞-algebras. We will see later that there are adjunctions comparing algebraic quantum field theories and prefactorization algebras, although they are usually not equal. However, as we will see in Section 10.7, there is one case where they coincide. When this happens, both the category of algebraic quantum field theories and the category of prefactorization algebras are canonically isomorphic to the category of C-diagrams of commutative monoids in M. Therefore, in this case homotopy algebraic quantum field theories and homotopy prefactorization algebras have the structure of homotopy coherent diagrams of E_∞-algebras.

7.2 Coherence Theorem

Recall from Definition 4.5.5 the concept of an algebra over a colored operad. In this section, we prove the following coherence result for algebras over the Boardman-Vogt construction of a colored operad.

Theorem 7.2.1. *Suppose* (O, γ^o) *is a* $\mathfrak{C}$-*colored operad in* M. *Then a* WO-*algebra is exactly a pair* (X, λ) *consisting of*

- *a* $\mathfrak{C}$-*colored object* X *in* M *and*

- *a structure morphism*

$$J[T] \otimes O[T] \otimes X_{\underline{c}} \xrightarrow{\lambda_T} X_d \in \mathsf{M} \tag{7.2.2}$$

for each $(\underline{c}; d) \in \mathsf{Prof}(\mathfrak{C}) \times \mathfrak{C}$ *and* $T \in \underline{\mathsf{Tree}}^{\mathfrak{C}}\binom{d}{\underline{c}}$

that satisfies the following four conditions.

Associativity *For* $\left(\underline{c} = (c_1, \ldots, c_n); d\right) \in \mathsf{Prof}(\mathfrak{C}) \times \mathfrak{C}$ *with* $n \geq 1$, $T \in \underline{\mathsf{Tree}}^{\mathfrak{C}}\binom{d}{\underline{c}}$, $T_j \in \underline{\mathsf{Tree}}^{\mathfrak{C}}\binom{c_j}{\underline{b}_j}$ *for* $1 \leq j \leq n$, $\underline{b} = (\underline{b}_1, \ldots, \underline{b}_n)$, *and*

$$G = \mathsf{Graft}(T; T_1, \ldots, T_n) \in \underline{\mathsf{Tree}}^{\mathfrak{C}}\binom{d}{\underline{b}}$$

the grafting (3.3.1), *the diagram*

$$\tag{7.2.3}$$

is commutative. Here $\pi = \otimes_S 1$ *is the morphism in Lemma 6.2.7 for the grafting* G.

Unity *For each* $c \in \mathfrak{C}$, *the composition*

$$X_c \xrightarrow{\cong} J[\uparrow_c] \otimes O[\uparrow_c] \otimes X_c \xrightarrow{\lambda_{\uparrow_c}} X_c \tag{7.2.4}$$

is the identity morphism of X_c.

Equivariance *For each* $T \in \underline{\mathsf{Tree}}^{\mathfrak{C}}\binom{d}{\underline{c}}$ *and permutation* $\sigma \in \Sigma_{|\underline{c}|}$, *the diagram*

$$\tag{7.2.5}$$

is commutative, in which $T\sigma \in \underline{\mathsf{Tree}}^{\mathfrak{C}}\binom{d}{\underline{c}\sigma}$ *is the same as* T *except that its ordering is* $\zeta_T\sigma$ *with* ζ_T *the ordering of* T. *The permutation* $\sigma^{-1} : X_{\underline{c}} \xrightarrow{\cong} X_{\underline{c}\sigma}$ *permutes the factors in* $X_{\underline{c}}$.

Wedge Condition *For $T \in \underline{\text{Tree}}^{\mathfrak{C}}(^d_{\underline{c}})$, $H_v \in \underline{\text{Tree}}^{\mathfrak{C}}(v)$ for each $v \in \text{Vt}(T)$, and $K = T(H_v)_{v \in T}$ the tree substitution, the diagram*

$$
\begin{array}{ccc}
J[T] \otimes O[K] \otimes X_{\underline{c}} & \xrightarrow{\left(\text{Id}, \otimes\gamma^o_{H_v}, \text{Id}\right)} & J[T] \otimes O[T] \otimes X_{\underline{c}} \\
\downarrow{\scriptstyle (J, \text{Id})} & & \downarrow{\scriptstyle \lambda_T} \\
J[K] \otimes O[K] \otimes X_{\underline{c}} & \xrightarrow{\lambda_K} & X_d
\end{array}
\tag{7.2.6}
$$

is commutative.

A morphism $f : (X, \lambda^X) \longrightarrow (Y, \lambda^Y)$ of WO-algebras is a morphism of the underlying $\mathfrak{C}$-colored objects that respects the structure morphisms in (7.2.2) in the obvious sense.

Proof. Given a WO-algebra (X, λ) in the sense of Definition 4.5.5, we define the structure morphism λ_T as the composition

$$
\begin{array}{ccc}
J[T] \otimes O[T] \otimes X_{\underline{c}} & \xrightarrow{\lambda_T} & X_d \\
\downarrow{\scriptstyle (\omega_T, \text{Id})} & & \| \\
\text{WO}(^d_{\underline{c}}) \otimes X_{\underline{c}} & \xrightarrow{\lambda} & X_d
\end{array}
\tag{7.2.7}
$$

for $T \in \underline{\text{Tree}}^{\mathfrak{C}}(^d_{\underline{c}})$. The wedge condition (7.2.6) is satisfied by the coend definition of $\text{WO}(^d_{\underline{c}})$ because

$$
(H_v)_{v \in T} : K = T(H_v)_{v \in T} \longrightarrow T
$$

is a morphism in $\underline{\text{Tree}}^{\mathfrak{C}}(^d_{\underline{c}})$. Using Proposition 6.3.14, we infer that the above associativity, unity, and equivariance conditions (7.2.3)-(7.2.5) follow from those in Definition 4.5.5.

Conversely, given a pair (X, λ) as in the statement above, we define the morphism

$$
\text{WO}(^d_{\underline{c}}) \otimes X_{\underline{c}} \xrightarrow{\lambda} X_d \in \text{M}
$$

for $(\underline{c}; d) \in \text{Prof}(\mathfrak{C}) \times \mathfrak{C}$ by insisting that the diagram (7.2.7) be commutative for all $T \in \underline{\text{Tree}}^{\mathfrak{C}}(^d_{\underline{c}})$. The wedge condition (7.2.6) guarantees that this morphism λ is entrywise well-defined. The associativity, unity, and equivariance axioms in Definition 4.5.5 now follow from the assumed associativity, unity, and equivariance conditions (7.2.3)-(7.2.5). $\qquad\square$

The following observation says that the colored units of O also act as the identity on a WO-algebra. We will use this result when we discuss homotopy inverses in homotopy algebraic quantum field theories and homotopy prefactorization algebras. Recall the linear graphs in Example 3.1.19.

Corollary 7.2.8. *Suppose* $(\mathsf{O}, \gamma^\mathsf{O})$ *is a* $\mathfrak{C}$*-colored operad in* M*, and* (X, λ) *is a* WO*-algebra. Then for each* $c \in \mathfrak{C}$*, the diagram*

$$
\begin{array}{ccc}
\mathsf{J}[\mathrm{Lin}_{(c,c)}] \otimes \mathsf{O}[\uparrow_c] \otimes X_c & \xrightarrow{(\mathrm{Id}, \gamma^\mathsf{O}_{\uparrow_c}, \mathrm{Id})} & \mathsf{J}[\mathrm{Lin}_{(c,c)}] \otimes \mathsf{O}[\mathrm{Lin}_{(c,c)}] \otimes X_c \\
{\scriptstyle \mathrm{Id}} \downarrow & & \downarrow {\scriptstyle \lambda_{\mathrm{Lin}_{(c,c)}}} \\
\mathbb{1} \otimes \mathbb{1} \otimes X_c & \xrightarrow{\ \cong\ } & X_c
\end{array}
$$

is commutative, in which $\mathrm{Lin}_{(c,c)}$ *is the linear graph with one vertex and profile* (c, c).

Proof. The diagram

$$
\begin{array}{ccc}
\mathsf{J}[\mathrm{Lin}_{(c,c)}] \otimes \mathsf{O}[\uparrow_c] \otimes X_c & \xrightarrow{(\mathrm{Id}, \gamma^\mathsf{O}_{\uparrow_c}, \mathrm{Id})} & \mathsf{J}[\mathrm{Lin}_{(c,c)}] \otimes \mathsf{O}[\mathrm{Lin}_{(c,c)}] \otimes X_c \\
{\scriptstyle \mathrm{Id}} \downarrow & & \downarrow {\scriptstyle \lambda_{\mathrm{Lin}_{(c,c)}}} \\
\mathsf{J}[\uparrow_c] \otimes \mathsf{O}[\uparrow_c] \otimes X_c & \xrightarrow{\ \lambda_{\uparrow_c}\ } & X_c
\end{array}
$$

is commutative by the wedge condition (7.2.6) because

$$
\mathrm{Lin}_{(c,c)}(\uparrow_c) = \ \uparrow_c \xrightarrow{\ (\uparrow_c)\ } \mathrm{Lin}_{(c,c)}
$$

is a morphism in $\underline{\mathrm{Tree}}^{\mathfrak{C}}(^c_c)$. By the unity condition (7.2.4), the bottom horizontal morphism $\lambda_{\uparrow_c}$ is the isomorphism $\mathbb{1} \otimes \mathbb{1} \otimes X_c \cong X_c$. $\square$

Recall from Corollary 6.4.7 that the augmentation $\eta : \mathsf{WO} \longrightarrow \mathsf{O}$ induces a change-of-operad adjunction

$$
\eta_! : \mathsf{Alg}_\mathsf{M}(\mathsf{WO}) \rightleftarrows \mathsf{Alg}_\mathsf{M}(\mathsf{O}) : \eta^*.
$$

The next observation describes the structure morphisms of a WO-algebra that is the pullback of an O-algebra.

Corollary 7.2.9. *Suppose* $(\mathsf{O}, \gamma^\mathsf{O})$ *is a* $\mathfrak{C}$*-colored operad in* M*, and* (X, λ) *is an* O*-algebra. For* $T \in \underline{\mathrm{Tree}}^{\mathfrak{C}}(^d_{\underline{c}})$*, the structure morphism* λ_T *in (7.2.2) for the* WO*-algebra* $\eta^*(X, \lambda)$ *is the composition*

$$
\begin{array}{ccc}
\mathsf{J}[T] \otimes \mathsf{O}[T] \otimes X_{\underline{c}} & \xrightarrow{\qquad \lambda_T \qquad} & X_d \\
{\scriptstyle (\otimes_{|T|} \epsilon, \mathrm{Id})} \downarrow & & \uparrow {\scriptstyle \lambda} \\
\mathbb{1}[T] \otimes \mathsf{O}[T] \otimes X_{\underline{c}} \xrightarrow{\ \cong\ } \mathsf{O}[T] \otimes X_{\underline{c}} & \xrightarrow{(\gamma^\mathsf{O}_T, \mathrm{Id})} & \mathsf{O}(^d_{\underline{c}}) \otimes X_{\underline{c}}
\end{array}
$$

in which $\epsilon : J \longrightarrow \mathbb{1}$ *is the counit of the commutative segment* J.

Proof. By Definition 5.1.5 and (7.2.7), λ_T is the composition

$$
\begin{array}{ccc}
\mathsf{J}[T] \otimes \mathsf{O}[T] \otimes X_{\underline{c}} & \xrightarrow{\ \lambda_T\ } & X_d \\
{\scriptstyle(\omega_T,\mathrm{Id})}\Big\downarrow & & \Big\uparrow{\scriptstyle\lambda} \\
\mathsf{WO}\binom{d}{\underline{c}} \otimes X_{\underline{c}} & \xrightarrow{\ (\eta,\mathrm{Id})\ } & \mathsf{O}\binom{d}{\underline{c}} \otimes X_{\underline{c}}.
\end{array}
$$

Now we observe that

$$
\eta \circ \omega_T = (\gamma_T^{\mathsf{O}})(\cong)\left(\bigotimes_{|T|} \epsilon, \mathrm{Id}\right)
$$

by the definition of the augmentation in Theorem 6.4.4. $\qquad\square$

Interpretation 7.2.10. When an O-algebra is regarded as a WO-algebra, the structure morphism λ_T is given by first forgetting the length of internal edges using the counit ϵ. Then one composes the elements in O using the operadic structure morphism γ_T^{O}, and follows that by the O-action structure morphism. $\qquad\diamond$

7.3 Homotopy Coherent Diagrams

For the next several sections, we will discuss some relevant examples of algebras over the Boardman-Vogt construction. Suppose C is a small category with object set $\mathfrak{C}$. In this section, we discuss algebras over the Boardman-Vogt construction of the colored operad for C-diagrams, called homotopy coherent C-diagrams. We will explain that these algebras are C-diagrams up to a family of coherent homotopies. Homotopy coherent diagrams of topological spaces have a long history; see, for example [Berger and Moerdijk (2007); Cordier and Porter (1986, 1997); Vogt (1973)].

Motivation 7.3.1. The physical relevance of homotopy coherent diagrams is that the isotony axiom in quantum field theory, sometimes called the locality axiom, is not always satisfied in relevant examples; see, for example, [Benini *et. al.* (2014); Becker *et. al.* (2017b)]. Instead, one should expect a homotopy version of functoriality, as suggested in [Benini and Schenkel (2017)] Section 5. Homotopy theory has taught us that when certain properties hold only up to homotopy (for example, homotopy associativity), there is usually a whole family of higher structure that encodes the specific homotopies and their relations. We will see in the following few sections that the Boardman-Vogt construction is very convenient for encoding such a family of higher structure. Homotopy coherent diagrams are also closely related to a homotopy version of the time-slice axiom, as we will explain in Section 7.4. $\diamond$

Recall from Example 4.5.21 that there is a $\mathfrak{C}$-colored operad $\mathsf{C}^{\mathrm{diag}}$ whose algebras are C-diagrams in M.

Definition 7.3.2. Objects in the category $\mathsf{Alg}_{\mathsf{M}}(\mathsf{WC}^{\mathrm{diag}})$ are called *homotopy coherent* C-*diagrams in* M, where $\mathsf{WC}^{\mathrm{diag}}$ is the Boardman-Vogt construction of the $\mathfrak{C}$-colored operad $\mathsf{C}^{\mathrm{diag}}$.

When applied to the $\mathfrak{C}$-colored operad $\mathsf{C}^{\mathsf{diag}}$, Corollary 6.4.7 and Corollary 6.5.10 yield the following adjunction.

Corollary 7.3.3. *The augmentation* $\eta : \mathsf{WC}^{\mathsf{diag}} \longrightarrow \mathsf{C}^{\mathsf{diag}}$ *induces an adjunction*

$$\mathsf{Alg_M}(\mathsf{WC}^{\mathsf{diag}}) \underset{\eta^*}{\overset{\eta_!}{\rightleftarrows}} \mathsf{Alg_M}(\mathsf{C}^{\mathsf{diag}})$$

that is a Quillen equivalence if $\mathsf{M} = \mathsf{Chain}_{\mathbb{K}}$ *with* $\mathbb{K}$ *a field of characteristic* 0.

Interpretation 7.3.4. This adjunction says that each C-diagram in M can be regarded as a homotopy coherent C-diagram in M via the augmentation η. The left adjoint $\eta_!$ rectifies each homotopy coherent C-diagram in M to a C-diagram in M.◇

Recall the linear graphs $\mathsf{Lin}_{\underline{c}}$ in Example 3.1.19 and the substitution category $\underline{\mathsf{Linear}}^{\mathfrak{C}}\binom{d}{c}$ for linear graphs in Definition 3.2.11. The objects in $\underline{\mathsf{Linear}}^{\mathfrak{C}}\binom{d}{c}$ are linear graphs with input color c and output color d. Its morphisms are given by tree substitution, but only for linear graphs. The following is the coherence theorem for homotopy coherent C-diagrams.

Theorem 7.3.5. *A homotopy coherent* C-*diagram in* M *is exactly a pair* (X, λ) *consisting of*

- *a* $\mathfrak{C}$-*colored object* X *in* M *and*
- *a structure morphism*

$$\mathsf{J}[\mathsf{Lin}_{\underline{c}}] \otimes X_{c_0} \xrightarrow{\lambda_{\underline{c}}^{f}} X_{c_n} \in \mathsf{M} \tag{7.3.6}$$

for

- *each profile* $\underline{c} = (c_0, \ldots, c_n) \in \mathsf{Prof}(\mathfrak{C})$ *with* $n \geq 0$;
- *each sequence of composable* C-*morphisms* $\underline{f} = (f_1, \ldots, f_n)$ *with* $f_j \in \mathsf{C}(c_{j-1}, c_j)$ *for* $1 \leq j \leq n$

that satisfies the following three conditions.

Associativity *Suppose* $0 \leq n \leq p$, $\underline{c} = (c_0, \ldots, c_n)$, *and* $\underline{c}' = (c_n, \ldots, c_p) \in \mathsf{Prof}(\mathfrak{C})$. *Suppose* $f_j \in \mathsf{C}(c_{j-1}, c_j)$ *for each* $1 \leq j \leq p$ *with* $\underline{f} = (f_1, \ldots, f_n)$ *and* $\underline{f}' = (f_{n+1}, \ldots, f_p)$. *Then the diagram*

$$
\begin{array}{ccc}
\mathsf{J}[\mathsf{Lin}_{\underline{c}'}] \otimes \mathsf{J}[\mathsf{Lin}_{\underline{c}}] \otimes X_{c_0} & \xrightarrow{(\pi, \mathrm{Id})} & \mathsf{J}[\mathsf{Lin}_{(c_0, \ldots, c_p)}] \otimes X_{c_0} \\
{\scriptstyle (\mathrm{Id}, \lambda_{\underline{c}}^{f})} \downarrow & & \downarrow {\scriptstyle \lambda_{(c_0, \ldots, c_p)}^{(f, f')}} \\
\mathsf{J}[\mathsf{Lin}_{\underline{c}'}] \otimes X_{c_n} & \xrightarrow{\lambda_{\underline{c}'}^{f'}} & X_{c_p}
\end{array}
\tag{7.3.7}
$$

is commutative, in which $\mathsf{Lin}_{(c_0, \ldots, c_p)}$ *is regarded as the grafting* (3.3.1) *of* $\mathsf{Lin}_{\underline{c}'}$ *and* $\mathsf{Lin}_{\underline{c}}$ *with* π *the morphism in Lemma 6.2.7.*

Unity *For each $c \in \mathfrak{C}$, the composition*

$$X_c \xrightarrow{\;\cong\;} J[\operatorname{Lin}_{(c)}] \otimes X_c \xrightarrow{\;\lambda_{(c)}^{\varnothing}\;} X_c \qquad (7.3.8)$$

is the identity morphism of X_c, where $\operatorname{Lin}_{(c)} = \uparrow_c$ is the c-colored exceptional edge.

Wedge Condition *Suppose*

$$\underline{c} = (c_0, \dots, c_n) \in \mathsf{Prof}(\mathfrak{C}) \text{ with } n \geq 1,$$
$$\underline{b}_j = (c_{j-1} = b_0^j, b_1^j, \dots, b_{k_j}^j = c_j) \in \mathsf{Prof}(\mathfrak{C}) \text{ with } k_j \geq 0 \text{ for } 1 \leq j \leq n, \text{ and}$$
$$\underline{b} = (\underline{b}_1, \dots, \underline{b}_n).$$

Suppose

$$f_i^j \in \mathsf{C}(b_{i-1}^j, b_i^j) \text{ for each } 1 \leq j \leq n \text{ and } 1 \leq i \leq k_j,$$
$$\underline{f}^j = (f_1^j, \dots, f_{k_j}^j),$$
$$\underline{f} = (\underline{f}^1, \dots, \underline{f}^n), \text{ and}$$
$$f^j = f_{k_j}^j \circ \dots \circ f_1^j \in \mathsf{C}(c_{j-1}, c_j).$$

Then the diagram

$$
\begin{array}{ccc}
J[\operatorname{Lin}_{\underline{c}}] \otimes X_{c_0} & \xrightarrow{\lambda_{\underline{c}}^{(f^1,\dots,f^n)}} & X_{c_n} \\
{\scriptstyle (J,\mathrm{Id})}\downarrow & & \| \\
J[\operatorname{Lin}_{\underline{b}}] \otimes X_{c_0} & \xrightarrow{\lambda_{\underline{b}}^{\underline{f}}} & X_{c_n}
\end{array}
\qquad (7.3.9)
$$

is commutative, in which $\operatorname{Lin}_{\underline{b}}$ is regarded as the tree substitution

$$\operatorname{Lin}_{\underline{b}} = \operatorname{Lin}_{\underline{c}}\left(\operatorname{Lin}_{\underline{b}_j}\right)_{j=1}^n.$$

Proof. This is the special case of the Coherence Theorem 7.2.1 for the $\mathfrak{C}$-colored operad $\mathsf{O} = \mathsf{C}^{\mathrm{diag}}$. Indeed, recall from Example 4.5.21 that the $\mathfrak{C}$-colored operad $\mathsf{C}^{\mathrm{diag}}$ is concentrated in unary entries:

$$\mathsf{C}^{\mathrm{diag}}\binom{d}{\underline{c}} = \begin{cases} \coprod_{\mathsf{C}(c,d)} \mathbb{1} & \text{if } \underline{c} = c \in \mathfrak{C}, \\ \varnothing & \text{if } |\underline{c}| \neq 1 \end{cases}$$

for $(\underline{c}; d) \in \mathsf{Prof}(\mathfrak{C}) \times \mathfrak{C}$. Its equivariant structure is trivial. Its colored units and operadic composition come from the identity morphisms and the categorical composition in C. Since $\mathsf{C}^{\mathrm{diag}}$ is concentrated in unary entries, if $T \in \underline{\mathsf{Tree}}^{\mathfrak{C}}\binom{d}{\underline{c}}$ is not a linear graph, then

$$\mathsf{C}^{\mathrm{diag}}[T] = \bigotimes_{v \in T} \mathsf{C}^{\mathrm{diag}}\binom{\mathrm{out}(v)}{\mathrm{in}(v)} = \varnothing.$$

So when T is not a linear graph, the structure morphism

$$J[T] \otimes \mathsf{C}^{\mathrm{diag}}[T] \otimes X_{\underline{c}} \xrightarrow{\lambda_T} X_d$$

in (7.2.2) for a $\mathsf{WC}^{\mathsf{diag}}$-algebra is the trivial morphism $\varnothing \longrightarrow X_d$. In particular, the equivariance condition (7.2.5) is trivial for $\mathsf{WC}^{\mathsf{diag}}$-algebras.

For $\underline{c} = (c_0, \ldots, c_n) \in \mathsf{Prof}(\mathfrak{C})$ with $n \geq 0$, there is a natural isomorphism

$$\mathsf{C}^{\mathsf{diag}}[\mathrm{Lin}_{\underline{c}}] = \bigotimes_{j=1}^n \mathsf{C}^{\mathsf{diag}}\binom{c_j}{c_{j-1}} = \bigotimes_{j=1}^n \left[\coprod_{\mathsf{C}(c_{j-1}, c_j)} \mathbb{1} \right] \cong \coprod_{\prod_{j=1}^n \mathsf{C}(c_{j-1}, c_j)} \mathbb{1}.$$

This implies that there is a natural isomorphism

$$\mathsf{J}[\mathrm{Lin}_{\underline{c}}] \otimes \mathsf{C}^{\mathsf{diag}}[\mathrm{Lin}_{\underline{c}}] \otimes X_{c_0} \cong \coprod_{\prod_{j=1}^n \mathsf{C}(c_{j-1}, c_j)} \mathsf{J}[\mathrm{Lin}_{\underline{c}}] \otimes X_{c_0}.$$

So the structure morphism

$$\mathsf{J}[\mathrm{Lin}_{\underline{c}}] \otimes \mathsf{C}^{\mathsf{diag}}[\mathrm{Lin}_{\underline{c}}] \otimes X_{c_0} \xrightarrow{\ \lambda_{\mathrm{Lin}_{\underline{c}}}\ } X_{c_n}$$

in (7.2.2) is uniquely determined by the restrictions $\lambda_{\underline{c}}^f$ as stated in (7.3.6). The associativity, unity, and wedge conditions (7.3.7)-(7.3.9) above are exactly those in the Coherence Theorem 7.2.1 for linear graphs. $\qquad\square$

Interpretation 7.3.10. Intuitively, one should think of the structure morphism $\lambda_{\underline{c}}^f$ in (7.3.6) as determined by the decorated linear graph

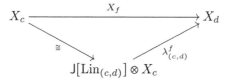

with n vertices decorated by the C-morphisms f_j. The colors c_j are the colors of the edges. If $n = 0$, then this is the c_0-colored exceptional edge $\uparrow_{c_0}$ with $\underline{c} = (c_0)$ and $(f_j) = \varnothing$. $\qquad\diamond$

Suppose (X, λ) is a $\mathsf{WC}^{\mathsf{diag}}$-algebra, i.e., a homotopy coherent C-diagram in M. In the next few examples, we will explain some of the structure on X that suggests that it is a C-diagram up to coherent higher homotopies.

Example 7.3.11 (Assignment on morphisms). For each morphism $f \in \mathsf{C}(c, d)$, the structure morphism in (7.3.6) yields the morphism

$$
\begin{array}{ccc}
X_c & \xrightarrow{\quad X_f \quad} & X_d \\
{\scriptstyle \cong}\searrow & & \nearrow{\scriptstyle \lambda_{(c,d)}^f} \\
& \mathsf{J}[\mathrm{Lin}_{(c,d)}] \otimes X_c &
\end{array}
$$

in M. If furthermore $f = \mathrm{Id}_c$, then X_{Id_c} is the identity morphism of X_c by Corollary 7.2.8. In X is an actual C-diagram, then it would preserve composition, i.e., $X_{fg} = X_f \circ X_g$ whenever fg is defined. For a homotopy coherent C-diagram, we will see in the next example that $X_{(-)}$ preserves composition up to a specified homotopy. $\diamond$

Example 7.3.12 (Homotopy preservation of composition). Suppose $(f,g) \in$ $\mathsf{C}(c,d) \times \mathsf{C}(b,c)$ is a pair of composable C-morphisms. Consider the diagram

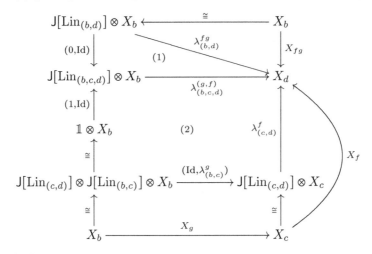

in M, in which

$$\mathsf{J}[\mathrm{Lin}_{(b,c,d)}] = J, \qquad \mathsf{J}[\mathrm{Lin}_{(b,d)}] = \mathsf{J}[\mathrm{Lin}_{(c,d)}] = \mathsf{J}[\mathrm{Lin}_{(b,c)}] = \mathbb{1},$$

and $0,1 : \mathbb{1} \longrightarrow J$ are part of the commutative segment J. This diagram is commutative:

- The upper right triangle is the definition of X_{fg}.
- The sub-diagram (1) is commutative by the wedge condition (7.3.9) with $n = 1$, $\underline{c} = (b,d)$, $\underline{b}_1 = (b,c,d)$, and $\underline{f}^1 = (g,f)$.
- The sub-diagram (2) is commutative by the associativity condition (7.3.7) with $n = 1$, $p = 2$, $\underline{c} = (b,c)$, $\underline{c}' = (c,d)$, $\underline{f} = (g)$, and $\underline{f}' = (f)$.
- The bottom rectangle is commutative by naturality and the definition of X_g.
- The lower right stripe is the definition of X_f.

We will call the morphisms $0, 1 : \mathbb{1} \longrightarrow J$ the 0-end and the 1-end of J, respectively. The above commutative diagram says that the structure morphism $\lambda_{(b,c,d)}^{(g,f)}$ is X_{fg} at the 0-end and the composition $X_f \circ X_g$ at the 1-end. So the structure morphism $\lambda_{(b,c,d)}^{(g,f)}$ is a homotopy from X_{fg} to the composition $X_f \circ X_g$. Therefore, a homotopy coherent C-diagram preserves composition up to a specified homotopy. It is important to observe that we are not just saying that the morphisms X_{fg} and $X_f \circ X_g$ are homotopic. Instead, a specific structure morphism $\lambda_{(b,c,d)}^{(g,f)}$ of a homotopy coherent C-diagram acts as the homotopy. There are higher homotopies for longer strings of composable C-morphisms, as we will see in the next example.   ⬦

Example 7.3.13 (Homotopy preservation of triple composition). Suppose given a triple of composable C-morphisms

$$(f,g,h) \in \mathsf{C}(c,d) \times \mathsf{C}(b,c) \times \mathsf{C}(a,b).$$

Consider the diagram

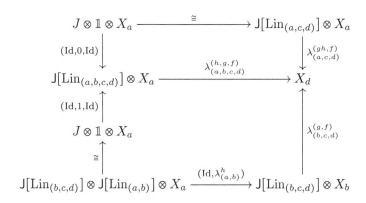

in M, in which

$$J[\mathrm{Lin}_{(a,b,c,d)}] \cong J \otimes J, \quad J[\mathrm{Lin}_{(a,b)}] = \mathbb{1},$$
$$J[\mathrm{Lin}_{(a,c,d)}] = J[\mathrm{Lin}_{(b,c,d)}] = J.$$

This diagram is commutative:

- The top rectangle is commutative by the wedge condition (7.3.9) with $n = 2$, $\underline{c} = (a,c,d)$, $\underline{b}_1 = (a,b,c)$, $\underline{b}_2 = (c,d)$, $\underline{f}^1 = (h,g)$, and $\underline{f}^2 = (f)$.
- The bottom square is commutative by the associativity condition (7.3.7) with $n = 1$, $p = 3$, $\underline{c} = (a,b)$, $\underline{c}' = (b,c,d)$, $\underline{f} = (h)$, and $\underline{f}' = (g,f)$.

This commutative diagram says that the structure morphism $\lambda^{(h,g,f)}_{(a,b,c,d)}$ yields a higher homotopy from $\lambda^{(gh,f)}_{(a,c,d)}$ to the composition $\lambda^{(g,f)}_{(b,c,d)} \circ (\mathrm{Id}, \lambda^{h}_{(a,b)})$.

Furthermore, as explained in Example 7.3.12:

- $\lambda^{(gh,f)}_{(a,c,d)}$ is a homotopy from $X_{fgh} : X_a \longrightarrow X_d$ to $X_f \circ X_{gh} : X_a \longrightarrow X_d$.
- $\lambda^{(g,f)}_{(b,c,d)}$ is a homotopy from $X_{fg} : X_b \longrightarrow X_d$ to $X_f \circ X_g : X_b \longrightarrow X_d$.

Altogether the above commutative diagram expresses a specific homotopy from X_{fgh} to $X_f \circ X_g \circ X_h$. $\diamond$

Example 7.3.14 (Homotopy preservation of triple composition). In Example 7.3.13 the commutative diagram only uses one copy of J to express a higher homotopy between the homotopies $\lambda^{(gh,f)}_{(a,c,d)}$ and $\lambda^{(g,f)}_{(b,c,d)} \circ (\mathrm{Id}, \lambda^{h}_{(a,b)})$. There is a similar

commutative diagram

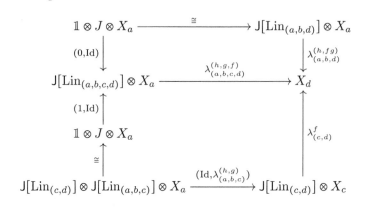

in M that makes use of the other copy of J in $\mathsf{J}[\mathrm{Lin}_{(a,b,c,d)}]$. Once again the top rectangle is commutative by the wedge condition (7.3.9), and the bottom square is commutative by the associativity condition (7.3.7).

This commutative diagram says that the structure morphism $\lambda^{(h,g,f)}_{(a,b,c,d)}$ yields:

- a higher homotopy from $\lambda^{(h,fg)}_{(a,b,d)}$ to the composition $\lambda^{f}_{(c,d)} \circ (\mathrm{Id}, \lambda^{(h,g)}_{(a,b,c)})$;
- another homotopy from X_{fgh} to $X_f \circ X_g \circ X_h$.

For longer strings of composable C-morphisms, there are similar commutative diagrams that express the structure morphisms $\lambda^{f}_{\underline{c}}$ as a family of coherent higher homotopies. The main point is that we are not trying to write down this infinite family of coherent homotopies from the ground up. Instead, all of them are neatly packaged in the Boardman-Vogt construction $\mathsf{WC}^{\mathrm{diag}}$ of the $\mathfrak{C}$-colored operad $\mathsf{C}^{\mathrm{diag}}$.
◇

7.4 Homotopy Inverses

In this section, we discuss a homotopy coherent version of an inverse using the Boardman-Vogt construction.

Motivation 7.4.1. Physically homotopy inverses are homotopy manifestations of the time-slice axiom in both homotopy algebraic quantum field theories and homotopy prefactorization algebras. The upshot of the time-slice axiom is that certain structure morphisms in algebraic quantum field theories are supposed to be invertible, e.g., if they correspond to Cauchy morphisms between oriented, time-oriented, and globally hyperbolic Lorentzian manifolds. The homotopy version of the time-slice axiom says that these structure morphisms are invertible up to specified homotopies.

As in Section 7.3, suppose C is a small category with object set $\mathfrak{C}$. If X is a C-diagram in M and if $f \in \mathsf{C}(c,d)$ is an isomorphism, then the morphism X_f :

$X_c \longrightarrow X_d$ in M is also an isomorphism with inverse $X_{f^{-1}}$, since

$$X_f \circ X_{f^{-1}} = X_{f \circ f^{-1}} = X_{\mathrm{Id}_d} = \mathrm{Id}_{X_d},$$
$$X_{f^{-1}} \circ X_f = X_{f^{-1} \circ f} = X_{\mathrm{Id}_c} = \mathrm{Id}_{X_c}.$$

If X is a homotopy coherent C-diagram, then we should replace the first equality in each line with a specified homotopy. In other words, X_f and $X_{f^{-1}}$ should be homotopy inverses of each other via specific structure morphisms. We will explain this in the following result. ◇

We will reuse the notation in Example 7.3.11.

Corollary 7.4.2. *Suppose* (X, λ) *is a homotopy coherent* C-*diagram in* M, *and* $f \in C(c, d)$ *is an isomorphism with inverse* $f^{-1} \in C(d, c)$. *Then the morphisms*

$$X_f : X_c \longrightarrow X_d \quad and \quad X_{f^{-1}} : X_d \longrightarrow X_c \in \mathsf{M}$$

are homotopy inverses of each other in the following sense.

(1) $X_{f^{-1}}$ *is a left homotopy inverse of* X_f *in the sense that the diagram*

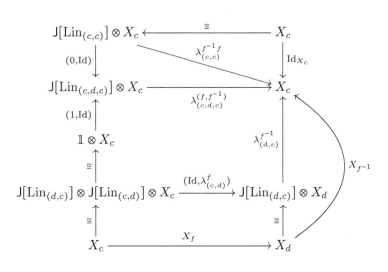

in M *is commutative, in which*

$$J[\mathrm{Lin}_{(c,d,c)}] = J \quad and \quad J[\mathrm{Lin}_{(c,c)}] = J[\mathrm{Lin}_{(d,c)}] = J[\mathrm{Lin}_{(c,d)}] = \mathbb{1}.$$

(2) $X_{f^{-1}}$ is a right homotopy inverse of X_f in the sense that the diagram

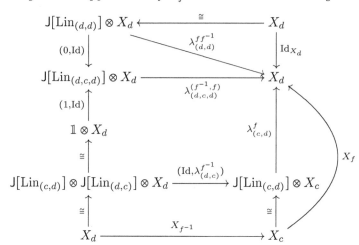

in M *is commutative, in which* $J[\mathrm{Lin}_{(d,c,d)}] = J$.

Proof. The first assertion is the special case of Example 7.3.12 for the composable pair of C-morphisms

$$(f^{-1}, f) \in C(d,c) \times C(c,d),$$

since by Example 7.3.11 X_{Id_c} is equal to Id_{X_c}. Similarly, the second assertion is the special case of Example 7.3.12 for the composable pair of C-morphisms

$$(f, f^{-1}) \in C(c,d) \times C(d,c).$$

$\square$

Interpretation 7.4.3. In a homotopy coherent C-diagram (X,λ), the structure morphism X_f for an invertible morphism f in C has the structure morphism $X_{f^{-1}}$ as a two-sided homotopy inverse. Moreover, the two homotopies are the structure morphisms $\lambda_{(c,d,c)}^{(f,f^{-1})}$ and $\lambda_{(d,c,d)}^{(f^{-1},f)}$. Therefore, a homotopy inverse and the homotopies are already encoded in the Boardman-Vogt construction $\mathsf{WC}^{\mathrm{diag}}$ of the colored operad $\mathsf{C}^{\mathrm{diag}}$. $\diamond$

7.5 A_∞-Algebras

In this section, we discuss a homotopy version of monoids, called strongly homotopy associative algebras or A_∞-algebras, as algebras over the Boardman-Vogt construction of the associative operad.

Motivation 7.5.1. Recall from Example 4.5.17 that the associative operad As is a 1-colored operad in M whose category of algebras is canonically isomorphic to the category of monoids in M (Definition 2.6.1). The physical relevance of A_∞-algebras is that an algebraic quantum field theory is a diagram of monoids satisfying the causality axiom and possibly the time-slice axiom. Strict associativity is not a homotopy invariant concept. Instead, the work of Stasheff [Stasheff (1963)] taught us that a homotopy version of a monoid is an A_∞-algebra. Therefore, A_∞-algebras will arise naturally in the study of homotopy algebraic quantum field theories.   ◇

Definition 7.5.2. Objects in the category $\mathsf{Alg_M}(\mathsf{WAs})$ are called A_∞-*algebras in* M, where WAs is the Boardman-Vogt construction of the associative operad As.

When applied to the associative operad, Corollary 6.4.7 and Corollary 6.5.10 yield the following adjunction.

Corollary 7.5.3. *The augmentation* $\eta : \mathsf{WAs} \longrightarrow \mathsf{As}$ *induces an adjunction*

$$\mathsf{Alg_M}(\mathsf{WAs}) \xrightarrow[\eta^*]{\eta_!} \mathsf{Alg_M}(\mathsf{As})$$

that is a Quillen equivalence if $\mathsf{M} = \mathsf{Chain}_{\mathbb{K}}$ *with* $\mathbb{K}$ *a field of characteristic zero.*

Interpretation 7.5.4. This adjunction says that each monoid in M can be regarded as an A_∞-algebra in M via the augmentation η. The left adjoint $\eta_!$ rectifies each A_∞-algebra in M to a monoid in M.   ◇

Since in this section we are discussing 1-colored operads, we will be using 1-colored trees. The substitution category, as in Definition 3.2.11, of 1-colored trees with n inputs is denoted by $\underline{\mathsf{Tree}}(n)$. Its objects are 1-colored trees with n inputs, and its morphisms are given by tree substitution. The following result is the coherence theorem for A_∞-algebras.

Theorem 7.5.5. *An* A_∞-*algebra in* M *is exactly a pair* (X, λ) *consisting of*

- *an object* $X \in \mathsf{M}$ *and*
- *a structure morphism*

$$\mathsf{J}[T] \otimes X^{\otimes n} \xrightarrow{\lambda_T^{\{\sigma_v\}_{v \in T}}} X \in \mathsf{M} \tag{7.5.6}$$

for

- *each* $T \in \underline{\mathsf{Tree}}(n)$ *with* $n \geq 0$ *and*
- *each* $\{\sigma_v\}_{v \in T} \in \prod_{v \in T} \Sigma_{|\mathsf{in}(v)|}$

that satisfies the following four conditions.

Associativity *Suppose* $T \in \underline{\mathsf{Tree}}(n)$ *with* $n \geq 1$, $T_j \in \underline{\mathsf{Tree}}(k_j)$ *with* $k_j \geq 0$ *and* $1 \leq j \leq n$, $k = k_1 + \cdots + k_n$,

$$G = \mathsf{Graft}(T; T_1, \ldots, T_n) \in \underline{\mathsf{Tree}}(k)$$

is the grafting, $\{\sigma_v\} \in \prod_{v \in T} \Sigma_{|\mathsf{in}(v)|}$, and $\{\sigma_u^j\} \in \prod_{u \in T_j} \Sigma_{|\mathsf{in}(u)|}$ for $1 \le j \le n$. Then the diagram

$$\mathsf{J}[T] \otimes \left(\overset{n}{\underset{j=1}{\bigotimes}} \mathsf{J}[T_j] \right) \otimes X^{\otimes k} \xrightarrow{\;(\pi, \mathrm{Id})\;} \mathsf{J}[G] \otimes X^{\otimes k} \qquad (7.5.7)$$

$$\text{permute} \downarrow \cong$$

$$\mathsf{J}[T] \otimes \overset{n}{\underset{j=1}{\bigotimes}} \left(\mathsf{J}[T_j] \otimes X^{\otimes k_j} \right) \qquad\qquad \lambda_G^{\{\sigma_v, \sigma_u^j\}}$$

$$\left(\mathrm{Id}, \otimes_j \lambda_{T_j}^{\{\sigma_u^j\}_{u \in T_j}} \right) \downarrow$$

$$\mathsf{J}[T] \otimes \overset{n}{\underset{j=1}{\bigotimes}} X \xrightarrow{\;\lambda_T^{\{\sigma_v\}_{v \in T}}\;} X$$

is commutative. Here $\pi = \otimes_S 1$ is the morphism in Lemma 6.2.7 for the grafting G. In the structure morphism $\lambda_G^{\{\sigma_v, \sigma_u^j\}}$, we have $v \in T$, $u \in T_j$, and $1 \le j \le n$.

Unity *The composition*

$$X \xrightarrow{\;\cong\;} \mathsf{J}[\uparrow] \otimes X \xrightarrow{\;\lambda_\uparrow^\varnothing\;} X \qquad (7.5.8)$$

is the identity morphism of X, where $\uparrow$ is the 1-colored exceptional edge.

Equivariance *For $T \in \underline{\mathsf{Tree}}(n)$, $\sigma \in \Sigma_n$, and $\{\sigma_v\} \in \prod_{v \in T} \Sigma_{|\mathsf{in}(v)|}$, the diagram*

$$\mathsf{J}[T] \otimes X^{\otimes n} \xrightarrow{\;\lambda_T^{\{\sigma_v\}}\;} X \qquad (7.5.9)$$

$$(\mathrm{Id}, \sigma^{-1}) \downarrow \qquad\qquad \|$$

$$\mathsf{J}[T\sigma] \otimes X^{\otimes n} \xrightarrow{\;\lambda_{T\sigma}^{\{\sigma_v\}}\;} X$$

is commutative, in which $T\sigma \in \underline{\mathsf{Tree}}(n)$ is the same as T except that its ordering is $\zeta_T \sigma$ with ζ_T the ordering of T.

Wedge Condition *Suppose $T \in \underline{\mathsf{Tree}}(n)$, $H_v \in \underline{\mathsf{Tree}}(|\mathsf{in}(v)|)$ for $v \in \mathsf{Vt}(T)$, $K = T(H_v)_{v \in T}$ is the tree substitution, and $\sigma_u^v \in \Sigma_{|\mathsf{in}(u)|}$ for each $v \in \mathsf{Vt}(T)$ and $u \in \mathsf{Vt}(H_v)$. Then the diagram*

$$\mathsf{J}[T] \otimes X^{\otimes n} \xrightarrow{\;\lambda_T^{\{\tau_v\}}\;} X \qquad (7.5.10)$$

$$(\mathsf{J}, \mathrm{Id}) \downarrow \qquad\qquad \|$$

$$\mathsf{J}[K] \otimes X^{\otimes n} \xrightarrow{\;\lambda_K^{\{\sigma_u^v\}}\;} X$$

is commutative. For each $v \in \mathsf{Vt}(T)$, τ_v is defined as

$$\tau_v = \gamma_{H_v}^{\mathsf{As}} \left(\{\sigma_u^v\}_{u \in H_v} \right) \in \Sigma_{|\mathsf{in}(v)|}$$

in which

$$\prod_{u \in H_v} \Sigma_{|\mathsf{in}(u)|} = \mathsf{As}[H_v] \xrightarrow{\;\gamma_{H_v}^{\mathsf{As}}\;} \mathsf{As}(|\mathsf{in}(v)|) = \Sigma_{|\mathsf{in}(v)|}$$

is the operadic structure morphism for H_v of the associative operad in Set, as in (4.4.10).

Proof. This is the special case of the Coherence Theorem 7.2.1 for the associative operad As in M. Indeed, recall that the entries of the associative operad in M are

$$\mathsf{As}(n) = \coprod_{\sigma \in \Sigma_n} \mathbb{1}$$

for $n \geq 0$. For a 1-colored tree $T \in \underline{\mathsf{Tree}}(n)$, there is a natural isomorphism

$$\mathsf{As}[T] = \bigotimes_{v \in T} \mathsf{As}(|\mathrm{in}(v)|) = \bigotimes_{v \in T} \left[\coprod_{\sigma \in \Sigma_{|\mathrm{in}(v)|}} \mathbb{1} \right] \cong \coprod_{\{\sigma_v\} \in \prod_{v \in T} \Sigma_{|\mathrm{in}(v)|}} \mathbb{1}.$$

It follows that there is a natural isomorphism

$$\mathsf{J}[T] \otimes \mathsf{As}[T] \otimes X^{\otimes n} \cong \coprod_{\{\sigma_v\} \in \prod_{v \in T} \Sigma_{|\mathrm{in}(v)|}} \mathsf{J}[T] \otimes X^{\otimes n}.$$

Therefore, the structure morphism λ_T in (7.2.2) is uniquely determined by the restrictions $\lambda_T^{\{\sigma_v\}_{v \in T}}$ as stated in (7.5.6). The associativity, unity, equivariance, and wedge conditions (7.5.7)-(7.5.10) above are exactly those in the Coherence Theorem 7.2.1 for 1-colored trees. □

Suppose (X, λ) is an A_∞-algebra in M. In the next few examples, we will explain some of the structure on X that suggests that it is a monoid up to coherent higher homotopies.

Example 7.5.11 (Multiplication). Suppose Cor_n is the 1-colored corolla with n legs; see Example 3.1.21 where corollas were defined. The structure morphism in (7.5.6) yields the composition

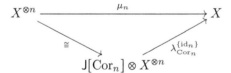

with $\mathsf{J}[\mathrm{Cor}_n] = \mathbb{1}$ and $\mathrm{id}_n \in \Sigma_n$ the identity permutation. By Corollary 7.2.8

$$\mu_1 : X \longrightarrow X$$

is the identity morphism on X, and so $\lambda_{\mathrm{Cor}_1}^{\{\mathrm{id}_1\}}$ is the isomorphism $\mathbb{1} \otimes X \cong X$. If X is a monoid, then we would expect

$$\mu_2 : X \otimes X \longrightarrow X$$

to be strictly associative with $\mu_0 : \mathbb{1} \longrightarrow X$ as a strict two-sided unit. For an A_∞-algebra, we expect homotopy associativity and a homotopy unit, as explained in the following examples. ◇

Example 7.5.12 (Left homotopy unit). Here we explain why $\mu_0 : \mathbb{1} \longrightarrow X$ is a left homotopy unit of μ_2. Consider the grafting

$$K = \mathsf{Graft}(\mathrm{Cor}_2; \mathrm{Cor}_0, \uparrow) \in \underline{\mathsf{Tree}}(1)$$

which may be visualized as follows.

Consider the diagram

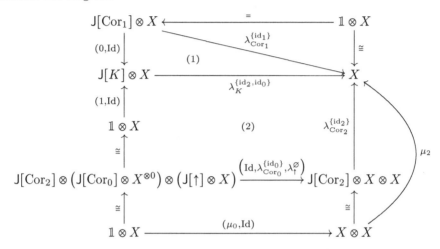

in M, in which

$$J[K] = J \quad \text{and} \quad J[\mathrm{Cor}_n] = J[\uparrow] = X^{\otimes 0} = \mathbb{1}.$$

This diagram is commutative:

- The top right triangle is commutative because $\lambda_{\mathrm{Cor}_1}^{\{\mathrm{id}_1\}}$ is the isomorphism $\mathbb{1} \otimes X \cong X$.
- The triangle (1) is commutative by the wedge condition (7.5.10) for the tree substitution $K = \mathrm{Cor}_1(K)$.
- The square (2) is commutative by the associativity condition (7.5.7) using the grafting definition of K.
- The bottom rectangle is commutative by the definition of μ_0 and the unity condition (7.5.8).
- The lower right stripe is the definition of μ_2.

The above commutative diagram says that the structure morphism $\lambda_K^{\{\mathrm{id}_2,\mathrm{id}_0\}}$ is a homotopy from the isomorphism $\mathbb{1} \otimes X \cong X$ to the composition $\mu_2 \circ (\mu_0, \mathrm{Id})$. So in an A_∞-algebra, μ_0 is a left homotopy unit of μ_2.   ◇

Example 7.5.13 (Right homotopy unit). Similarly, consider the grafting

$$G = \mathsf{Graft}\big(\mathrm{Cor}_2; \uparrow, \mathrm{Cor}_0\big) \in \underline{\mathsf{Tree}}(1)$$

which may be visualized as follows.

As above there is a commutative diagram

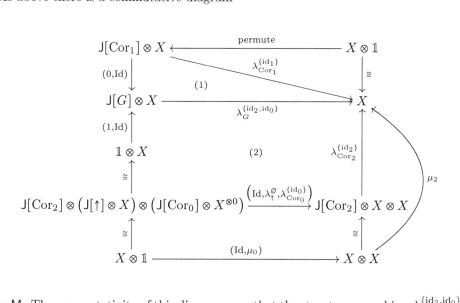

in M. The commutativity of this diagram says that the structure morphism $\lambda_G^{\{id_2, id_0\}}$ is a homotopy from the isomorphism $X \otimes \mathbb{1} \cong X$ to the composition $\mu_2 \circ (\mathrm{Id}, \mu_0)$. So in an A_∞-algebra, μ_0 is also a right homotopy unit of μ_2. ◇

Example 7.5.14 (Homotopy associativity). Let us now observe that the structure morphism $\mu_2 : X^{\otimes 2} \longrightarrow X$ is homotopy associative in the following sense. Consider the 1-colored trees

$$K = \mathsf{Graft}\big(\mathsf{Cor}_2; \mathsf{Cor}_2, \uparrow\big) \quad \text{and} \quad G = \mathsf{Graft}\big(\mathsf{Cor}_2; \uparrow, \mathsf{Cor}_2\big) \in \underline{\mathsf{Tree}}(3)$$

which may be visualized as follows.

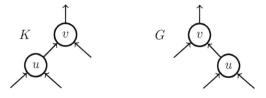

Consider the diagram

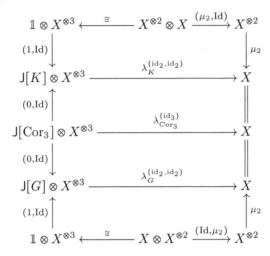

in M, in which

$$\mathsf{J}[K] = \mathsf{J}[G] = J \quad \text{and} \quad \mathsf{J}[\mathrm{Cor}_3] = \mathbb{1}.$$

This diagram is commutative:

- The top rectangle is commutative by (i) the definition of μ_2, (ii) the unity condition (7.5.8), and (iii) the associativity condition (7.5.7) using the grafting definition of K.
- The second rectangle from the top is commutative by the wedge condition (7.5.10) for the tree substitution $K = \mathrm{Cor}_3(K)$.
- The third rectangle from the top is commutative by the wedge condition (7.5.10) for the tree substitution $G = \mathrm{Cor}_3(G)$.
- The bottom rectangle is commutative by (i) the definition of μ_2, (ii) the unity condition (7.5.8), and (iii) the associativity condition (7.5.7) using the grafting definition of G.

The top half of the commutative diagram says that the structure morphism $\lambda_K^{\{\mathrm{id}_2,\mathrm{id}_2\}}$ is a homotopy from $\lambda_{\mathrm{Cor}_3}^{\{\mathrm{id}_3\}}$ to the composition $\mu_2 \circ (\mu_2, \mathrm{Id})$. The bottom half of the commutative diagram says that the structure morphism $\lambda_G^{\{\mathrm{id}_2,\mathrm{id}_2\}}$ is a homotopy from $\lambda_{\mathrm{Cor}_3}^{\{\mathrm{id}_3\}}$ to the composition $\mu_2 \circ (\mathrm{Id}, \mu_2)$. The entire commutative diagram together exhibits a homotopy between the compositions $\mu_2 \circ (\mu_2, \mathrm{Id})$ and $\mu_2 \circ (\mathrm{Id}, \mu_2)$. So in an A_∞-algebra, the morphism μ_2 is homotopy associative.

This is only the first layer of the higher homotopy associative structure in an A_∞-algebra. For example, similar to the discussion above, one can consider 1-colored trees with more than one internal edges. Any iterated composition of the various μ_k's as represented by a 1-colored tree $T \in \underline{\mathsf{Tree}}(n)$ is homotopic to the structure morphism $\lambda_{\mathrm{Cor}_n}^{\{\mathrm{id}_n\}}$ via the homotopy $\lambda_T^{\{\mathrm{id}_v\}}$, where $\mathrm{id}_v \in \Sigma_{|\mathrm{in}(v)|}$ is the identity permutation for each $v \in \mathsf{Vt}(T)$. The point is that we do not need to write these relations down one-by-one from the ground up. Instead, all of the higher homotopy associative structure is neatly packed in the Boardman-Vogt construction WAs of the associative operad. ◇

7.6 E_∞-Algebras

In this section, we discuss a homotopy version of commutative monoids, called strongly homotopy commutative algebras or E_∞-algebras, as algebras over the Boardman-Vogt construction of the commutative operad.

Motivation 7.6.1. Recall from Example 4.5.19 that the commutative operad Com is a 1-colored operad in M whose category of algebras is canonically isomorphic to the category of commutative monoids in M (Definition 2.6.1). The physical relevance of E_∞-algebras is that a prefactorization algebra includes a commutative monoid in its structure. Strict commutativity is not a homotopy invariant concept. A homotopy coherent version of a commutative monoid is an E_∞-algebra. Therefore, E_∞-algebras will arise naturally in the study of homotopy prefactorization algebras. ◇

Definition 7.6.2. Objects in the category $\mathsf{Alg_M(WCom)}$ are called E_∞-*algebras in* M, where WCom is the Boardman-Vogt construction of the commutative operad Com.

When applied to the commutative operad, Corollary 6.4.7 and Corollary 6.5.10 yield the following adjunction.

Corollary 7.6.3. *The augmentation* $\eta : \mathsf{WCom} \longrightarrow \mathsf{Com}$ *induces an adjunction*

$$\mathsf{Alg_M(WCom)} \underset{\eta^*}{\overset{\eta_!}{\rightleftarrows}} \mathsf{Alg_M(Com)}$$

that is a Quillen equivalence if $\mathsf{M} = \mathsf{Chain}_\mathbb{K}$ *with* $\mathbb{K}$ *a field of characteristic zero*

Interpretation 7.6.4. This adjunction says that each commutative monoid in M can be regarded as an E_∞-algebra in M via the augmentation η. The left adjoint $\eta_!$ rectifies each E_∞-algebra in M to a commutative monoid in M. ◇

When applied to the operad morphism $f : \mathsf{As} \longrightarrow \mathsf{Com}$ in Example 5.1.10, Corollary 6.4.9 yields the following result.

Corollary 7.6.5. *There is a diagram of adjunctions*

$$
\begin{array}{ccc}
\mathsf{Alg_M(WAs)} & \underset{(\mathsf{W}f)^*}{\overset{(\mathsf{W}f)_!}{\rightleftarrows}} & \mathsf{Alg_M(WCom)} \\
\eta_! \big\updownarrow \eta^* & & \eta_! \big\updownarrow \eta^* \\
\mathsf{Alg_M(As)} & \underset{f^*}{\overset{f_!}{\rightleftarrows}} & \mathsf{Alg_M(Com)}
\end{array}
$$

in which

$$f_! \circ \eta_! = \eta_! \circ (\mathsf{W}f)_! \quad and \quad \eta^* \circ f^* = (\mathsf{W}f)^* \circ \eta^*.$$

Remark 7.6.6. The bottom adjunction, the left adjunction, and the right adjunction are the ones in Example 5.1.10, Corollary 7.5.3, and Corollary 7.6.3, respectively. In the top adjunction, the right adjoint $(\mathsf{W}f)^*$ sends each E_∞-algebra to its underlying A_∞-algebra. The left adjoint $(\mathsf{W}f)_!$ sends each A_∞-algebra to an E_∞-algebra. ◇

The following result is the coherence theorem for E_∞-algebras.

Theorem 7.6.7. *An E_∞-algebra in* M *is exactly a pair* (X, λ) *consisting of*

- *an object $X \in$ M and*
- *a structure morphism*

$$\mathsf{J}[T] \otimes X^{\otimes n} \xrightarrow{\lambda_T} X \tag{7.6.8}$$

for each $n \geq 0$ and $T \in \underline{\mathsf{Tree}}(n)$

that satisfies the following four conditions.

Associativity *Suppose $T \in \underline{\mathsf{Tree}}(n)$ with $n \geq 1$, $T_j \in \underline{\mathsf{Tree}}(k_j)$ with $k_j \geq 0$ and $1 \leq j \leq n$, $k = k_1 + \cdots + k_n$, and*

$$G = \mathsf{Graft}(T; T_1, \ldots, T_n) \in \underline{\mathsf{Tree}}(k)$$

is the grafting. Then the diagram

$$
\begin{array}{ccc}
\mathsf{J}[T] \otimes \left(\overset{n}{\underset{j=1}{\bigotimes}} \mathsf{J}[T_j] \right) \otimes X^{\otimes k} & \xrightarrow{(\pi,\mathrm{Id})} & \mathsf{J}[G] \otimes X^{\otimes k} \\
\text{permute} \downarrow \cong & & \\
\mathsf{J}[T] \otimes \overset{n}{\underset{j=1}{\bigotimes}} \left(\mathsf{J}[T_j] \otimes X^{\otimes k_j} \right) & & \downarrow \lambda_G \\
\left(\mathrm{Id}, \otimes_j \lambda_{T_j} \right) \downarrow & & \\
\mathsf{J}[T] \otimes \overset{n}{\underset{j=1}{\bigotimes}} X & \xrightarrow{\lambda_T} & X
\end{array}
\tag{7.6.9}
$$

is commutative. Here $\pi = \otimes_S 1$ is the morphism in Lemma 6.2.7 for the grafting G.

Unity *The composition*

$$X \xrightarrow{\cong} \mathsf{J}[\uparrow] \otimes X \xrightarrow{\lambda_\uparrow} X \tag{7.6.10}$$

is the identity morphism of X, where $\uparrow$ is the 1-colored exceptional edge.

Equivariance *For $T \in \underline{\mathsf{Tree}}(n)$ and $\sigma \in \Sigma_n$, the diagram*

$$
\begin{array}{ccc}
\mathsf{J}[T] \otimes X^{\otimes n} & \xrightarrow{\ \lambda_T\ } & X \\
{\scriptstyle (\mathrm{Id},\sigma^{-1})}\big\downarrow & & \big\| \\
\mathsf{J}[T\sigma] \otimes X^{\otimes n} & \xrightarrow{\ \lambda_{T\sigma}\ } & X
\end{array}
\tag{7.6.11}
$$

is commutative, in which $T\sigma \in \underline{\mathsf{Tree}}(n)$ is the same as T except that its ordering is $\zeta_T \sigma$ with ζ_T the ordering of T.

Wedge Condition *Suppose $T \in \underline{\mathsf{Tree}}(n)$, $H_v \in \underline{\mathsf{Tree}}(|\mathrm{in}(v)|)$ for $v \in \mathsf{Vt}(T)$, and $K = T(H_v)_{v\in T}$ is the tree substitution. Then the diagram*

$$
\begin{array}{ccc}
\mathsf{J}[T] \otimes X^{\otimes n} & \xrightarrow{\ \lambda_T\ } & X \\
{\scriptstyle (\mathsf{J},\mathrm{Id})}\big\downarrow & & \big\| \\
\mathsf{J}[K] \otimes X^{\otimes n} & \xrightarrow{\ \lambda_K\ } & X
\end{array}
\tag{7.6.12}
$$

is commutative.

Proof. This is the special case of the Coherence Theorem 7.2.1 for the commutative operad Com in M. Indeed, recall that the entries of the commutative operad in M are

$$
\mathsf{Com}(n) = \mathbb{1} \quad \text{for} \quad n \geq 0.
$$

For a 1-colored tree $T \in \underline{\mathsf{Tree}}(n)$, there is a natural isomorphism

$$
\mathsf{Com}[T] = \bigotimes_{v\in T} \mathsf{Com}(|\mathrm{in}(v)|) = \bigotimes_{v\in T} \mathbb{1} \cong \mathbb{1}.
$$

It follows that there is a natural isomorphism

$$
\mathsf{J}[T] \otimes \mathsf{Com}[T] \otimes X^{\otimes n} \cong \mathsf{J}[T] \otimes X^{\otimes n}.
$$

Therefore, the structure morphism λ_T in (7.2.2) becomes the morphism λ_T in (7.6.8). The associativity, unity, equivariance, and wedge conditions (7.6.9)-(7.6.12) above are exactly those in the Coherence Theorem 7.2.1 for 1-colored trees. $\qquad \square$

Suppose (X, λ) is an E_∞-algebra in M. In the next few examples, we will explain part of the structure on X.

Example 7.6.13 (A_∞-structure). The right adjoint $(\mathsf{W}f)^*$ in Corollary 7.6.5 pulls (X, λ) back to an A_∞-algebra. More explicitly, as an A_∞-algebra, its structure morphisms $\lambda_T^{\{\sigma_v\}_{v\in T}}$ in (7.5.6) are equal to the structure morphism λ_T in (7.6.8) for all choices of permutations $\{\sigma_v\} \in \prod_{v\in T} \Sigma_{|\mathrm{in}(v)|}$. In particular, the discussion in Example 7.5.11 to Example 7.5.14 also applies to an E_∞-algebra. $\qquad \diamond$

Example 7.6.14 (Strict commutativity). Consider the tree substitution
$$\mathrm{Cor}_n \sigma = \mathrm{Cor}_n(\mathrm{Cor}_n \sigma)$$
for $n \geq 0$, $\sigma \in \Sigma_n$, $\mathrm{Cor}_n \in \underline{\mathsf{Tree}}(n)$ the corolla with n inputs, and $\mathrm{Cor}_n \sigma$ a permuted corolla. Then the wedge condition (7.6.12) yields the equality
$$\lambda_{\mathrm{Cor}_n} = \lambda_{\mathrm{Cor}_n \sigma} : \mathbb{1} \otimes X^{\otimes n} \longrightarrow X.$$
The equivariance condition (7.6.11) with $T = \mathrm{Cor}_n$ is the following commutative diagram.

$$
\begin{array}{ccc}
X^{\otimes n} \cong \mathsf{J}[\mathrm{Cor}_n] \otimes X^{\otimes n} & \xrightarrow{\ \lambda_{\mathrm{Cor}_n}\ } & X \\
{\scriptstyle (\mathrm{Id},\sigma^{-1})}\big\downarrow & & \big\| \\
X^{\otimes n} \cong \mathsf{J}[\mathrm{Cor}_n \sigma] \otimes X^{\otimes n} & \xrightarrow{\ \lambda_{\mathrm{Cor}_n \sigma}\ } & X
\end{array}
$$

Since $\lambda_{\mathrm{Cor}_n} = \lambda_{\mathrm{Cor}_n \sigma}$, we conclude that the structure morphism λ_{Cor_n} is invariant under permutations of the X factors in its domain. For more general 1-colored trees, the structure morphism is invariant under permutations of its domain factors up to a specified homotopy, as we will see in the next example.   ◇

Example 7.6.15 (Homotopy commutativity). Suppose $K \in \underline{\mathsf{Tree}}(n)$ is not an exceptional edge, and $\sigma \in \Sigma_n$. Recall that $|K|$ denotes the number of internal edges in K. Consider the diagram

$$
\begin{array}{ccc}
\mathsf{J}[K] \otimes X^{\otimes n} & \xrightarrow{\ \lambda_K\ } & X \\
{\scriptstyle \left(0^{\otimes |K|}\circ\cong,\mathrm{Id}\right)}\big\uparrow & & \big\| \\
\mathsf{J}[\mathrm{Cor}_n] \otimes X^{\otimes n} & \xrightarrow{\ \lambda_{\mathrm{Cor}_n}\ } & X \\
{\scriptstyle \left(0^{\otimes |K|}\circ\cong,\mathrm{Id}\right)}\big\downarrow & & \big\| \\
\mathsf{J}[K\sigma] \otimes X^{\otimes n} & \xrightarrow{\ \lambda_{K\sigma}\ } & X \\
{\scriptstyle (\mathrm{Id},\sigma)}\big\downarrow & & \big\| \\
\mathsf{J}[K] \otimes X^{\otimes n} & \xrightarrow{\ \lambda_K\ } & X
\end{array}
$$

in M. This diagram is commutative:

- The top square is commutative by the wedge condition (7.6.12) for the tree substitution $K = \mathrm{Cor}_n(K)$, in which $0^{\otimes |K|} \circ \cong$ is the composition
$$\mathsf{J}[\mathrm{Cor}_n] = \mathbb{1} \xrightarrow{\ \cong\ } \mathbb{1}^{\otimes |K|} \xrightarrow{\ 0^{\otimes |K|}\ } J^{\otimes |K|} \cong \mathsf{J}[K]$$
with $0 : \mathbb{1} \longrightarrow J$ a part of the commutative segment J.
- The middle square is commutative for the same reason for the tree substitution $K\sigma = \mathrm{Cor}_n(K\sigma)$.
- The bottom square is commutative by the equivariance condition (7.6.11).

The entire commutative diagram together says that the structure morphism λ_K is homotopic to the composition $\lambda_K \circ (\mathrm{Id}, \sigma)$. So the structure morphism λ_K is commutative in its domain X factors up to a specified homotopy.   ◇

7.7 Homotopy Coherent Diagrams of A_∞-Algebras

In this section, we discuss a homotopy coherent version of a diagram of monoids using the Boardman-Vogt construction. Fix a small category C with object set $\mathfrak{C}$.

Motivation 7.7.1. An algebraic quantum field theory on an orthogonal category $\overline{\mathsf{C}}$ is, first of all, a functor $\mathfrak{A} : \mathsf{C} \longrightarrow \mathsf{Mon}(\mathsf{M})$ from C to monoids in M. Therefore, we should expect a homotopy algebraic quantum field theory to have the structure of a homotopy coherent C-diagram of A_∞-algebras. This is a combination of the structures in Section 7.3 and Section 7.5, in the sense that it forgets to a homotopy coherent C-diagram in M and that entrywise it is an A_∞-algebra. We saw in Example 4.5.22 that C-diagrams of monoids in M are exactly algebras over the colored operad $\mathsf{O}_{\mathsf{C}}^{\mathsf{M}}$. Their homotopy coherent analogues should therefore be algebras over the Boardman-Vogt construction of $\mathsf{O}_{\mathsf{C}}^{\mathsf{M}}$. ◇

Definition 7.7.2. Objects in the category $\mathsf{Alg}_{\mathsf{M}}(\mathsf{WO}_{\mathsf{C}}^{\mathsf{M}})$ are called *homotopy coherent C-diagrams of A_∞-algebras in M*, where $\mathsf{WO}_{\mathsf{C}}^{\mathsf{M}}$ is the Boardman-Vogt construction of the $\mathfrak{C}$-colored operad $\mathsf{O}_{\mathsf{C}}^{\mathsf{M}}$ in Example 4.5.22.

When applied to the colored operad $\mathsf{O}_{\mathsf{C}}^{\mathsf{M}}$, Corollary 6.4.7 and Corollary 6.5.10 yield the following adjunction.

Corollary 7.7.3. *The augmentation $\eta : \mathsf{WO}_{\mathsf{C}}^{\mathsf{M}} \longrightarrow \mathsf{O}_{\mathsf{C}}^{\mathsf{M}}$ induces an adjunction*

$$\mathsf{Alg}_{\mathsf{M}}(\mathsf{WO}_{\mathsf{C}}^{\mathsf{M}}) \underset{\eta^*}{\overset{\eta_!}{\rightleftarrows}} \mathsf{Alg}_{\mathsf{M}}(\mathsf{O}_{\mathsf{C}}^{\mathsf{M}}) \cong \mathsf{Mon}(\mathsf{M})^{\mathsf{C}}$$

that is a Quillen equivalence if $\mathsf{M} = \mathsf{Chain}_{\mathbb{K}}$ with $\mathbb{K}$ a field of characteristic zero

Interpretation 7.7.4. Each C-diagram of monoids in M can be regarded as a $\mathsf{WO}_{\mathsf{C}}^{\mathsf{M}}$-algebra via the augmentation η. The left adjoint $\eta_!$ rectifies each homotopy coherent C-diagram of A_∞-algebras in M to a C-diagram of monoids in M. ◇

The colored operad $\mathsf{O}_{\mathsf{C}}^{\mathsf{M}}$ for C-diagrams of monoids in M is related to the associative operad As in Example 4.5.17 and the C-diagram operad $\mathsf{C}^{\mathsf{diag}}$ in Example 4.5.21 as follows. We will denote the unique color in As by $*$. A copy of $\mathbb{1}$ corresponding to an element x will be denoted by $\mathbb{1}_x$. The following observation is proved by a direct inspection.

Lemma 7.7.5. *Consider the $\mathfrak{C}$-colored operad $\mathsf{O}_{\mathsf{C}}^{\mathsf{M}}$ in Example 4.5.22.*

(1) For each $c \in \mathfrak{C}$, there is an operad morphism

$$\mathsf{As} \xrightarrow{\iota_c} \mathsf{O}_{\mathsf{C}}^{\mathsf{M}}$$

that sends ∗ to c and is entrywise defined by the commutative diagrams

$$
\begin{array}{ccc}
\mathbb{1}_\sigma & \xrightarrow{\quad = \quad} & \mathbb{1}_{(\sigma,\{\mathrm{Id}_c\}_{j=1}^n)} \\
{\scriptstyle \text{inclusion}}\big\downarrow & & \big\downarrow{\scriptstyle \text{inclusion}} \\
\mathsf{As}(n) = \coprod\limits_{\sigma\in\Sigma_n}\mathbb{1} & \xrightarrow{\;\iota_c\;} & \coprod\limits_{\Sigma_n\times\prod\limits_{j=1}^n \mathsf{C}(c,c)}\mathbb{1} = \mathsf{O}^{\mathsf{M}}_{\mathsf{C}}\!\left(\begin{smallmatrix} c \\ c,\dots,c \end{smallmatrix}\right)
\end{array}
$$

for $n \geq 0$ and $\sigma \in \Sigma_n$.

(2) *There is a morphism of $\mathfrak{C}$-colored operads*

$$
\mathsf{C}^{\mathrm{diag}} \xrightarrow{\quad i \quad} \mathsf{O}^{\mathsf{M}}_{\mathsf{C}}
$$

that is entrywise defined by the commutative diagrams

$$
\begin{array}{ccc}
\mathbb{1}_f & \xrightarrow{\quad = \quad} & \mathbb{1}_{(\mathrm{id}_1,f)} \\
{\scriptstyle \text{inclusion}}\big\downarrow & & \big\downarrow{\scriptstyle \text{inclusion}} \\
\mathsf{C}^{\mathrm{diag}}\!\left(\begin{smallmatrix} d \\ c \end{smallmatrix}\right) = \coprod\limits_{f\in\mathsf{C}(c,d)}\mathbb{1} & \xrightarrow{\quad i \quad} & \coprod\limits_{\Sigma_1\times\mathsf{C}(c,d)}\mathbb{1} = \mathsf{O}^{\mathsf{M}}_{\mathsf{C}}\!\left(\begin{smallmatrix} d \\ c \end{smallmatrix}\right)
\end{array}
$$

for $c,d \in \mathfrak{C}$ and $f \in \mathsf{C}(c,d)$. In all other entries $\left(\begin{smallmatrix} d \\ \underline{c} \end{smallmatrix}\right)$, i is the unique morphism from the initial object to $\mathsf{O}^{\mathsf{M}}_{\mathsf{C}}\left(\begin{smallmatrix} d \\ \underline{c} \end{smallmatrix}\right)$.

Applying Corollary 6.4.7 to the operad morphisms in Lemma 7.7.5, we obtain the following result.

Corollary 7.7.6. *There is a diagram of change-of-operad adjunctions*

$$
\begin{array}{ccccc}
\mathsf{Alg}_{\mathsf{M}}(\mathsf{WAs}) & \underset{\mathsf{W}\iota_c^*}{\overset{(\mathsf{W}\iota_c)_!}{\rightleftarrows}} & \mathsf{Alg}_{\mathsf{M}}(\mathsf{WO}^{\mathsf{M}}_{\mathsf{C}}) & \underset{\mathsf{W}i^*}{\overset{\mathsf{W}i_!}{\rightleftarrows}} & \mathsf{Alg}_{\mathsf{M}}(\mathsf{WC}^{\mathrm{diag}}) \\[4pt]
{\scriptstyle \eta_!}\big\updownarrow{\scriptstyle \eta^*} & & {\scriptstyle \eta_!}\big\updownarrow{\scriptstyle \eta^*} & & {\scriptstyle \eta_!}\big\updownarrow{\scriptstyle \eta^*} \\[4pt]
\mathsf{Mon}(\mathsf{M}) \cong \mathsf{Alg}_{\mathsf{M}}(\mathsf{As}) & \underset{\iota_c^*}{\overset{(\iota_c)_!}{\rightleftarrows}} & \mathsf{Alg}_{\mathsf{M}}(\mathsf{O}^{\mathsf{M}}_{\mathsf{C}}) \cong \mathsf{Mon}(\mathsf{M})^{\mathsf{C}} & \underset{i^*}{\overset{i_!}{\rightleftarrows}} & \mathsf{Alg}_{\mathsf{M}}(\mathsf{C}^{\mathrm{diag}}) \cong \mathsf{M}^{\mathsf{C}}
\end{array}
$$

with commuting left adjoint diagrams and commuting right adjoint diagrams, where the left half is defined for each $c \in \mathfrak{C}$.

Interpretation 7.7.7. In the adjunction $(\iota_c)_! \dashv \iota_c^*$, the right adjoint ι_c^* remembers only the monoid at the c-colored entry. In the adjunction $i_! \dashv i^*$, the right adjoint i^* remembers only the underlying C-diagram in M. The right adjoint $\mathsf{W}\iota_c^*$ remembers only the A_∞-algebra at the c-colored entry, while $\mathsf{W}i^*$ remembers only the underlying homotopy coherent C-diagram in M.   ◇

The following result is the coherence theorem for homotopy coherent C-diagrams of A_∞-algebras in M. If the base category is Set, then we will denote $\mathsf{O}^{\mathsf{Set}}_{\mathsf{C}}$ by O_{C}.

Theorem 7.7.8. *A* $\mathsf{WO}_\mathsf{C}^\mathsf{M}$-*algebra is exactly a pair* (X,λ) *consisting of*

- *a* $\mathfrak{C}$-*colored object* X *in* M *and*
- *a structure morphism*

$$\mathsf{J}[T] \otimes X_{\underline{c}} \xrightarrow{\;\lambda_T\big\{(\sigma^v,\underline{f}^v)\big\}_{v\in T}\;} X_d \;\in \mathsf{M} \tag{7.7.9}$$

for each $T \in \underline{\mathsf{Tree}}^{\mathfrak{C}}\big(\begin{smallmatrix}d\\\underline{c}\end{smallmatrix}\big)$ *with* $(\underline{c};d) \in \mathsf{Prof}(\mathfrak{C}) \times \mathfrak{C}$ *and each*

$$\big\{(\sigma^v,\underline{f}^v)\big\}_{v\in\mathsf{Vt}(T)} \in \prod_{v\in\mathsf{Vt}(T)}\left[\Sigma_{|\mathsf{in}(v)|} \times \prod_{j=1}^{|\mathsf{in}(v)|} \mathsf{C}\big(\mathsf{in}(v)_j,\mathsf{out}(v)\big)\right] = \prod_{v\in\mathsf{Vt}(T)} \mathsf{O}_\mathsf{C}(v)$$

that satisfies the following four conditions.

Associativity *Suppose* $\big(\underline{c}=(c_1,\ldots,c_n);d\big) \in \mathsf{Prof}(\mathfrak{C}) \times \mathfrak{C}$ *with* $n \geq 1$, $T \in \underline{\mathsf{Tree}}^{\mathfrak{C}}\big(\begin{smallmatrix}d\\\underline{c}\end{smallmatrix}\big)$, $T_j \in \underline{\mathsf{Tree}}^{\mathfrak{C}}\big(\begin{smallmatrix}c_j\\\underline{b_j}\end{smallmatrix}\big)$ *for* $1 \leq j \leq n$, $\underline{b}=(\underline{b_1},\ldots,\underline{b_n})$,

$$G = \mathsf{Graft}(T;T_1,\ldots,T_n) \in \underline{\mathsf{Tree}}^{\mathfrak{C}}\big(\begin{smallmatrix}d\\\underline{b}\end{smallmatrix}\big)$$

is the grafting (3.3.1), $\big\{(\sigma^v,\underline{f}^v)\big\}$ *is as above, and*

$$\big\{(\sigma^u,\underline{f}^u)\big\}_{u\in\mathsf{Vt}(T_j)} \in \prod_{u\in\mathsf{Vt}(T_j)}\left[\Sigma_{|\mathsf{in}(u)|} \times \prod_{k=1}^{|\mathsf{in}(u)|} \mathsf{C}\big(\mathsf{in}(u)_k,\mathsf{out}(u)\big)\right] = \prod_{u\in\mathsf{Vt}(T_j)} \mathsf{O}_\mathsf{C}(u)$$

for each $1 \leq j \leq n$. *Then the diagram*

$$\tag{7.7.10}$$

is commutative. Here $\pi = \otimes_S 1$ *is the morphism in Lemma 6.2.7 for the grafting* G.

Unity *For each* $c \in \mathfrak{C}$, *the composition*

$$X_c \xrightarrow{\;\cong\;} \mathsf{J}[\uparrow_c] \otimes X_c \xrightarrow{\;\lambda_{\uparrow_c}\{\varnothing\}\;} X_c \tag{7.7.11}$$

is the identity morphism of X_c.

Equivariance *For each* $T \in \underline{\mathsf{Tree}}^{\mathfrak{C}}\binom{d}{\underline{c}}$, $\{(\sigma^v, \underline{f}^v)\}$ *as above, and permutation* $\sigma \in \Sigma_{|\underline{c}|}$, *the diagram*

$$
\begin{array}{ccc}
\mathsf{J}[T] \otimes X_{\underline{c}} & \xrightarrow{\lambda_T\{(\sigma^v, \underline{f}^v)\}_{v \in T}} & X_d \\
{\scriptstyle (\mathrm{Id}, \sigma^{-1})} \downarrow & & \| \\
\mathsf{J}[T\sigma] \otimes X_{\underline{c}\sigma} & \xrightarrow{\lambda_{T\sigma}\{(\sigma^v, \underline{f}^v)\}_{v \in T\sigma}} & X_d
\end{array}
\tag{7.7.12}
$$

is commutative, in which $T\sigma \in \underline{\mathsf{Tree}}^{\mathfrak{C}}\binom{d}{\underline{c}\sigma}$ *is the same as* T *except that its ordering is* $\zeta_T\sigma$ *with* ζ_T *the ordering of* T. *The permutation* $\sigma^{-1} : X_{\underline{c}} \xrightarrow{\cong} X_{\underline{c}\sigma}$ *permutes the factors in* $X_{\underline{c}}$.

Wedge Condition *Suppose* $T \in \underline{\mathsf{Tree}}^{\mathfrak{C}}\binom{d}{\underline{c}}$, $H_v \in \underline{\mathsf{Tree}}^{\mathfrak{C}}(v)$ *for each* $v \in \mathsf{Vt}(T)$, $K = T(H_v)_{v \in T}$ *is the tree substitution, and*

$$
\{(\sigma^u, \underline{f}^u)\}_{u \in \mathsf{Vt}(H_v)} \in \prod_{u \in \mathsf{Vt}(H_v)} \left[\Sigma_{|\mathrm{in}(u)|} \times \prod_{j=1}^{|\mathrm{in}(u)|} \mathsf{C}\big(\mathrm{in}(u)_j, \mathrm{out}(u)\big) \right] = \prod_{u \in \mathsf{Vt}(H_v)} \mathsf{O}_{\mathsf{C}}(u)
$$

for each $v \in \mathsf{Vt}(T)$. *Then the diagram*

$$
\begin{array}{ccc}
\mathsf{J}[T] \otimes X_{\underline{c}} & \xrightarrow{\lambda_T\{(\tau^v, \underline{g}^v)\}_{v \in T}} & X_d \\
{\scriptstyle (\mathsf{J}, \mathrm{Id})} \downarrow & & \| \\
\mathsf{J}[K] \otimes X_{\underline{c}} & \xrightarrow{\lambda_K\{(\sigma^u, \underline{f}^u)\}_{u \in K}} & X_d
\end{array}
\tag{7.7.13}
$$

is commutative. Here for each $v \in \mathsf{Vt}(T)$,

$$
(\tau^v, \underline{g}^v) = \gamma_{H_v}^{\mathsf{O}_{\mathsf{C}}}\Big(\{(\sigma^u, \underline{f}^u)\}_{u \in H_v}\Big) \in \mathsf{O}_{\mathsf{C}}(v) = \Sigma_{|\mathrm{in}(v)|} \times \prod_{j=1}^{|\mathrm{in}(v)|} \mathsf{C}\big(\mathrm{in}(v)_j, \mathrm{out}(v)\big)
$$

with

$$
\mathsf{O}_{\mathsf{C}}[H_v] = \prod_{u \in H_v} \mathsf{O}_{\mathsf{C}}(u) \xrightarrow{\gamma_{H_v}^{\mathsf{O}_{\mathsf{C}}}} \mathsf{O}_{\mathsf{C}}(v)
$$

the operadic structure morphism of O_{C} *for* H_v *in* (4.4.10).

A morphism $f : (X, \lambda^X) \longrightarrow (Y, \lambda^Y)$ *of* $\mathsf{WO}_{\mathsf{C}}^{\mathsf{M}}$-*algebras is a morphism of the underlying* $\mathfrak{C}$-*colored objects that respects the structure morphisms in* (7.7.9) *in the obvious sense.*

Proof. This is the special case of the Coherence Theorem 7.2.1 applied to the $\mathfrak{C}$-colored operad $\mathsf{O}_{\mathsf{C}}^{\mathsf{M}}$. Indeed, since

$$
\mathsf{O}_{\mathsf{C}}^{\mathsf{M}}\binom{d}{\underline{c}} = \coprod_{\Sigma_n \times \prod_{j=1}^{n} \mathsf{C}(c_j, d)} \mathbb{1} = \coprod_{\mathsf{O}_{\mathsf{C}}\binom{d}{\underline{c}}} \mathbb{1},
$$

for each $T \in \underline{\mathsf{Tree}}^{\mathfrak{C}}\binom{d}{\underline{c}}$ there is a canonical isomorphism

$$O_{\mathsf{C}}^{\mathsf{M}}[T] = \bigotimes_{v \in T} O_{\mathsf{C}}^{\mathsf{M}}(v) = \bigotimes_{v \in T} \left(\coprod_{O_{\mathsf{C}}(v)} \mathbb{1} \right) \cong \coprod_{\prod_{v \in T} O_{\mathsf{C}}(v)} \mathbb{1}.$$

This implies that there is a canonical isomorphism

$$J[T] \otimes O_{\mathsf{C}}^{\mathsf{M}}[T] \otimes X_{\underline{c}} \cong \coprod_{\prod_{v \in T} O_{\mathsf{C}}(v)} J[T] \otimes X_{\underline{c}}.$$

Therefore, the structure morphism λ_T in (7.2.2) is uniquely determined by the restricted structure morphisms $\lambda_T \{ (\sigma^v, \underline{f}^v) \}_{v \in T}$ in (7.7.9). The above associativity, unity, equivariance, and wedge conditions are those in Theorem 7.2.1. □

Example 7.7.14 (Objectwise A_∞-algebra). Suppose (X, λ) is a $\mathsf{WO}_{\mathsf{C}}^{\mathsf{M}}$-algebra, and $c \in \mathfrak{C}$. Under the right adjoint $\mathsf{W}\iota_c^*$ in Corollary 7.7.6, we have that

$$\mathsf{W}\iota_c^*(X, \lambda) \in \mathsf{Alg}_{\mathsf{M}}(\mathsf{WAs}),$$

i.e., an A_∞-algebra. Explicit, its underlying object is $X_c \in \mathsf{M}$. For $T \in \underline{\mathsf{Tree}}(n)$ and $\{\sigma_v\}_{v \in T} \in \prod_{v \in T} \Sigma_{|\mathsf{in}(v)|}$, the A_∞-algebra structure morphism

$$J[T] \otimes X_c^{\otimes n} \xrightarrow{\lambda_T^{\{\sigma_v\}_{v \in T}}} X_c \in \mathsf{M}$$

in (7.5.6) is the structure morphism

$$\lambda_{T_c} \left\{ \left(\sigma_v, \{\mathrm{Id}_c\}_{j=1}^{|\mathsf{in}(v)|} \right) \right\}_{v \in T}$$

in (7.7.9), where $T_c \in \underline{\mathsf{Tree}}^{\mathfrak{C}}\binom{c}{c,\dots,c}$ is the c-colored tree obtained from T by replacing every edge color by c. ◇

Example 7.7.15 (Underlying homotopy coherent C-diagram). Suppose (X, λ) is a $\mathsf{WO}_{\mathsf{C}}^{\mathsf{M}}$-algebra. Under the right adjoint $\mathsf{W}i^*$ in Corollary 7.7.6, we have that

$$\mathsf{W}i^*(X, \lambda) \in \mathsf{Alg}_{\mathsf{M}}(\mathsf{WC}^{\mathsf{diag}}),$$

i.e., a homotopy coherent C-diagram in M. Explicitly, the homotopy coherent C-diagram structure morphism

$$J[\mathsf{Lin}_{\underline{c}}] \otimes X_{c_0} \xrightarrow{\lambda_{\underline{c}}^{\underline{f}}} X_{c_n} \in \mathsf{M}$$

in (7.3.6) is the structure morphism

$$\lambda_{\mathsf{Lin}_{\underline{c}}} \left\{ (\mathrm{id}_1, f_j) \right\}_{1 \le j \le n}$$

in (7.7.9). ◇

There is also a homotopy coherent compatibility between the homotopy coherent C-diagram structure and the objectwise A_∞-algebra structure in a $\mathsf{WO}_{\mathsf{C}}^{\mathsf{M}}$-algebra. We will explain it in details in Section 9.9 in the context that we actually care about, namely homotopy algebraic quantum field theories.

7.8 Homotopy Coherent Diagrams of E_∞-Algebras

In this section, we discuss a homotopy coherent version of a diagram of commutative monoids using the Boardman-Vogt construction. Fix a small category C with object set $\mathfrak{C}$.

Motivation 7.8.1. As we will see in Section 10.7, there is one situation where prefactorization algebras coincide with algebraic quantum field theories. In this case, both categories are the categories of C-diagrams of commutative monoids in M. Homotopy prefactorization algebras, which coincide with homotopy algebraic quantum field theories, are therefore homotopy coherent C-diagrams of E_∞-algebras in M.   ⋄

Definition 7.8.2. Objects in the category $\mathsf{Alg}_\mathsf{M}\big(\mathsf{WCom}^\mathsf{C}\big)$ are called *homotopy coherent* C-*diagrams of* E_∞-*algebras in* M, where WCom^C is the Boardman-Vogt construction of the $\mathfrak{C}$-colored operad Com^C in Example 4.5.23.

When applied to the colored operad Com^C, Corollary 6.4.7 and Corollary 6.5.10 yield the following adjunction.

Corollary 7.8.3. *The augmentation* $\eta : \mathsf{WCom}^\mathsf{C} \longrightarrow \mathsf{Com}^\mathsf{C}$ *induces an adjunction*

$$\mathsf{Alg}_\mathsf{M}\big(\mathsf{WCom}^\mathsf{C}\big) \underset{\eta^*}{\overset{\eta_!}{\rightleftarrows}} \mathsf{Alg}_\mathsf{M}\big(\mathsf{Com}^\mathsf{C}\big) \cong \mathsf{Com}(\mathsf{M})^\mathsf{C}$$

that is a Quillen equivalence if $\mathsf{M} = \mathsf{Chain}_\mathbb{K}$ *with* $\mathbb{K}$ *a field of characteristic zero*

Interpretation 7.8.4. Each C-diagram of commutative monoids in M can be regarded as a WCom^C-algebra via the augmentation η. The left adjoint $\eta_!$ rectifies each homotopy coherent C-diagram of E_∞-algebras in M to a C-diagram of commutative monoids in M.   ⋄

The colored operad Com^C for C-diagrams of commutative monoids in M is related to the commutative operad Com in Example 4.5.19 and the colored operad $\mathsf{O}_\mathsf{C}^\mathsf{M}$ for C-diagrams of monoids in M in Example 4.5.22 as follows. We will denote the unique color in Com by $*$. The following observation is proved by a direct inspection.

Lemma 7.8.5. *Consider the* $\mathfrak{C}$-*colored operad* Com^C *in Example 4.5.23.*

(1) For each $c \in \mathfrak{C}$, *there is an operad morphism*

$$\mathsf{Com} \xrightarrow{\iota_c} \mathsf{Com}^\mathsf{C}$$

that sends $*$ *to* c *and is entrywise defined by the commutative diagram*

$$\begin{array}{ccc}
\mathbb{1} & \xrightarrow{\ =\ } & \mathbb{1}_{\{\mathrm{Id}_c\}_{j=1}^n} \\
{\scriptstyle =}\Big\downarrow & & \Big\downarrow{\scriptstyle \text{inclusion}} \\
\mathsf{Com}(n) & \xrightarrow{\ \iota_c\ } & \underset{\overset{n}{\underset{j=1}{\prod}} \mathsf{C}(c,c)}{\coprod} \mathbb{1} = \mathsf{Com}^\mathsf{C}\big(\begin{smallmatrix} c \\ c,\ldots,c \end{smallmatrix}\big)
\end{array}$$

for $n \geq 0$.

(2) There is a morphism of $\mathfrak{C}$-colored operads

$$O_{\mathsf{C}}^{\mathsf{M}} \xrightarrow{\ p\ } \mathsf{Com}^{\mathsf{C}}$$

that is entrywise defined by the commutative diagrams

for $(\underline{c}; d) \in \mathsf{Prof}(\mathfrak{C}) \times \mathfrak{C}$ with $\underline{c} = (c_1, \ldots, c_n)$, $\sigma \in \Sigma_n$, and $\underline{f} \in \prod_{j=1}^{n} \mathsf{C}(c_j, d)$.

Applying Corollary 6.4.7 to the operad morphisms in Lemma 7.8.5, we obtain the following result.

Corollary 7.8.6. *There is a diagram of change-of-operad adjunctions*

with commuting left adjoint diagrams and commuting right adjoint diagrams, where the left half is defined for each $c \in \mathfrak{C}$.

Interpretation 7.8.7. In the adjunction $(\iota_c)_! \dashv \iota_c^*$, the right adjoint ι_c^* remembers only the commutative monoid at the c-colored entry. In the adjunction $p_! \dashv p^*$, the right adjoint p^* sends a C-diagram of commutative monoids in M to its underlying C-diagram of monoids in M; i.e., it forgets about the commutativity. The right adjoint $W\iota_c^*$ remembers only the E_∞-algebra at the c-colored entry. The right adjoint Wp^* sends a homotopy coherent C-diagram of E_∞-algebras in M to the underlying homotopy coherent C-diagram of A_∞-algebras in M. Combined with Corollary 7.7.6, one can forget further down to the underlying homotopy coherent C-diagram in M. ◇

The following result is the coherence theorem for homotopy coherent C-diagrams of E_∞-algebras in M. If the base category is Set, then we will write $\mathsf{Com}^{\mathsf{C}}$ as $\mathsf{Com}_{\mathsf{Set}}^{\mathsf{C}}$.

Theorem 7.8.8. *A $\mathsf{WCom}^{\mathsf{C}}$-algebra is exactly a pair (X, λ) consisting of*

- *a $\mathfrak{C}$-colored object X in M and*

- *a structure morphism*

$$J[T] \otimes X_{\underline{c}} \xrightarrow{\ \lambda_T\{\underline{f}^v\}_{v \in T}\ } X_d \in \mathsf{M} \tag{7.8.9}$$

for each $T \in \underline{\mathsf{Tree}}^{\mathfrak{C}}\binom{d}{\underline{c}}$ with $(\underline{c}; d) \in \mathsf{Prof}(\mathfrak{C}) \times \mathfrak{C}$ and each

$$\{\underline{f}^v\}_{v \in \mathsf{Vt}(T)} \in \prod_{v \in \mathsf{Vt}(T)} \prod_{j=1}^{|\mathrm{in}(v)|} \mathsf{C}\big(\mathrm{in}(v)_j, \mathrm{out}(v)\big) = \prod_{v \in \mathsf{Vt}(T)} \mathsf{Com}_{\mathsf{Set}}^{\mathsf{C}}(v)$$

that satisfies the following four conditions.

Associativity *Suppose* $\big(\underline{c} = (c_1,\dots,c_n); d\big) \in \mathsf{Prof}(\mathfrak{C}) \times \mathfrak{C}$ *with* $n \geq 1$, $T \in \underline{\mathsf{Tree}}^{\mathfrak{C}}\binom{d}{\underline{c}}$, $T_j \in \underline{\mathsf{Tree}}^{\mathfrak{C}}\binom{c_j}{\underline{b}_j}$ *for* $1 \leq j \leq n$, $\underline{b} = (\underline{b}_1,\dots,\underline{b}_n)$,

$$G = \mathsf{Graft}(T; T_1,\dots,T_n) \in \underline{\mathsf{Tree}}^{\mathfrak{C}}\binom{d}{\underline{b}}$$

is the grafting (3.3.1), $\{\underline{f}^v\}$ *is as above, and*

$$\{\underline{f}^u\}_{u \in \mathsf{Vt}(T_j)} \in \prod_{u \in \mathsf{Vt}(T_j)} \prod_{k=1}^{|\mathrm{in}(u)|} \mathsf{C}\big(\mathrm{in}(u)_k, \mathrm{out}(u)\big) = \prod_{u \in \mathsf{Vt}(T_j)} \mathsf{Com}_{\mathsf{Set}}^{\mathsf{C}}(u)$$

for each $1 \leq j \leq n$. *Then the diagram*

$$J[T] \otimes \Big(\bigotimes_{j=1}^{n} J[T_j] \Big) \otimes X_{\underline{b}} \tag{7.8.10}$$

is commutative. Here $\pi = \otimes_S 1$ *is the morphism in Lemma 6.2.7 for the grafting* G.

Unity *For each* $c \in \mathfrak{C}$, *the composition*

$$X_c \xrightarrow{\ \cong\ } J[\uparrow_c] \otimes X_c \xrightarrow{\ \lambda_{\uparrow_c}\{\varnothing\}\ } X_c \tag{7.8.11}$$

is the identity morphism of X_c.

Equivariance *For each* $T \in \underline{\mathsf{Tree}}^{\mathfrak{C}}\binom{d}{\underline{c}}$, $\{\underline{f}^v\}$ *as above, and permutation* $\sigma \in \Sigma_{|\underline{c}|}$, *the diagram*

$$\begin{array}{ccc}
J[T] \otimes X_{\underline{c}} & \xrightarrow{\ \lambda_T\{\underline{f}^v\}_{v \in T}\ } & X_d \\
{\scriptstyle (\mathrm{Id},\sigma^{-1})} \Big\downarrow & & \Big\| \\
J[T\sigma] \otimes X_{\underline{c}\sigma} & \xrightarrow{\ \lambda_{T\sigma}\{\underline{f}^v\}_{v \in T\sigma}\ } & X_d
\end{array} \tag{7.8.12}$$

is commutative, in which $T\sigma \in \underline{\text{Tree}}^{\mathfrak{C}}\binom{d}{c\sigma}$ *is the same as* T *except that its ordering is* $\zeta_T\sigma$ *with* ζ_T *the ordering of* T. *The permutation* $\sigma^{-1} : X_{\underline{c}} \xrightarrow{\cong} X_{\underline{c}\sigma}$ *permutes the factors in* $X_{\underline{c}}$.

Wedge Condition *Suppose* $T \in \underline{\text{Tree}}^{\mathfrak{C}}\binom{d}{\underline{c}}$, $H_v \in \underline{\text{Tree}}^{\mathfrak{C}}(v)$ *for each* $v \in \text{Vt}(T)$, $K = T(H_v)_{v \in T}$ *is the tree substitution, and*

$$\left\{\underline{f}^u\right\}_{u \in \text{Vt}(H_v)} \in \prod_{u \in \text{Vt}(H_v)} \prod_{j=1}^{|\text{in}(u)|} \mathsf{C}\big(\text{in}(u)_j, \text{out}(u)\big) = \prod_{u \in \text{Vt}(H_v)} \mathsf{Com}^{\mathsf{C}}_{\mathsf{Set}}(u)$$

for each $v \in \text{Vt}(T)$. *Then the diagram*

$$\begin{array}{ccc} \mathsf{J}[T] \otimes X_{\underline{c}} & \xrightarrow{\lambda_T\{\underline{g}^v\}_{v \in T}} & X_d \\ {\scriptstyle (\mathsf{J},\text{Id})}\Big\downarrow & & \Big\| \\ \mathsf{J}[K] \otimes X_{\underline{c}} & \xrightarrow{\lambda_K\{\underline{f}^u\}_{u \in K}} & X_d \end{array} \tag{7.8.13}$$

is commutative. Here for each $v \in \text{Vt}(T)$,

$$\underline{g}^v = \gamma^{\mathsf{Com}^{\mathsf{C}}_{\mathsf{Set}}}_{H_v}\left(\{\underline{f}^u\}_{u \in H_v}\right) \in \mathsf{Com}^{\mathsf{C}}_{\mathsf{Set}}(v) = \prod_{j=1}^{|\text{in}(v)|} \mathsf{C}\big(\text{in}(v)_j, \text{out}(v)\big)$$

with

$$\mathsf{Com}^{\mathsf{C}}_{\mathsf{Set}}[H_v] = \prod_{u \in H_v} \mathsf{Com}^{\mathsf{C}}_{\mathsf{Set}}(u) \xrightarrow{\gamma^{\mathsf{Com}^{\mathsf{C}}_{\mathsf{Set}}}_{H_v}} \mathsf{Com}^{\mathsf{C}}_{\mathsf{Set}}(v)$$

the operadic structure morphism of $\mathsf{Com}^{\mathsf{C}}_{\mathsf{Set}}$ *for* H_v *in* (4.4.10).

A morphism $f : (X, \lambda^X) \longrightarrow (Y, \lambda^Y)$ *of* $\mathsf{WCom}^{\mathsf{C}}$-*algebras is a morphism of the underlying* $\mathfrak{C}$-*colored objects that respects the structure morphisms in* (7.8.9) *in the obvious sense.*

Proof. This is the special case of the Coherence Theorem 7.2.1 applied to the $\mathfrak{C}$-colored operad $\mathsf{Com}^{\mathsf{C}}$. Indeed, since

$$\mathsf{Com}^{\mathsf{C}}\binom{d}{\underline{c}} = \coprod_{\prod\limits_{j=1}^{n} \mathsf{C}(c_j, d)} \mathbb{1} = \coprod_{\mathsf{Com}^{\mathsf{C}}_{\mathsf{Set}}\binom{d}{\underline{c}}} \mathbb{1},$$

for each $T \in \underline{\text{Tree}}^{\mathfrak{C}}\binom{d}{\underline{c}}$ there is a canonical isomorphism

$$\mathsf{Com}^{\mathsf{C}}[T] = \bigotimes_{v \in T} \mathsf{Com}^{\mathsf{C}}(v) = \bigotimes_{v \in T}\left(\coprod_{\mathsf{Com}^{\mathsf{C}}_{\mathsf{Set}}(v)} \mathbb{1}\right) \cong \coprod_{\prod\limits_{v \in T} \mathsf{Com}^{\mathsf{C}}_{\mathsf{Set}}(v)} \mathbb{1}.$$

This implies that there is a canonical isomorphism

$$\mathsf{J}[T] \otimes \mathsf{Com}^{\mathsf{C}}[T] \otimes X_{\underline{c}} \cong \coprod_{\prod\limits_{v \in T} \mathsf{Com}^{\mathsf{C}}_{\mathsf{Set}}(v)} \mathsf{J}[T] \otimes X_{\underline{c}}.$$

Therefore, the structure morphism λ_T in (7.2.2) is uniquely determined by the restricted structure morphisms $\lambda_T\{\underline{f}^v\}_{v \in T}$ in (7.8.9). The above associativity, unity, equivariance, and wedge conditions are those in Theorem 7.2.1. $\square$

Example 7.8.14 (Objectwise E_∞-algebra). Suppose (X, λ) is a WCom^C-algebra, and $c \in \mathfrak{C}$. Under the right adjoint $\mathsf{W}\iota_c^*$ in Corollary 7.8.6, we have that

$$\mathsf{W}\iota_c^*(X, \lambda) \in \mathsf{Alg}_\mathsf{M}(\mathsf{WCom}),$$

i.e., an E_∞-algebra. Explicit, its underlying object is $X_c \in \mathsf{M}$. For $T \in \underline{\mathsf{Tree}}(n)$, the E_∞-algebra structure morphism

$$J[T] \otimes X_c^{\otimes n} \xrightarrow{\ \lambda_T\ } X_c \in \mathsf{M}$$

in (7.6.8) is the structure morphism

$$\lambda_{T_c}\left\{\{\mathrm{Id}_c\}_{j=1}^{|\mathrm{in}(v)|}\right\}_{v \in T}$$

in (7.8.9), where $T_c \in \underline{\mathsf{Tree}}^{\mathfrak{C}}\binom{c}{c,\dots,c}$ is the c-colored tree obtained from T by replacing every edge color by c.   ◇

Example 7.8.15 (Underlying homotopy coherent C-diagram). Suppose (X, λ) is a WCom^C-algebra. Under the right adjoints

$$\mathsf{Alg}_\mathsf{M}\left(\mathsf{WCom}^\mathsf{C}\right) \xrightarrow{\ \mathsf{W}p^*\ } \mathsf{Alg}_\mathsf{M}(\mathsf{WO}_\mathsf{C}^\mathsf{M}) \xrightarrow{\ \mathsf{W}i^*\ } \mathsf{Alg}_\mathsf{M}(\mathsf{WC}^{\mathrm{diag}})$$

in Corollary 7.8.6 and Corollary 7.7.6, we have that

$$(\mathsf{W}i^*)(\mathsf{W}p)^*(X, \lambda) \in \mathsf{Alg}_\mathsf{M}(\mathsf{WC}^{\mathrm{diag}}),$$

i.e., a homotopy coherent C-diagram in M. Explicitly, the homotopy coherent C-diagram structure morphism

$$J[\underline{\mathrm{Lin}}_{\underline{c}}] \otimes X_{c_0} \xrightarrow{\ \lambda_{\underline{c}}^{f}\ } X_{c_n} \in \mathsf{M}$$

in (7.3.6) is the structure morphism $\lambda_{\underline{\mathrm{Lin}}_{\underline{c}}}\{f_j\}_{1 \le j \le n}$ in (7.8.9).   ◇

Chapter 8

Algebraic Quantum Field Theories

This chapter is about algebraic quantum field theories in the operadic framework of [Benini *et. al.* (2017)]. In Section 8.1 we provide a brief description of the traditional Haag-Kastler approach to algebraic quantum field theories and how it may be generalized to an operadic framework. In Section 8.2 we discuss orthogonal categories and algebraic quantum field theories defined on them. In Section 8.3 we discuss the colored operads in [Benini *et. al.* (2017)] whose algebras are algebraic quantum field theories. Many examples are discussed in Section 8.4, including diagrams of (commutative) monoids, chiral conformal, Euclidean, and locally covariant quantum field theories, various flavors of quantum gauge theories, and quantum field theories on spacetimes with timelike boundary. In Section 8.5 we study homotopical properties of the category of algebraic quantum field theories.

As in previous chapters, $(\mathsf{M}, \otimes, \mathbb{1})$ is a fixed cocomplete symmetric monoidal closed category, such as $\mathsf{Vect}_{\mathbb{K}}$ and $\mathsf{Chain}_{\mathbb{K}}$, and $\mathfrak{C}$ is a non-empty set.

8.1 From Haag-Kastler Axioms to Operads

In this section, we provide a brief overview of the traditional approach to algebraic quantum field theories due to Haag and Kastler. Then we review how the Haag-Kastler approach is modified to the operadic viewpoint in [Benini *et. al.* (2017)], which is what the rest of this chapter is about.

Haag and Kastler [Haag and Kastler (1964)] defined an algebraic quantum field theory on a fixed Lorentzian spacetime X as a rule $\mathfrak{A}$ that assigns

- to each suitable spacetime region $U \subseteq X$ a unital associative algebra $\mathfrak{A}(U)$ and
- to each inclusion $f : U \subseteq V$ an injective algebra homomorphism $\mathfrak{A}(f) : \mathfrak{A}(U) \longrightarrow \mathfrak{A}(V)$.

The algebra $\mathfrak{A}(U)$ is the algebra of quantum observables in U. The homomorphism $\mathfrak{A}(f)$ sends each observable in U to an observable in the larger region V. The condition that each $\mathfrak{A}(f)$ be injective is called the *isotony axiom*. Moreover, it is assumed that the following axioms are satisfied.

Causality Axiom If $U_1 \subseteq V$ and $U_2 \subseteq V$ are causally disjoint regions in V, then each element in $\mathfrak{A}(U_1)$ and each element in $\mathfrak{A}(U_2)$ commute in $\mathfrak{A}(V)$.

Time-Slice Axiom If $U \subseteq V$ contains a Cauchy surface of V, then the homomorphism $\mathfrak{A}(U) \xrightarrow{\cong} \mathfrak{A}(V)$ is an isomorphism.

The causality axiom corresponds to the physical principle that effects do not travel faster than the speed of light. The time-slice axiom says that observables in a small time interval determine all observables.

The Haag-Kastler approach is generalized in [Brunetti *et. al.* (2003)] to the category of all oriented, time-oriented, and globally hyperbolic Lorentzian manifolds. To obtain other flavors of quantum field theories, such as chiral conformal and Euclidean quantum field theories, the above framework is abstracted one step further in [Benini *et. al.* (2017)] by replacing the category of spacetimes with an abstract small category C equipped with a set $\perp$ of pairs of morphisms $(g_1 : a \longrightarrow c, g_2 : b \longrightarrow c)$ with the same codomain. Physically one interprets the objects in C as the spacetimes of interest and the morphisms as inclusions of smaller regions into larger regions. A pair $(g_1, g_2) \in \perp$ means that their domains a and b are suitably disjoint regions in the common codomain c. The pair $\overline{\mathsf{C}} = (\mathsf{C}, \perp)$ is called an orthogonal category.

The causality axiom is implemented using the set $\perp$ of orthogonality relations. The time-slice axiom may be implemented by choosing a suitable set S of morphisms in C, corresponding to the Cauchy morphisms in the Lorentzian case. In addition to the domain category, the target category can also be replaced by the category of monoids in a symmetric monoidal category M, with $\mathsf{M} = \mathsf{Vect}_{\mathbb{K}}$ being the traditional case. So now an algebraic quantum field theory on $\overline{\mathsf{C}} = (\mathsf{C}, \perp)$ is a functor $\mathfrak{A} : \mathsf{C} \longrightarrow \mathsf{Mon}(\mathsf{M})$ that satisfies the causality axiom and, if a set S of morphisms is given, the time-slice axiom.

In Example 4.5.22 we saw that there is an $\mathsf{Ob}(\mathsf{C})$-colored operad $\mathsf{O}_{\mathsf{C}}^{\mathsf{M}}$ whose algebras are exactly C-diagrams of monoids in M. With a bit more work, one can build the causality axiom into the colored operad. So there is an $\mathsf{Ob}(\mathsf{C})$-colored operad $\mathsf{O}_{\overline{\mathsf{C}}}^{\mathsf{M}}$ whose category of algebras is exactly the category of algebraic quantum field theories on $\overline{\mathsf{C}} = (\mathsf{C}, \perp)$. To implement the time-slice axiom, one first replaces the small category C with its S-localization $\mathsf{C}[S^{-1}]$ and the orthogonality relation $\perp$ by a suitable pushforward. In Section 8.3 we will discuss this operadic framework for algebraic quantum field theories.

One might wonder what happened to the isotony axiom, which requires that each homomorphism $\mathfrak{A}(f)$ be injective. Various models of quantum gauge theories actually do not satisfy the isotony axiom; see for example [Becker *et. al.* (2017,b); Benini *et. al.* (2014,b); Benini and Schenkel (2017); Dappiaggi and Lang (2012); Sanders *et. al.* (2014)]. Since the operadic framework is general enough to include some flavors of quantum gauge theories, it is reasonable to drop the isotony axiom.

8.2 AQFT as Functors

In this section, we review the functor definition of algebraic quantum field theories as discussed in [Benini *et. al.* (2017)]. All the assertions in this section are from [Benini *et. al.* (2017)], where the reader may find more details.

The causality axiom says that certain elements from separated regions should commute. The following concept of an orthogonality relation is used to formalize the idea of separated regions.

Definition 8.2.1. Suppose C is a small category. An *orthogonality relation* on C is a subset $\perp$ of pairs of morphisms in C such that if $(f, g) \in \perp$, then f and g have the same codomain. Furthermore, it is required to satisfy the following three axioms.

Symmetry If $(f, g) \in \perp$, then $(g, f) \in \perp$.

Post-Composition If $(g_1, g_2) \in \perp$, then $(fg_1, fg_2) \in \perp$ for all composable morphisms f in C.

Pre-Composition If $(g_1, g_2) \in \perp$, then $(g_1 h_1, g_2 h_2) \in \perp$ for all composable morphisms h_1 and h_2 in C.

If $(f, g) \in \perp$, then we also write $f \perp g$ and say that f and g are *orthogonal*.

- An *orthogonal category* is a small category equipped with an orthogonality relation.
- An *orthogonal functor*

$$F : \overline{\mathsf{C}} = (\mathsf{C}, \perp^{\mathsf{C}}) \longrightarrow (\mathsf{D}, \perp^{\mathsf{D}}) = \overline{\mathsf{D}}$$

  between orthogonal categories is a functor $F : \mathsf{C} \longrightarrow \mathsf{D}$ that preserves the orthogonality relations in the sense that $(f, g) \in \perp^{\mathsf{C}}$ implies $(Ff, Fg) \in \perp^{\mathsf{D}}$.
- The category of orthogonal categories and orthogonal functors is denoted by $\mathsf{OrthCat}$.

Orthogonality relations can be pulled back and pushed forward via any functor. The following observation is [Benini *et. al.* (2017)] Lemma 4.29, which follows directly from Definition 8.2.1.

Lemma 8.2.2. *Suppose $F : \mathsf{C} \longrightarrow \mathsf{D}$ is a functor between small categories.*

(1) If $\perp^{\mathsf{C}}$ is an orthogonality relation on C, then

$$F_*(\perp^{\mathsf{C}}) = \left\{ \left(fF(g_1)h_1, fF(g_2)h_2 \right) : (g_1, g_2) \in \perp^{\mathsf{C}}, f, h_1, h_2 \in \mathsf{D} \right\}$$

is an orthogonality relation on D, called the pushforward of $\perp^{\mathsf{C}}$ along F. Moreover,

$$F : (\mathsf{C}, \perp^{\mathsf{C}}) \longrightarrow \left(\mathsf{D}, F_*(\perp^{\mathsf{C}}) \right)$$

is an orthogonal functor.

(2) If $\perp^D$ is an orthogonality relation on D, then

$$F^*(\perp^D) = \left\{(f_1, f_2) : \mathrm{codomain}(f_1) = \mathrm{codomain}(f_2), \, (Ff_1, Ff_2) \in \perp^D\right\}$$

is an orthogonality relation on C, called the pullback of $\perp^D$ along F. Moreover,

$$F : \left(\mathsf{C}, F^*(\perp^D)\right) \longrightarrow (\mathsf{D}, \perp^D)$$

is an orthogonal functor.

Recall from Section 2.6 that Mon(M) is the category of monoids in M. The isotony axiom in the Haag-Kastler setting [Haag and Kastler (1964)] says that for each inclusion of regions there is a corresponding inclusion of algebras. In the categorical setting, instead of algebra inclusions, we ask for a functor from the category of regions to the category of monoids.

Definition 8.2.3. Suppose $\overline{\mathsf{C}} = (\mathsf{C}, \perp)$ is an orthogonal category, and S is a set of morphisms in C.

(1) A functor $\mathfrak{A} : \mathsf{C} \longrightarrow \mathsf{Mon}(\mathsf{M})$ satisfies the *causality axiom* if for each orthogonal pair $(g_1 : a \to c, g_2 : b \to c) \in \perp$, the diagram

$$
\begin{array}{ccccc}
\mathfrak{A}(a) \otimes \mathfrak{A}(b) & \xrightarrow{\left(\mathfrak{A}(g_1), \mathfrak{A}(g_2)\right)} & \mathfrak{A}(c) \otimes \mathfrak{A}(c) & \xrightarrow[\cong]{\text{permute}} & \mathfrak{A}(c) \otimes (c) \\
{\scriptstyle \left(\mathfrak{A}(g_1), \mathfrak{A}(g_2)\right)} \Big\downarrow & & & & \Big\downarrow {\scriptstyle \mu_c} \\
\mathfrak{A}(c) \otimes \mathfrak{A}(c) & & \xrightarrow{\hspace{3cm} \mu_c \hspace{3cm}} & & \mathfrak{A}(c)
\end{array}
\tag{8.2.4}
$$

in M is commutative, where μ_c is the monoid multiplication in $\mathfrak{A}(c)$.

(2) An *algebraic quantum field theory on $\overline{\mathsf{C}}$* is a functor $\mathfrak{A} : \mathsf{C} \longrightarrow \mathsf{Mon}(\mathsf{M})$ that satisfies the causality axiom.

(3) The full subcategory of the diagram category $\mathsf{Mon}(\mathsf{M})^{\mathsf{C}}$ whose objects are algebraic quantum field theories on $\overline{\mathsf{C}}$ is denoted by $\mathsf{QFT}(\overline{\mathsf{C}})$.

(4) A functor $\mathfrak{A} : \mathsf{C} \longrightarrow \mathsf{Mon}(\mathsf{M})$ satisfies the *time-slice axiom* with respect to S if for each $s : a \longrightarrow b \in S$, the morphism

$$\mathfrak{A}(s) : \mathfrak{A}(a) \xrightarrow{\ \cong\ } \mathfrak{A}(b) \in \mathsf{Mon}(\mathsf{M})$$

is an isomorphism.

(5) The full subcategory of $\mathsf{QFT}(\overline{\mathsf{C}})$ consisting of algebraic quantum field theories on $\overline{\mathsf{C}}$ that satisfy the time-slice axiom with respect to S is denoted by $\mathsf{QFT}(\overline{\mathsf{C}}, S)$.

Interpretation 8.2.5. Physically the objects in the orthogonal category $\overline{\mathsf{C}}$ are the spacetime regions of interest. The functor $\mathfrak{A}$ assigns a monoid of quantum observables $\mathfrak{A}(c)$ to each region c. The orthogonality relation $\perp$ specifies the disjoint regions. The causality axiom says that, if a and b are disjoint regions in c, then an observable from a and an observable from b commute in c. The set S specifies the Cauchy morphisms. The time-slice axiom says that Cauchy morphisms are sent to isomorphisms of monoids of quantum observables. For an orthogonal category $\overline{\mathsf{C}}$, $\mathsf{QFT}(\overline{\mathsf{C}})$ is the category of all the quantum field theories associated to the spacetime regions in C.   ⋄

Remark 8.2.6. In [Benini *et. al.* (2017)] the causality axiom, the time-slice axiom, and an algebraic quantum field theory are called $\perp$-commutativity, W-constancy, and a $\perp$-commutative functor $\mathsf{C} \longrightarrow \mathsf{Mon}(\mathsf{M})$, respectively. Moreover, $\mathsf{QFT}(\overline{\mathsf{C}})$ is denoted by $\mathsf{Mon}_{\mathsf{M}}^{\overline{\mathsf{C}}}$ in [Benini *et. al.* (2017)]. ◇

Recall the concept of a localization of a category in Section 2.8. The following result is [Benini *et. al.* (2017)] Lemma 4.30, which uses the pushforward orthogonality relation in Lemma 8.2.2.

Lemma 8.2.7. *Suppose* $\overline{\mathsf{C}} = (\mathsf{C}, \perp)$ *is an orthogonal category, and* S *is a set of morphisms in* C *with* S-*localization* $\ell : \mathsf{C} \longrightarrow \mathsf{C}[S^{-1}]$. *Suppose*

$$\overline{\mathsf{C}[S^{-1}]} = \Big(\mathsf{C}[S^{-1}], \ell_*(\perp)\Big)$$

is the orthogonal category equipped with the pushforward of $\perp$ *along* ℓ. *Then there is a canonical isomorphism*

$$\mathsf{QFT}(\overline{\mathsf{C}}, S) \cong \mathsf{QFT}(\overline{\mathsf{C}[S^{-1}]}).$$

Using this isomorphism, we will regard $\mathsf{QFT}(\overline{\mathsf{C}[S^{-1}]})$ *as a full subcategory of* $\mathsf{QFT}(\overline{\mathsf{C}})$.

Proof. Let us first indicate the correspondence between objects. First, an object on the right side yields an object on the left side by pre-composition with the S-localization ℓ.

On the other hand, suppose $\mathfrak{A} : \mathsf{C} \longrightarrow \mathsf{Mon}(\mathsf{M})$ is a functor that satisfies the causality axiom and the time-slice axiom with respect to S. Then by the universal property of the S-localization, there is a unique functor

$$\mathfrak{B} : \mathsf{C}[S^{-1}] \longrightarrow \mathsf{Mon}(\mathsf{M}) \quad \text{such that} \quad \mathfrak{A} = \mathfrak{B}\ell.$$

To see that $\mathfrak{B}$ satisfies the causality axiom, recall that each orthogonal pair in the pushforward $\ell_*(\perp)$ has the form $\big(f\ell(g_1)h_1, f\ell(g_2)h_2\big)$ with

- $(g_1 : a \to c) \perp (g_2 : b \to c)$ and
- $f : \ell c \to d, h_1 : x \to \ell a, h_2 : y \to \ell b \in \mathsf{C}[S^{-1}]$.

We want to know that the outermost diagram in

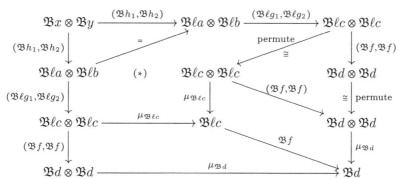

is commutative. The upper left triangle and the upper right triangle are commutative by definition and the symmetry in M, respectively. The bottom trapezoid is equal to the adjacent parallelogram, which is commutative because $\mathfrak{B}f$ is a morphism of monoids in M. Since $\mathfrak{B}\ell = \mathfrak{A}$, the sub-diagram $(*)$ is the commutative diagram (8.2.4).

To see the correspondence between morphisms, we simply use the description in Theorem 2.8.3 of the morphisms in the localization.   □

Interpretation 8.2.8. Lemma 8.2.7 says that the time-slice axiom in algebraic quantum field theories can be implemented by replacing the orthogonal category with its localization along with the pushforward orthogonality relation. Therefore, algebraic quantum field theories and those that satisfy the time-slice axiom can be studied in the same setting.   ◇

8.3   AQFT as Operad Algebras

In this section, following [Benini *et. al.* (2017)] we describe a colored operad whose algebras are algebraic quantum field theories on a given orthogonal category.

Motivation 8.3.1. From the previous section, an algebraic quantum field theory on an orthogonal category $(C, \perp)$ is a functor $\mathfrak{A} : C \longrightarrow \mathsf{Mon}(M)$ that satisfies the causality axiom and, if a set S of morphisms is given, the time-slice axiom with respect to S. The time-slice axiom says that certain structure morphisms are invertible, which by Lemma 8.2.7 can be implemented by using the S-localization of C. The functor $\mathfrak{A}$ itself is a C-diagram of monoids in M, while the causality axiom is a form of commutativity. As we saw in Examples 4.5.19 and 4.5.22, commutative monoids and diagrams of monoids can all be modeled using (colored) operads. Therefore, it is natural to expect a colored operad whose algebras are algebraic quantum field theories. Recall from Definition 4.2.1 the description of a colored operad in terms of generating operations and generating axioms.   ◇

Definition 8.3.2. Suppose $\overline{C} = (C, \perp)$ is an orthogonal category with $\mathsf{Ob}(C) = \mathfrak{C}$. Define the following sets and functions.

Entries Define the object $O_{\overline{C}} \in \mathsf{Set}^{\mathsf{Prof}(\mathfrak{C}) \times \mathfrak{C}}$ entrywise as the quotient set

$$O_{\overline{C}}\binom{d}{\underline{c}} = \left(\Sigma_n \times \prod_{j=1}^{n} C(c_j, d) \right) \Big/ \sim \quad \text{for } \binom{d}{\underline{c}} = \binom{d}{c_1, \ldots, c_n} \in \mathsf{Prof}(\mathfrak{C}) \times \mathfrak{C}$$

in which the equivalence relation $\sim$ is defined as follows. For $(\sigma, \underline{f})$ and $(\sigma', \underline{f}')$ in $\Sigma_n \times \prod_{j=1}^{n} C(c_j, d)$, we define

$$(\sigma, \underline{f}) \sim (\sigma', \underline{f}')$$

if and only if the following two conditions hold:

- $\underline{f} = \underline{f}'$ in $\prod_{j=1}^{n} C(c_j, d)$.

- $\sigma\sigma'^{-1}$ factors as a product $\tau_1\cdots\tau_r$ of transpositions in Σ_n such that, for each $1 \le k \le r$, the right permutation

$$\tau_k : \underline{f}\sigma^{-1}\tau_1\cdots\tau_{k-1} \longrightarrow \underline{f}\sigma^{-1}\tau_1\ldots\tau_k$$

is a transposition of two morphisms in C that are adjacent and orthogonal in $\underline{f}\sigma^{-1}\tau_1\cdots\tau_{k-1}$.

The equivalence class of $(\sigma, \underline{f})$ is denoted by $[\sigma, \underline{f}]$.

Equivariance For $\tau \in \Sigma_{|\underline{c}|}$, define the map

$$\mathsf{O}_{\overline{\mathsf{C}}}\binom{d}{\underline{c}} \overset{\tau}{\longrightarrow} \mathsf{O}_{\overline{\mathsf{C}}}\binom{d}{\underline{c}\tau}$$

by $[\sigma, \underline{f}]\tau = [\sigma\tau, \underline{f}\tau]$.

Colored Units For $c \in \mathfrak{C}$, the c-colored unit in $\mathsf{O}_{\overline{\mathsf{C}}}\binom{c}{c}$ is $[\mathrm{id}_1, \mathrm{Id}_c]$.

Operadic Composition For $(\underline{c}; d) \in \mathsf{Prof}(\mathfrak{C}) \times \mathfrak{C}$ with $|\underline{c}| = n \ge 1$, $\underline{b}_j = (b_{j1}, \ldots, b_{jk_j}) \in \mathsf{Prof}(\mathfrak{C})$ for $1 \le j \le n$ with $|\underline{b}_j| = k_j \ge 0$, and $\underline{b} = (\underline{b}_1, \ldots, \underline{b}_n)$, define the map

$$\mathsf{O}_{\overline{\mathsf{C}}}\binom{d}{\underline{c}} \times \prod_{j=1}^{n} \mathsf{O}_{\overline{\mathsf{C}}}\binom{c_j}{\underline{b}_j} \overset{\gamma}{\longrightarrow} \mathsf{O}_{\overline{\mathsf{C}}}\binom{d}{\underline{b}}$$

by

$$\gamma\left([\sigma, \underline{f}]; \{[\tau_j, \underline{g}_j]\}_{j=1}^{n}\right) = \left[\sigma\langle\tau_1, \ldots, \tau_n\rangle, \left(f_1\underline{g}_1, \ldots, f_n\underline{g}_n\right)\right]$$

where

$$f_j\underline{g}_j = \left(f_j g_{j1}, \ldots, f_j g_{jk_j}\right) \in \prod_{i=1}^{k_j} \mathsf{C}(b_{ji}, d) \quad \text{if} \quad \underline{g}_j = \left(g_{j1}, \ldots, g_{jk_j}\right) \in \prod_{i=1}^{k_j} \mathsf{C}(b_{ji}, c_j)$$

and

$$\sigma\langle\tau_1, \ldots, \tau_n\rangle = \underbrace{\sigma\langle k_1, \ldots, k_n\rangle}_{\text{block permutation}} \circ \underbrace{(\tau_1 \oplus \cdots \oplus \tau_n)}_{\text{block sum}} \in \Sigma_{k_1+\cdots+k_n}$$

as in (4.5.18).

Interpretation 8.3.3. In the previous definition, one should think of $[\sigma, f]$ as a three-step operation:

(1) Apply the morphisms in $\underline{f}$ to observables in $(c_1, \ldots, c_n)$.
(2) Permute the result from the left by σ.
(3) Multiply the observables in d.

The equivalence relation $\sim$ is generated by transpositions of adjacent orthogonal pairs. ◇

Example 8.3.4. The equivalence relation $\sim$ only has an effect when the sequence $\underline{f}$ has length > 1. So for any colors $c, d \in \mathfrak{C}$, there is a canonical bijection

$$\mathsf{C}(c, d) \overset{\cong}{\longrightarrow} \Sigma_1 \times \mathsf{C}(c, d) = \mathsf{O}_{\overline{\mathsf{C}}}\binom{d}{c},$$

sending $f \in \mathsf{C}(c, d)$ to $[\mathrm{id}_1, f] \in \mathsf{O}_{\overline{\mathsf{C}}}\binom{d}{c}$. ◇

The following concept is the orthogonal version of an equivalence of categories in Definition 2.1.8.

Definition 8.3.5. An *orthogonal equivalence* is an orthogonal functor $F : \overline{C} \longrightarrow \overline{D}$ such that

- $F : C \longrightarrow D$ is an equivalence of categories, and
- $\perp^C = F^*(\perp^D)$.

Recall from Example 5.3.3 (i) the strong symmetric monoidal functor $\mathsf{Set} \longrightarrow \mathsf{M}$ that sends a set S to $\coprod_S \mathbb{1}$ and (ii) the corresponding functor

$$(-)^M : \mathsf{Operad}(\mathsf{Set}) \longrightarrow \mathsf{Operad}(\mathsf{M})$$

between categories of operads. The following observations are the main categorical properties of the above construction. They are from [Benini *et. al.* (2017)] Proposition 4.11, Proposition 4.16, Theorem 4.27, Proposition 5.4, and Theorem 5.11.

Theorem 8.3.6. *Suppose* $\overline{C} = (C, \perp)$ *is an orthogonal category with* $\mathsf{Ob}(C) = \mathfrak{C}$.

(1) With the structure in Definition 8.3.2, $O_{\overline{C}}$ is a $\mathfrak{C}$-colored operad in Set.

(2) The construction $O_{(-)}$ defines a functor

$$O_{(-)} : \mathsf{OrthCat} \longrightarrow \mathsf{Operad}(\mathsf{Set}).$$

(3) There is a canonical isomorphism

$$\mathsf{Alg}_\mathsf{M}\left(O_{\overline{C}}^\mathsf{M}\right) \cong \mathsf{QFT}(\overline{C}). \tag{8.3.7}$$

(4) For each set S of morphisms in C, the S-localization functor ℓ induces a change-of-operad adjunction

$$\mathsf{QFT}(\overline{C}) \cong \mathsf{Alg}_\mathsf{M}\left(O_{\overline{C}}^\mathsf{M}\right) \underset{(O_\ell^\mathsf{M})^*}{\overset{(O_\ell^\mathsf{M})_!}{\rightleftarrows}} \mathsf{Alg}_\mathsf{M}\left(O_{\overline{C[S^{-1}]}}^\mathsf{M}\right) \cong \mathsf{QFT}(\overline{C}, S) \tag{8.3.8}$$

whose counit

$$\epsilon : (O_\ell^\mathsf{M})_!(O_\ell^\mathsf{M})^* \overset{\cong}{\longrightarrow} \mathrm{Id}_{\mathsf{Alg}_\mathsf{M}\left(O_{\overline{C[S^{-1}]}}^\mathsf{M}\right)}$$

is a natural isomorphism.

(5) Each orthogonal functor $F : \overline{C} \longrightarrow \overline{D}$ induces a change-of-operad adjunction

$$\mathsf{QFT}(\overline{C}) \cong \mathsf{Alg}_\mathsf{M}\left(O_{\overline{C}}^\mathsf{M}\right) \underset{(O_F^\mathsf{M})^*}{\overset{(O_F^\mathsf{M})_!}{\rightleftarrows}} \mathsf{Alg}_\mathsf{M}\left(O_{\overline{D}}^\mathsf{M}\right) \cong \mathsf{QFT}(\overline{D}) . \tag{8.3.9}$$

(6) If $F : \overline{C} \longrightarrow \overline{D}$ is an orthogonal equivalence, then the change-of-operad adjunction in (8.3.9) is an adjoint equivalence.

Proof. For the first assertion, one checks directly that the structure morphisms for $O_{\overline{C}}$ are well-defined and that they satisfy the axioms in Definition 4.2.1. The second assertion also follows from a direct inspection.

For the third assertion, let us describe the correspondence between objects. From the left side, by Definition 4.5.5 an $O_{\overline{C}}^{\mathsf{M}}$-algebra consists of a $\mathfrak{C}$-colored object $X = \{X_c\}_{c \in \mathfrak{C}}$ in M together with a structure morphism

$$\coprod_{O_{\overline{C}}\binom{d}{\underline{c}}} X_{\underline{c}} \cong O_{\overline{C}}^{\mathsf{M}}\binom{d}{\underline{c}} \otimes X_{\underline{c}} \xrightarrow{\ \lambda\ } X_d \in \mathsf{M}$$

for each $(\underline{c}; d) \in \mathsf{Prof}(\mathfrak{C}) \times \mathfrak{C}$ that satisfies the associativity, unity, and equivariance axioms. The restriction of λ to a copy of $X_{\underline{c}}$ corresponding to an element $x \in O_{\overline{C}}\binom{d}{\underline{c}}$,

$$X_{\underline{c}} \xrightarrow[\text{summand}]{\ x\ } O_{\overline{C}}^{\mathsf{M}}\binom{d}{\underline{c}} \otimes X_{\underline{c}} \xrightarrow{\ \lambda\ } X_d \ ,$$

will be denoted by λ_x. Define a functor $\mathfrak{A}_X : \mathsf{C} \longrightarrow \mathsf{M}$ by setting

$$\mathfrak{A}_X(c) = X_c \quad \text{for} \quad c \in \mathfrak{C},$$
$$\mathfrak{A}_X(f) = \lambda_{[\mathrm{id}_1, f]} : X_c \longrightarrow X_d \quad \text{for} \quad f \in \mathsf{C}(c, d).$$

One checks that $\mathfrak{A}_X$ is well-defined. Moreover, it extends to a functor

$$\mathfrak{A}_X : \mathsf{C} \longrightarrow \mathsf{Mon}(\mathsf{M})$$

such that, for each $c \in \mathfrak{C}$, $\mathfrak{A}_X(c) = X_c$ has monoid multiplication

$$\lambda_{[\mathrm{id}_2, (\mathrm{Id}_c, \mathrm{Id}_c)]} : X_c \otimes X_c \longrightarrow X_c$$

and unit

$$\lambda_{[\mathrm{id}_0, \varnothing]} : \mathbb{1} \longrightarrow X_c.$$

That $\mathfrak{A}_X$ satisfies the causality axiom is a consequence of the equivalence relation $\sim$ that defines each entry of $O_{\overline{C}}$. So $\mathfrak{A}_X$ is an algebraic quantum field theory on $\overline{C}$.

For the converse, the key point is that the $\mathfrak{C}$-colored operad $O_{\overline{C}}$ is generated by the elements

- $\mu_c = [\mathrm{id}_2, (\mathrm{Id}_c, \mathrm{Id}_c)] \in O_{\overline{C}}\binom{c}{c,c}$,
- $[\mathrm{id}_1, f] \in O_{\overline{C}}\binom{d}{c}$, and
- $1_c = [\mathrm{id}_0, \varnothing] \in O_{\overline{C}}\binom{c}{\varnothing}$

for all $c, d \in \mathfrak{C}$ and $f \in \mathsf{C}(c, d)$, and permutations. Indeed, for each $m \geq 3$ and $c \in \mathsf{C}$, the element

$$\mu_m = [\mathrm{id}_m, \underbrace{(\mathrm{Id}_c, \ldots, \mathrm{Id}_c)}_{m}] \in O_{\overline{C}}\binom{c}{c,\ldots,c}$$

is equal to

$$\gamma\big(\mu_2; [\mathrm{id}_1, \mathrm{Id}_c], \mu_{m-1}\big),$$

so by induction all the μ_m are generated by μ_2 and the c-colored unit. For $n \geq 2$ and $f_j \in C(c_j, d)$ for $1 \leq j \leq n$, we have

$$\left[\mathrm{id}_n, (f_1, \ldots, f_n)\right] = \gamma\left(\mu_n; \left\{[\mathrm{id}_1, f_j]\right\}_{j=1}^n\right) \in O_{\overline{C}}\binom{d}{c_1, \ldots, c_n}.$$

So together with permutations the above elements generate all of $O_{\overline{C}}$. Furthermore, one checks that all the generating relations among these generators are already reflected in the properties of an algebraic quantum field theory. Therefore, using the previous paragraph and the axioms in Definition 4.5.5, an algebraic quantum field theory on $\overline{C}$ determines an $O_{\overline{C}}^{\mathsf{M}}$-algebra.

The change-of-operad adjunction (8.3.8) in the fourth assertion is a consequence of Theorem 5.1.8 applied to the morphism

$$O_{\ell}^{\mathsf{M}} : O_{\overline{C}}^{\mathsf{M}} \longrightarrow O_{\overline{C[S^{-1}]}}^{\mathsf{M}}$$

of $\mathfrak{C}$-colored operads, the previous two assertions, and Lemma 8.2.7. The counit is a natural isomorphism by [Mac Lane (1998)] VI.3 Theorem 1 because the right adjoint $(O_{\ell}^{\mathsf{M}})^*$ is full and faithful, which in turn is true because on both sides a morphism is a natural assignment of a monoid morphism to each object in C.

The change-of-operad adjunction (8.3.9) in assertion (5) is a consequence of Theorem 5.1.8 applied to the operad morphism

$$O_F^{\mathsf{M}} : O_{\overline{C}}^{\mathsf{M}} \longrightarrow O_{\overline{D}}^{\mathsf{M}}$$

and of assertions (2) and (3).

For assertion (6), since a left adjoint is unique up to a unique isomorphism, it is enough to show that the right adjoint $(O_F^{\mathsf{M}})^*$ is an equivalence of categories, i.e., full, faithful, and essentially surjective. By the isomorphism (8.3.7) in assertion (3), it is enough to show that the functor

$$\mathsf{QFT}(\overline{D}) \xrightarrow{\ F^*\ } \mathsf{QFT}(\overline{C})$$

is an equivalence of categories. For $\mathfrak{A} \in \mathsf{QFT}(\overline{D})$, this functor is defined as

$$F^*\mathfrak{A} = \mathfrak{A}F,$$

i.e., pre-composition with $F : \mathsf{C} \longrightarrow \mathsf{D}$. Similarly, for $\mathfrak{A}, \mathfrak{B} \in \mathsf{QFT}(\overline{D})$, the function on morphisms

$$\mathsf{QFT}(\overline{D})(\mathfrak{A}, \mathfrak{B}) \xrightarrow{\ F^*\ } \mathsf{QFT}(\overline{C})(F^*\mathfrak{A}, F^*\mathfrak{B})$$
$$\| \qquad\qquad\qquad\qquad \|$$
$$\mathsf{Mon}(\mathsf{M})^{\mathsf{D}}(\mathfrak{A}, \mathfrak{B}) \qquad \mathsf{Mon}(\mathsf{M})^{\mathsf{C}}(\mathfrak{A}F, \mathfrak{B}F)$$

is given by pre-composition with F. Using that $F : \mathsf{C} \longrightarrow \mathsf{D}$ is an equivalence of categories, one checks that the function on morphisms F^* is a bijection. Therefore, the functor F^* is full and faithful.

To see that the functor F^* is essentially surjective, suppose $\mathfrak{A} \in \mathsf{QFT}(\overline{C})$. We must show that there exist $\mathfrak{B} \in \mathsf{QFT}(\overline{D})$ and an isomorphism $F^*\mathfrak{B} \cong \mathfrak{A}$. Define a functor $\mathfrak{B} : \mathsf{D} \longrightarrow \mathsf{Mon}(\mathsf{M})$ as follows.

- For each object $d \in \mathsf{D}$, since F is an equivalence of categories, we can choose
  - an object $d' \in \mathsf{C}$ and
  - an isomorphism $\rho_d : d \xrightarrow{\cong} Fd'$.

  We can furthermore insist that, if d is in the image of F, then d' is chosen from the F-pre-image of d and that $\rho_d = \mathrm{Id}_d$. Define
  $$\mathfrak{B}(d) = \mathfrak{A}(d') \in \mathsf{Mon}(\mathsf{M}).$$

- Suppose given a morphism $f \in \mathsf{D}(d_1, d_2)$. In the previous step, we have chosen objects $d_1', d_2' \in \mathsf{C}$ and isomorphisms $d_1 \xrightarrow{\cong} Fd_1'$ and $d_2 \xrightarrow{\cong} Fd_2'$. These choices yield bijections
  $$\mathsf{C}(d_1', d_2') \xrightarrow[\cong]{F} \mathsf{D}(Fd_1', Fd_2') \xrightarrow{\cong} \mathsf{D}(d_1, d_2),$$

  so f has a unique pre-image $f' \in \mathsf{C}(d_1', d_2')$. Define
  $$\mathfrak{B}(f) = \mathfrak{A}(f') : \mathfrak{B}(d_1) = \mathfrak{A}(d_1') \longrightarrow \mathfrak{A}(d_2') = \mathfrak{B}(d_2) \in \mathsf{Mon}(\mathsf{M}).$$

Using that F is an orthogonal equivalence, one can check that this actually defines a functor $\mathfrak{B}$ that satisfies the causality axiom, i.e., $\mathfrak{B} \in \mathsf{QFT}(\overline{\mathsf{D}})$. Furthermore, by construction $\mathfrak{A}$ and $F^*\mathfrak{B}$ are naturally isomorphic as functors $\mathsf{C} \longrightarrow \mathsf{Mon}(\mathsf{M})$. Therefore, F^* is essentially surjective. $\qquad\square$

Interpretation 8.3.10. Consider Theorem 8.3.6.

(1) Via the isomorphism (8.3.7), the causality axiom of algebraic quantum field theories are built into the $\mathfrak{C}$-colored operad $\mathsf{O}_{\overline{\mathfrak{C}}}$ via the equivalence relation $\sim$ in Definition 8.3.2. In particular, from the operadic viewpoint, the causality axiom is not an extra property that a functor $\mathsf{C} \longrightarrow \mathsf{Mon}(\mathsf{M})$ may or may not satisfy. Instead, every $\mathsf{O}_{\overline{\mathsf{C}}}^{\mathsf{M}}$-algebra already satisfies the causality axiom. Using this isomorphism, we will identify algebraic quantum field theories on $\overline{\mathsf{C}}$ with $\mathsf{O}_{\overline{\mathsf{C}}}^{\mathsf{M}}$-algebras.

(2) Similarly, via the isomorphism
$$\mathsf{Alg}_{\mathsf{M}}\left(\mathsf{O}_{\overline{\mathsf{C}[S^{-1}]}}^{\mathsf{M}}\right) \cong \mathsf{QFT}(\overline{\mathsf{C}}, S)$$

the time-slice axiom with respect to S is built into the $\mathfrak{C}$-colored operad $\mathsf{O}_{\overline{\mathsf{C}[S^{-1}]}}$. So every $\mathsf{O}_{\overline{\mathsf{C}[S^{-1}]}}^{\mathsf{M}}$-algebra already satisfies the time-slice axiom. Using this isomorphism, we will identify algebraic quantum field theories on $\overline{\mathsf{C}}$ that satisfy the time-slice axiom with $\mathsf{O}_{\overline{\mathsf{C}[S^{-1}]}}^{\mathsf{M}}$-algebras.

(3) The right adjoint $(\mathsf{O}_\ell^{\mathsf{M}})^*$ in the change-of-operad adjunction (8.3.8) says that each algebraic quantum field theory on $\overline{\mathsf{C}}$ that satisfies the time-slice axiom is in particular an algebraic quantum field theory on $\overline{\mathsf{C}}$. The left adjoint $(\mathsf{O}_\ell^{\mathsf{M}})_!$ assigns to each algebraic quantum field theory on $\overline{\mathsf{C}}$ another one that satisfies the time-slice axiom.

(4) The change-of-operad adjunction in (8.3.9) allows one to go back and forth between algebraic quantum field theories of different flavors, i.e., those on $\overline{\mathsf{C}}$ and those on $\overline{\mathsf{D}}$. ◇

Remark 8.3.11. In Theorem 8.3.6 assertion (6), we observed that the change-of-operad adjunction associated to an orthogonal equivalence is an adjoint equivalence. The proof given above uses (i) the canonical isomorphism (8.3.7) and (ii) elementary facts about an orthogonal equivalence. This line of argument is very similar to the well-known proof of Theorem 2.4.9 that characterizes equivalences of categories. On the other hand, the proof of this adjoint equivalence given in [Benini *et. al.* (2017)] Theorem 5.11 directly deals with the algebra categories $\mathsf{Alg}_{\mathsf{M}}(\mathsf{O}_{\overline{\mathsf{C}}}^{\mathsf{M}})$ and $\mathsf{Alg}_{\mathsf{M}}(\mathsf{O}_{\overline{\mathsf{D}}}^{\mathsf{M}})$, and uses more sophisticated techniques. Furthermore, Theorem 8.3.6 assertion (6) has a homotopy version, given below in Theorem 8.5.1. ◇

8.4 Examples of AQFT

In this section we provide examples of orthogonal categories and algebraic quantum field theories. The first two examples are the two extreme cases for the orthogonality relation.

Example 8.4.1 (Diagrams of monoids). Suppose C is a small category equipped with the empty orthogonality relation (i.e., $\perp\ =\varnothing$). Since the commutative diagram in the causality axiom (8.2.4) never happens, an algebraic quantum field theory on the *minimal orthogonal category*

$$\overline{\mathsf{C}_{\mathsf{min}}} = (\mathsf{C},\varnothing)$$

is exactly a functor $\mathsf{C} \longrightarrow \mathsf{Mon}(\mathsf{M})$. Therefore, there is an equality

$$\mathsf{QFT}(\overline{\mathsf{C}_{\mathsf{min}}}) = \mathsf{Mon}(\mathsf{M})^{\mathsf{C}},$$

the category of C-diagrams of monoids in M. ◇

Example 8.4.2 (Diagrams of commutative monoids). Suppose C is a small category, and suppose $\perp_{\mathsf{max}}$ is the set of all pairs of morphisms in C with the same codomain. In particular, for each object $c \in \mathsf{C}$, we have $\mathrm{Id}_c \perp_{\mathsf{max}} \mathrm{Id}_c$, so the causality axiom (8.2.4) says that $\mathfrak{A}(c)$ is a commutative monoid in M. For the *maximal orthogonal category*

$$\overline{\mathsf{C}_{\mathsf{max}}} = (\mathsf{C}, \perp_{\mathsf{max}}),$$

each $\mathfrak{A} \in \mathsf{QFT}(\overline{\mathsf{C}_{\mathsf{max}}})$ is in particular a C-diagram of commutative monoids in M. Conversely, each C-diagram of commutative monoids satisfies the causality axiom because the multiplication μ_c is commutative. Therefore, in this case we have

$$\mathsf{QFT}(\overline{\mathsf{C}_{\mathsf{max}}}) = \mathsf{Com}(\mathsf{M})^{\mathsf{C}},$$

the category of C-diagrams of commutative monoids in M. We interpret this equality as saying that, when observables always commute, algebraic quantum field theories reduce to the classical case. ◇

Example 8.4.3 (Underlying diagrams of monoids). For each orthogonal category $\overline{\mathsf{C}} = (\mathsf{C}, \perp)$, there are orthogonal functors

$$\overline{\mathsf{C}_{\min}} = (\mathsf{C}, \varnothing) \xrightarrow{\ i_0\ } \overline{\mathsf{C}} \xrightarrow{\ i_1\ } \overline{\mathsf{C}_{\max}} = (\mathsf{C}, \perp_{\max})$$

whose underlying functors are the identity functors on C. By Theorem 8.3.6 they induce the following two change-of-operad adjunctions.

$$
\mathsf{Alg}_{\mathsf{M}}\!\left(\mathsf{O}^{\mathsf{M}}_{\overline{\mathsf{C}_{\min}}}\right)
\underset{(\mathsf{O}^{\mathsf{M}}_{i_0})^*}{\overset{(\mathsf{O}^{\mathsf{M}}_{i_0})_!}{\rightleftarrows}}
\mathsf{Alg}_{\mathsf{M}}\!\left(\mathsf{O}^{\mathsf{M}}_{\overline{\mathsf{C}}}\right)
\underset{(\mathsf{O}^{\mathsf{M}}_{i_1})^*}{\overset{(\mathsf{O}^{\mathsf{M}}_{i_1})_!}{\rightleftarrows}}
\mathsf{Alg}_{\mathsf{M}}\!\left(\mathsf{O}^{\mathsf{M}}_{\overline{\mathsf{C}_{\max}}}\right)
$$

$$\downarrow\cong \qquad\qquad\qquad \downarrow\cong \qquad\qquad\qquad \downarrow\cong$$

$$\mathsf{Mon}(\mathsf{M})^{\mathsf{C}} = \mathsf{QFT}\!\left(\overline{\mathsf{C}_{\min}}\right) \qquad \mathsf{QFT}(\overline{\mathsf{C}}) \qquad \mathsf{QFT}\!\left(\overline{\mathsf{C}_{\max}}\right) = \mathsf{Com}(\mathsf{M})^{\mathsf{C}}$$

The right adjoint $(\mathsf{O}^{\mathsf{M}}_{i_0})^*$ sends each algebraic quantum field theory on $\overline{\mathsf{C}}$ to its underlying C-diagram of monoids. The other right adjoint $(\mathsf{O}^{\mathsf{M}}_{i_1})^*$ says that each C-diagram of commutative monoids is in particular an algebraic quantum field theory on $\overline{\mathsf{C}}$. $\diamond$

The next three examples are about bounded lattices and (equivariant) topological spaces.

Example 8.4.4 (Quantum field theories on bounded lattices). Suppose $(L, \leq)$ is a bounded lattice as in Example 2.2.12, also regarded as a small category. Two morphisms $g_1 : a \longrightarrow c$ and $g_2 : b \longrightarrow c$ in L are orthogonal if and only if $a \wedge b = 0$, which is the least element in L. This defines an orthogonal category $(L, \perp)$ and algebraic quantum field theories on it. $\diamond$

Example 8.4.5 (Quantum field theories on topological spaces). For each topological space X, recall from Example 2.2.13 that $\mathsf{Open}(X)$ is a bounded lattice. By Example 8.4.4 there is an orthogonal category $\overline{\mathsf{Open}(X)}$, where $U_1 \subset V$ and $U_2 \subset V$ are orthogonal if and only if U_1 and U_2 are disjoint. Corresponding to the orthogonal category $\overline{\mathsf{Open}(X)}$ is the category of algebraic quantum field theories on it. $\diamond$

Example 8.4.6 (Quantum field theories on equivariant topological spaces). Suppose G is a group, and X is a topological space in which G acts on the left by homeomorphisms. Suppose $\mathsf{Open}(X)_G$ is the category in Example 2.2.14. Define $\perp$ as the set of pairs of morphisms $U_1 \xrightarrow{\ i_1 g_1\ } V \xleftarrow{\ i_2 g_2\ } U_2$ in $\mathsf{Open}(X)_G$ of the form

$$U_1 \xrightarrow{\ g_1\ } g_1 U_1 \xrightarrow{\ i_1\ } V \xleftarrow{\ i_2\ } g_2 U_2 \xleftarrow{\ g_2\ } U_2$$

with $g_1, g_2 \in G$ and i_1, i_2 both inclusions such that $g_1 U_1$ and $g_2 U_2$ are disjoint. This defines an orthogonal category $\overline{\mathsf{Open}(X)_G}$. If G is the trivial group, then we recover the orthogonal category $\overline{\mathsf{Open}(X)}$ in Example 8.4.5. Corresponding to the orthogonal category $\overline{\mathsf{Open}(X)_G}$ is the category of algebraic quantum field theories on it. $\diamond$

Examples 8.4.7 to 8.4.12 below are from [Benini *et. al.* (2017)] and are about quantum field theories defined on spacetimes without additional geometric structure. The upshot is that the operadic framework in Section 8.2 includes many quantum field theories in the literature, including various flavors of chiral conformal, Euclidean, and locally covariant quantum field theories. To specify a particular flavor of quantum field theories, we simply choose the right orthogonal category $\overline{\mathsf{C}} = (\mathsf{C}, \perp)$ and, if there is a version of the time-slice axiom, a suitable set of morphisms $S \subset \mathrm{Mor}(\mathsf{C})$ to be localized.

Example 8.4.7 (Chiral conformal quantum field theories). In the context of Example 2.2.15, there is an orthogonal category

$$\overline{\mathsf{Man}^d} = (\mathsf{Man}^d, \perp)$$

in which Man^d is the category of d-dimensional oriented manifolds with orientation-preserving open embeddings as morphisms. Two morphisms $g_1 : X_1 \longrightarrow X$ and $g_2 : X_2 \longrightarrow X$ in Man^d are orthogonal if and only if their images are disjoint subsets in X. By Theorem 8.3.6 there is a canonical isomorphism

$$\mathsf{Alg}_\mathsf{M}\big(\mathsf{O}^\mathsf{M}_{\overline{\mathsf{Man}^d}}\big) \cong \mathsf{QFT}\big(\overline{\mathsf{Man}^d}\big).$$

When $\mathsf{M} = \mathsf{Vect}_\mathbb{K}$ and $d = 1$, the objects in $\mathsf{QFT}\big(\overline{\mathsf{Man}^d}\big)$ are coordinate-free chiral conformal nets of $\mathbb{K}$-algebras that satisfy the commutativity axiom for observables localized in disjoint regions [Bartels *et al.* (2015)]. ◇

Example 8.4.8 (Chiral conformal quantum field theories on discs). In the context of Example 2.2.16, there is an orthogonal category

$$\overline{\mathsf{Disc}^d} = (\mathsf{Disc}^d, \perp)$$

in which Disc^d is the full subcategory of Man^d of d-dimensional oriented manifolds diffeomorphic to $\mathbb{R}^d$. The orthogonality relation is the pullback of that on Man^d along the full subcategory inclusion

$$j : \mathsf{Disc}^d \longrightarrow \mathsf{Man}^d.$$

In other words, two morphisms $g_1 : X_1 \longrightarrow X$ and $g_2 : X_2 \longrightarrow X$ in Disc^d are orthogonal if and only if their images are disjoint subsets in X. By Theorem 8.3.6 there is a canonical isomorphism

$$\mathsf{Alg}_\mathsf{M}\big(\mathsf{O}^\mathsf{M}_{\overline{\mathsf{Disc}^d}}\big) \cong \mathsf{QFT}\big(\overline{\mathsf{Disc}^d}\big).$$

When $\mathsf{M} = \mathsf{Vect}_\mathbb{K}$ and $d = 1$, the objects in $\mathsf{QFT}\big(\overline{\mathsf{Disc}^d}\big)$ are coordinate-free chiral conformal nets of $\mathbb{K}$-algebras defined on intervals that satisfy the commutativity axiom for observables localized in disjoint intervals [Bartels *et al.* (2015)].

The full subcategory inclusion $j : \mathsf{Disc}^d \longrightarrow \mathsf{Man}^d$ induces an orthogonal functor by Lemma 8.2.2. So by Theorem 8.3.6, it induces a change-of-operad adjunction

$$\mathsf{QFT}\big(\overline{\mathsf{Disc}^d}\big) \cong \mathsf{Alg}_\mathsf{M}\big(\mathsf{O}^\mathsf{M}_{\overline{\mathsf{Disc}^d}}\big) \underset{(\mathsf{O}^\mathsf{M}_j)^*}{\overset{(\mathsf{O}^\mathsf{M}_j)_!}{\rightleftarrows}} \mathsf{Alg}_\mathsf{M}\big(\mathsf{O}^\mathsf{M}_{\overline{\mathsf{Man}^d}}\big) \cong \mathsf{QFT}\big(\overline{\mathsf{Man}^d}\big).$$

When $M = \mathsf{Vect}_{\mathbb{K}}$ and $d = 1$, this adjunction allows us to go back and forth between (i) coordinate-free chiral conformal nets of $\mathbb{K}$-algebras defined on intervals that satisfy the commutativity axiom for observables localized in disjoint intervals and (ii) those defined on all 1-dimensional oriented manifolds. $\diamond$

Example 8.4.9 (Chiral conformal quantum field theories on a fixed manifold). For a fixed oriented manifold $X \in \mathsf{Man}^d$, recall from Example 2.2.13 the category $\mathsf{Open}(X)$ whose objects are open subsets of X and whose morphisms are subset inclusions. Denote the induced functor by

$$\iota : \mathsf{Open}(X) \longrightarrow \mathsf{Man}^d,$$

and equip $\mathsf{Open}(X)$ with the pullback orthogonality relation along ι. In other words, for open subsets $U_1, U_2 \subseteq V \subseteq X$, the inclusions $U_1 \subseteq V$ and $U_2 \subseteq V$ are orthogonal if and only if U_1 and U_2 are disjoint subsets of V. This is a special case of Example 8.4.5. By Theorem 8.3.6 there is a canonical isomorphism

$$\mathsf{Alg}_{\mathsf{M}}\left(\mathsf{O}^{\mathsf{M}}_{\overline{\mathsf{Open}(X)}}\right) \cong \mathsf{QFT}\left(\overline{\mathsf{Open}(X)}\right).$$

When $M = \mathsf{Vect}_{\mathbb{K}}$, $d = 1$, and $X = S^1$, the objects in $\mathsf{QFT}\left(\overline{\mathsf{Open}(S^1)}\right)$ are chiral conformal nets of $\mathbb{K}$-algebras on the circle [Kawahigashi (2015); Rehren (2015)].

The subcategory inclusion $\iota : \mathsf{Open}(X) \longrightarrow \mathsf{Man}^d$ induces an orthogonal functor by Lemma 8.2.2. So by Theorem 8.3.6, it induces a change-of-operad adjunction

$$\mathsf{Alg}_{\mathsf{M}}\left(\mathsf{O}^{\mathsf{M}}_{\overline{\mathsf{Open}(X)}}\right) \underset{(\mathsf{O}^{\mathsf{M}}_{\iota})^*}{\overset{(\mathsf{O}^{\mathsf{M}}_{\iota})_!}{\rightleftarrows}} \mathsf{Alg}_{\mathsf{M}}\left(\mathsf{O}^{\mathsf{M}}_{\overline{\mathsf{Man}^d}}\right).$$

$$\mathsf{QFT}\left(\overline{\mathsf{Open}(X)}\right) \qquad\qquad \mathsf{QFT}\left(\overline{\mathsf{Man}^d}\right)$$

(with vertical isomorphisms $\cong$)

$\diamond$

Example 8.4.10 (Euclidean quantum field theories). In the context of Example 2.2.17, there is an orthogonal category

$$\overline{\mathsf{Riem}^d} = \left(\mathsf{Riem}^d, \perp\right)$$

in which Riem^d is the category with d-dimensional oriented Riemannian manifolds as objects and orientation-preserving isometric open embeddings as morphisms. Two morphisms $g_1 : X_1 \longrightarrow X$ and $g_2 : X_2 \longrightarrow X$ in Riem^d are orthogonal if and only if their images are disjoint subsets in X. By Theorem 8.3.6 there is a canonical isomorphism

$$\mathsf{Alg}_{\mathsf{M}}\left(\mathsf{O}^{\mathsf{M}}_{\overline{\mathsf{Riem}^d}}\right) \cong \mathsf{QFT}\left(\overline{\mathsf{Riem}^d}\right).$$

When $M = \mathsf{Vect}_{\mathbb{K}}$ the objects in $\mathsf{QFT}\left(\overline{\mathsf{Riem}^d}\right)$ are locally covariant versions of Euclidean quantum field theories that satisfy the commutativity axiom for observables localized in disjoint regions [Schlingemann (1999)]. As in Example 8.4.9, we may also restrict to a fixed oriented Riemannian manifold X and consider algebraic quantum field theories on $\overline{\mathsf{Open}(X)}$ as objects in $\mathsf{QFT}\left(\overline{\mathsf{Open}(X)}\right)$. $\diamond$

Example 8.4.11 (Locally covariant quantum field theories). In the context of Example 2.2.18, there is an orthogonal category

$$\overline{\mathsf{Loc}^d} = (\mathsf{Loc}^d, \perp)$$

in which Loc^d is the category of d-dimensional oriented, time-oriented, and globally hyperbolic Lorentzian manifolds. A morphism is an isometric embedding that preserves the orientations and time-orientations whose image is causally compatible and open. Two morphisms $g_1 : X_1 \longrightarrow X$ and $g_2 : X_2 \longrightarrow X$ in Loc^d are orthogonal if and only if their images are causally disjoint subsets in X. By Theorem 8.3.6 there is a canonical isomorphism

$$\mathsf{Alg}_\mathsf{M}\left(\mathsf{O}^\mathsf{M}_{\overline{\mathsf{Loc}^d}}\right) \cong \mathsf{QFT}\left(\overline{\mathsf{Loc}^d}\right).$$

When $\mathsf{M} = \mathsf{Vect}_\mathbb{K}$ the objects in $\mathsf{QFT}\left(\overline{\mathsf{Loc}^d}\right)$ are casual locally covariant quantum field theories that do not necessarily satisfy the isotony axiom [Brunetti *et. al.* (2003); Fewster (2013); Fewster and Verch (2015)].

To implement the time-slice axiom, recall that a morphism $f : X \longrightarrow Y$ in Loc^d is a *Cauchy morphism* if its image contains a Cauchy surface of Y. The set of Cauchy morphisms is denoted by S. By Theorem 8.3.6 there is a change-of-operad adjunction

$$\mathsf{QFT}\left(\overline{\mathsf{Loc}^d}\right) \cong \mathsf{Alg}_\mathsf{M}\left(\mathsf{O}^\mathsf{M}_{\overline{\mathsf{Loc}^d}}\right) \underset{(\mathsf{O}^\mathsf{M}_\ell)^*}{\overset{(\mathsf{O}^\mathsf{M}_\ell)_!}{\rightleftarrows}} \mathsf{Alg}_\mathsf{M}\left(\mathsf{O}^\mathsf{M}_{\overline{\mathsf{Loc}^d[S^{-1}]}}\right) \cong \mathsf{QFT}\left(\overline{\mathsf{Loc}^d}, S\right).$$

When M is the category $\mathsf{Vect}_\mathbb{K}$, the objects on the right side are causal locally covariant quantum field theories satisfying the time-slice axiom but not necessarily the isotony axiom.   ◇

Example 8.4.12 (Locally covariant quantum field theories on a fixed spacetime). In the context of Example 2.2.19, for each Lorentzian manifold $X \in \mathsf{Loc}^d$ consider the category $\mathsf{Gh}(X)$ of globally hyperbolic open subsets of X with subset inclusions as morphisms. As in Example 8.4.9, $\mathsf{Gh}(X)$ may be equipped with the pullback orthogonality relation along the subcategory inclusion $i : \mathsf{Gh}(X) \longrightarrow \mathsf{Loc}^d$. By Theorem 8.3.6 there is a canonical isomorphism

$$\mathsf{Alg}_\mathsf{M}\left(\mathsf{O}^\mathsf{M}_{\overline{\mathsf{Gh}(X)}}\right) \cong \mathsf{QFT}\left(\overline{\mathsf{Gh}(X)}\right).$$

When $\mathsf{M} = \mathsf{Vect}_\mathbb{K}$ the objects in $\mathsf{QFT}\left(\overline{\mathsf{Gh}(X)}\right)$ are locally covariant quantum field theories on X that do not necessarily satisfy the time-slice axiom and the isotony axiom.

As in Example 8.4.11, to implement the time-slice axiom, suppose S is the set of morphisms $U \subseteq V \subseteq X$ such that U contains a Cauchy surface of $i(V)$. By Theorem 8.3.6 the S-localization functor

$$\ell : \mathsf{Gh}(X) \longrightarrow \mathsf{Gh}(X)[S^{-1}]$$

induces a change-of-operad adjunction

$$\mathsf{Alg}_{\mathsf{M}}\left(\mathsf{O}_{\overline{\mathsf{Gh}(X)}}^{\mathsf{M}}\right) \underset{(\mathsf{O}_{\ell}^{\mathsf{M}})^{*}}{\overset{(\mathsf{O}_{\ell}^{\mathsf{M}})_{!}}{\rightleftarrows}} \mathsf{Alg}_{\mathsf{M}}\left(\mathsf{O}_{\overline{\mathsf{Gh}(X)[S^{-1}]}}^{\mathsf{M}}\right).$$

$$\Big\downarrow \cong \qquad\qquad\qquad \Big\downarrow \cong$$

$$\mathsf{QFT}\big(\overline{\mathsf{Gh}(X)}\big) \qquad\qquad \mathsf{QFT}\big(\overline{\mathsf{Gh}(X)}, S\big)$$

When $\mathsf{M} = \mathsf{Vect}_{\mathbb{K}}$ the objects on the right side are causal nets of $\mathbb{K}$-algebras satisfying the time-slice axiom but not necessarily the isotony axiom [Haag and Kastler (1964)]. ◇

Examples 8.4.13 to 8.4.16 below are from [Benini and Schenkel (2017)] and are about quantum field theories defined on spacetimes with additional geometric structures such as principal bundles, connections, and spin structure. A common feature is that the isotony axiom–which asks that each structure morphism $\mathfrak{A}(f) : \mathfrak{A}(X) \longrightarrow \mathfrak{A}(Y) \in \mathsf{Mon}(\mathsf{M})$ be a monomorphism–is usually not satisfied.

Example 8.4.13 (Dynamical quantum gauge theories on principal bundles). In the context of Example 2.2.20, recall that for each Lie group G there is a forgetful functor

$$\pi : \mathsf{Loc}_G^d \longrightarrow \mathsf{Loc}^d$$

that forgets about the bundle structure, where Loc_G^d is the category of d-dimensional oriented, time-oriented, and globally hyperbolic Lorentzian manifolds equipped with a principal G-bundle. Suppose:

- $S_G = \pi^{-1}(S) \subset \mathsf{Mor}(\mathsf{Loc}_G^d)$ is the π-pre-image of the set S of Cauchy morphisms in Loc^d.
- $\pi^*(\perp)$ is the pullback of the orthogonality relation $\perp$ in Loc^d in Example 8.4.11 along π.

The forgetful functor π and the universal property of localization induce a commutative diagram

$$\overline{\mathsf{Loc}_G^d} = (\mathsf{Loc}_G^d, \pi^*(\perp)) \xrightarrow{\quad\pi\quad} (\mathsf{Loc}^d, \perp) = \overline{\mathsf{Loc}^d}$$

$$\ell\Big\downarrow \qquad\qquad\qquad\qquad \Big\downarrow\ell$$

$$\overline{\mathsf{Loc}_G^d[S_G^{-1}]} = \big(\mathsf{Loc}_G^d[S_G^{-1}], \ell_*\pi^*(\perp)\big) \xrightarrow{\quad\pi'\quad} (\mathsf{Loc}^d[S^{-1}], \ell_*(\perp)) = \overline{\mathsf{Loc}^d[S^{-1}]}$$

in $\mathsf{OrthCat}$. The right vertical morphism is the S-localization functor on Loc^d, and $\ell_*(\perp)$ is the pushforward orthogonality relation along ℓ. The left vertical morphism is the S_G-localization functor on Loc_G^d, and $\ell_*\pi^*(\perp)$ is the pushforward orthogonality relation of $\pi^*(\perp)$ along ℓ. Since

$$\ell\pi(S_G) \subseteq \ell(S)$$

are all isomorphisms in $\mathsf{Loc}^d[S^{-1}]$, by the universal property of S_G-localization, there is a unique functor

$$\mathsf{Loc}^d_G[S_G^{-1}] \xrightarrow{\ \pi'\ } \mathsf{Loc}^d[S^{-1}] \quad \text{such that} \quad \ell\pi = \pi'\ell.$$

A direct inspection shows that π' is an orthogonal functor.

By Theorem 8.3.6 the functor π' induces a change-of-operad adjunction

$$\mathsf{Alg}_\mathsf{M}\Big(\mathsf{O}^\mathsf{M}_{\overline{\mathsf{Loc}^d_G[S_G^{-1}]}}\Big) \underset{(\mathsf{O}^\mathsf{M}_{\pi'})^*}{\overset{(\mathsf{O}^\mathsf{M}_{\pi'})_!}{\rightleftarrows}} \mathsf{Alg}_\mathsf{M}\Big(\mathsf{O}^\mathsf{M}_{\overline{\mathsf{Loc}^d[S^{-1}]}}\Big).$$

$$\Big\downarrow{\scriptstyle\cong} \qquad\qquad\qquad\qquad \Big\downarrow{\scriptstyle\cong}$$

$$\mathsf{QFT}\big(\overline{\mathsf{Loc}^d_G}, S_G\big) \qquad\qquad \mathsf{QFT}\big(\overline{\mathsf{Loc}^d}, S\big)$$

When $\mathsf{M} = \mathsf{Vect}_\mathbb{K}$ objects on the left side include dynamical quantum gauge theories on principal G-bundles that do not necessarily satisfy the isotony axiom [Benini et. al. (2014,b)]. ◇

Example 8.4.14 (Charged matter quantum field theories on background gauge fields). In the context of Example 2.2.21, recall that for each Lie group G there is a forgetful functor

$$\pi p : \mathsf{Loc}^d_{G,\mathrm{con}} \longrightarrow \mathsf{Loc}^d$$

that forgets about the bundle structure and the connection, where $\mathsf{Loc}^d_{G,\mathrm{con}}$ is the category of triples (X, P, C) with $(X, P) \in \mathsf{Loc}^d_G$ and C a connection on P. Suppose:

- $S_G = (\pi p)^{-1}(S) \subset \mathrm{Mor}(\mathsf{Loc}^d_{G,\mathrm{con}})$ is the (πp)-pre-image of the set S of Cauchy morphisms in Loc^d.
- $(\pi p)^*(\perp)$ is the pullback of the orthogonality relation $\perp$ in Loc^d in Example 8.4.11 along πp.

Exactly as in Example 8.4.13, the forgetful functor πp and the universal property of localization induce a commutative diagram

$$\overline{\mathsf{Loc}^d_{G,\mathrm{con}}} = \big(\mathsf{Loc}^d_{G,\mathrm{con}}, (\pi p)^*(\perp)\big) \xrightarrow{\ \pi p\ } \big(\mathsf{Loc}^d, \perp\big) = \overline{\mathsf{Loc}^d}$$

$$\ell\Big\downarrow \qquad\qquad\qquad\qquad\qquad\qquad \Big\downarrow\ell$$

$$\overline{\mathsf{Loc}^d_{G,\mathrm{con}}[S_G^{-1}]} = \big(\mathsf{Loc}^d_{G,\mathrm{con}}[S_G^{-1}], \ell_*(\pi p)^*(\perp)\big) \xrightarrow{\ \pi'\ } \big(\mathsf{Loc}^d[S^{-1}], \ell_*(\perp)\big) = \overline{\mathsf{Loc}^d[S^{-1}]}$$

in $\mathsf{OrthCat}$.

By Theorem 8.3.6 the functor π' induces a change-of-operad adjunction

$$\mathsf{Alg}_\mathsf{M}\Big(\mathsf{O}^\mathsf{M}_{\overline{\mathsf{Loc}^d_{G,\mathrm{con}}[S_G^{-1}]}}\Big) \underset{(\mathsf{O}^\mathsf{M}_{\pi'})^*}{\overset{(\mathsf{O}^\mathsf{M}_{\pi'})_!}{\rightleftarrows}} \mathsf{Alg}_\mathsf{M}\Big(\mathsf{O}^\mathsf{M}_{\overline{\mathsf{Loc}^d[S^{-1}]}}\Big).$$

$$\Big\downarrow{\scriptstyle\cong} \qquad\qquad\qquad\qquad \Big\downarrow{\scriptstyle\cong}$$

$$\mathsf{QFT}\big(\overline{\mathsf{Loc}^d_{G,\mathrm{con}}}, S_G\big) \qquad\qquad \mathsf{QFT}\big(\overline{\mathsf{Loc}^d}, S\big)$$

When M = Vect$_\mathbb{K}$ objects on the left side include charged matter quantum field theories on background gauge fields that do not necessarily satisfy the isotony axiom [Schenkel and Zahn (2017); Zahn (2014)]. ◇

Example 8.4.15 (Dirac and fermionic quantum field theories). In the context of Example 2.2.22, recall that there is a forgetful functor

$$\pi : \mathsf{SLoc}^d \longrightarrow \mathsf{Loc}^d,$$

where SLoc^d is the category of d-dimensional oriented, time-oriented, and globally hyperbolic Lorentzian spin manifolds. Suppose:

- $S_\pi \subset \mathsf{Mor}(\mathsf{SLoc}^d)$ is the π-pre-image of the set S of Cauchy morphisms in Loc^d.
- $\pi^*(\perp)$ is the pullback of the orthogonality relation $\perp$ in Loc^d in Example 8.4.11 along π.

Exactly, as in Example 8.4.13, the forgetful functor π and the universal property of localization induce a commutative diagram

$$
\begin{array}{ccc}
\overline{\mathsf{SLoc}^d} = (\mathsf{SLoc}^d, \pi^*(\perp)) & \xrightarrow{\quad\quad\pi\quad\quad} & (\mathsf{Loc}^d, \perp) = \overline{\mathsf{Loc}^d} \\
{\scriptstyle\ell}\downarrow & & \downarrow{\scriptstyle\ell} \\
\overline{\mathsf{SLoc}^d[S_\pi^{-1}]} = \left(\mathsf{SLoc}^d[S_\pi^{-1}], \ell_* \pi^*(\perp)\right) & \xrightarrow{\quad\pi'\quad} & \left(\mathsf{Loc}^d[S^{-1}], \ell_*(\perp)\right) = \overline{\mathsf{Loc}^d[S^{-1}]}
\end{array}
$$

in $\mathsf{OrthCat}$.

By Theorem 8.3.6 the functor π' induces a change-of-operad adjunction

$$
\begin{array}{ccc}
\mathsf{Alg}_\mathsf{M}\left(\mathsf{O}^\mathsf{M}_{\overline{\mathsf{SLoc}^d[S_\pi^{-1}]}}\right) & \underset{(\mathsf{O}^\mathsf{M}_{\pi'})^*}{\overset{(\mathsf{O}^\mathsf{M}_{\pi'})_!}{\rightleftarrows}} & \mathsf{Alg}_\mathsf{M}\left(\mathsf{O}^\mathsf{M}_{\overline{\mathsf{Loc}^d[S^{-1}]}}\right) . \\
{\scriptstyle\cong}\downarrow & & \downarrow{\scriptstyle\cong} \\
\mathsf{QFT}\left(\overline{\mathsf{SLoc}^d}, S_\pi\right) & & \mathsf{QFT}\left(\overline{\mathsf{Loc}^d}, S\right)
\end{array}
$$

When M = Vect$_\mathbb{K}$ objects on the left side include Dirac quantum fields that do not necessarily satisfy the isotony axiom [Dappiaggi *et. al.* (2009); Sanders (2010); Verch (2001)]. Furthermore, when M is the symmetric monoidal category of $\mathbb{K}$-supermodules, its monoids are $\mathbb{K}$-superalgebras. In this case, objects on the left side include fermionic quantum field theories that do not necessarily satisfy the isotony axiom [Bär and Ginoux (2011)]. ◇

Example 8.4.16 (Quantum field theories on structured spacetimes). Examples 8.4.13, 8.4.14, and 8.4.15 are subsumed by the following more general setting from [Benini and Schenkel (2017)]. Suppose given a functor $\pi : \mathsf{Str} \longrightarrow \mathsf{Loc}^d$ between small categories. One regards Str as the category of spacetimes with additional geometric structures with π the forgetful functor that forgets about the additional structures. Suppose:

- $S_\pi \subset \mathsf{Mor}(\mathsf{Str})$ is the π-pre-image of the set S of Cauchy morphisms in Loc^d.

- $\pi^*(\perp)$ is the pullback of the orthogonality relation $\perp$ in Loc^d in Example 8.4.11 along π.

Exactly as in Example 8.4.13, the forgetful functor π and the universal property of localization induce a commutative diagram

$$\begin{array}{ccc}
\overline{\mathsf{Str}} = (\mathsf{Str}, \pi^*(\perp)) & \xrightarrow{\quad\quad\pi\quad\quad} & (\mathsf{Loc}^d, \perp) = \overline{\mathsf{Loc}^d} \\
{\scriptstyle \ell}\downarrow & & \downarrow{\scriptstyle \ell} \\
\overline{\mathsf{Str}[S_\pi^{-1}]} = \big(\mathsf{Str}[S_\pi^{-1}], \ell_*\pi^*(\perp)\big) & \xrightarrow{\quad\pi'\quad} & (\mathsf{Loc}^d[S^{-1}], \ell_*(\perp)) = \overline{\mathsf{Loc}^d[S^{-1}]}
\end{array}$$

in $\mathsf{OrthCat}$.

By Theorem 8.3.6 the functor π' induces a change-of-operad adjunction

$$\begin{array}{ccc}
\mathsf{Alg}_{\mathsf{M}}\big(\mathsf{O}^{\mathsf{M}}_{\overline{\mathsf{Str}[S_\pi^{-1}]}}\big) & \underset{(\mathsf{O}^{\mathsf{M}}_{\pi'})^*}{\overset{(\mathsf{O}^{\mathsf{M}}_{\pi'})_!}{\rightleftarrows}} & \mathsf{Alg}_{\mathsf{M}}\big(\mathsf{O}^{\mathsf{M}}_{\overline{\mathsf{Loc}^d[S^{-1}]}}\big)\,. \\
{\scriptstyle \cong}\downarrow & & \downarrow{\scriptstyle \cong} \\
\mathsf{QFT}\big(\overline{\mathsf{Str}}, S_\pi\big) & & \mathsf{QFT}\big(\overline{\mathsf{Loc}^d}, S\big)
\end{array}$$

Objects on the left side are quantum field theories on $\pi : \mathsf{Str} \longrightarrow \mathsf{Loc}^d$ in the sense of [Benini and Schenkel (2017)] that do not necessarily satisfy the isotony axiom.⋄

The next example is from [Benini *et. al.* (2018)] and is about quantum field theories defined on spacetimes with timelike boundary.

Example 8.4.17 (Algebraic quantum field theories on spacetime with timelike boundary). Suppose X is a spacetime with timelike boundary as in Example 2.2.23. There is an orthogonal category

$$\overline{\mathsf{Reg}(X)} = (\mathsf{Reg}(X), \perp)$$

in which $\mathsf{Reg}(X)$ is the category of regions in X.

- Two morphisms $g_1 : U_1 \longrightarrow V$ and $g_2 : U_2 \longrightarrow V$ in $\mathsf{Reg}(X)$ are orthogonal if and only if U_1 and U_2 are causally disjoint in V.
- A *Cauchy morphism* $i : U \longrightarrow V$ in $\mathsf{Reg}(X)$ is a morphism such that $D(U) = D(V)$, where $D(U)$ is the set of points $x \in X$ such that every inextensible piecewise smooth future directed causal curve from x meets U.
- The set of all Cauchy morphisms in $\mathsf{Reg}(X)$ is denoted by S_X.

By Lemma 8.2.7 and Theorem 8.3.6, there are canonical isomorphisms

$$\mathsf{Alg}_{\mathsf{M}}\big(\mathsf{O}^{\mathsf{M}}_{\overline{\mathsf{Reg}(X)[S_X^{-1}]}}\big) \cong \mathsf{QFT}\big(\overline{\mathsf{Reg}(X)[S_X^{-1}]}\big) \cong \mathsf{QFT}\big(\overline{\mathsf{Reg}(X)}, S_X\big).$$

Objects in $\mathsf{QFT}(\overline{\mathsf{Reg}(X)}, S_X)$ are exactly the algebraic quantum field theories on X as in [Benini *et. al.* (2018)] Definition 3.1.

There is a full subcategory inclusion

$$j : \mathsf{Reg}(X_0) \longrightarrow \mathsf{Reg}(X)$$

in which $\mathsf{Reg}(X_0)$ is the category of regions in the interior X_0 of X. Suppose:

- $S_{X_0} = j^{-1}(S_X) \subset \mathsf{Mor}(\mathsf{Reg}(X_0))$ is the j-pre-image of the set S_X of Cauchy morphisms in $\mathsf{Reg}(X)$.
- $j^*(\perp)$ is the pullback of the orthogonality relation $\perp$ in $\mathsf{Reg}(X)$ along j.

Similar to Example 8.4.13, the full subcategory inclusion j and the universal property of localization induce a commutative diagram

$$
\begin{array}{ccc}
\overline{\mathsf{Reg}(X_0)} = (\mathsf{Reg}(X_0), j^*(\perp)) & \xrightarrow{\quad j \quad} & (\mathsf{Reg}(X), \perp) = \overline{\mathsf{Reg}(X)} \\
\ell \downarrow & & \downarrow \ell \\
\overline{\mathsf{Reg}(X_0)[S_{X_0}^{-1}]} = \left(\mathsf{Reg}(X_0)[S_{X_0}^{-1}], \ell_* j^*(\perp)\right) & \xrightarrow{\ j' \ } & \left(\mathsf{Reg}(X)[S_X^{-1}], \ell_*(\perp)\right) = \overline{\mathsf{Reg}(X)[S_X^{-1}]}
\end{array}
$$

in OrthCat.

By Theorem 8.3.6 the functor j' induces a change-of-operad adjunction

$$
\begin{array}{ccc}
\mathsf{Alg}_{\mathsf{M}}\left(\mathsf{O}^{\mathsf{M}}_{\overline{\mathsf{Reg}(X_0)[S_{X_0}^{-1}]}}\right) & \underset{(\mathsf{O}^{\mathsf{M}}_{j'})^*}{\overset{(\mathsf{O}^{\mathsf{M}}_{j'})_!}{\rightleftarrows}} & \mathsf{Alg}_{\mathsf{M}}\left(\mathsf{O}^{\mathsf{M}}_{\overline{\mathsf{Reg}(X)[S_X^{-1}]}}\right) \\
\cong \downarrow & & \downarrow \cong \\
\mathsf{QFT}\left(\overline{\mathsf{Reg}(X_0)}, S_{X_0}\right) & & \mathsf{QFT}\left(\overline{\mathsf{Reg}(X)}, S_X\right)
\end{array}
$$

The right adjoint is the restriction functor, while the left adjoint is called the *universal extension functor* in [Benini *et. al.* (2018)]. ◇

8.5 Homotopical Properties

In this section we study homotopical properties of the category $\mathsf{QFT}(\overline{\mathsf{C}})$ of algebraic quantum field theories on an orthogonal category $\overline{\mathsf{C}}$. For this to make sense, the base category M in this section is assumed to be a monoidal model category in which the colored operads under consideration are admissible in the sense of Definition 5.2.5. For example, one can take M to be Top, SSet, Cat, or $\mathsf{Chain}_{\mathbb{K}}$ with $\mathbb{K}$ a field of characteristic zero, in which all colored operads are admissible.

In Theorem 8.3.6(6) we noted that the change-of-operad adjunction

$$
\mathsf{QFT}(\overline{\mathsf{C}}) \cong \mathsf{Alg}_{\mathsf{M}}\left(\mathsf{O}^{\mathsf{M}}_{\overline{\mathsf{C}}}\right) \underset{(\mathsf{O}^{\mathsf{M}}_F)^*}{\overset{(\mathsf{O}^{\mathsf{M}}_F)_!}{\rightleftarrows}} \mathsf{Alg}_{\mathsf{M}}\left(\mathsf{O}^{\mathsf{M}}_{\overline{\mathsf{D}}}\right) \cong \mathsf{QFT}(\overline{\mathsf{D}})
$$

induced by an orthogonal equivalence $F : \overline{\mathsf{C}} \longrightarrow \overline{\mathsf{D}}$ is an adjoint equivalence. The following observation says that this is a Quillen equivalence as well.

Theorem 8.5.1. *Suppose $F : \overline{\mathsf{C}} \longrightarrow \overline{\mathsf{D}}$ is an orthogonal functor, and M is a monoidal model category in which the colored operads $\mathsf{O}^{\mathsf{M}}_{\overline{\mathsf{C}}}$ and $\mathsf{O}^{\mathsf{M}}_{\overline{\mathsf{D}}}$ are admissible.*

(1) The change-of-operad adjunction $(\mathsf{O}^{\mathsf{M}}_F)_! \dashv (\mathsf{O}^{\mathsf{M}}_F)^*$ is a Quillen adjunction.

(2) If F is an orthogonal equivalence, then the operad morphism

$$\mathsf{O}^{\mathsf{M}}_F : \mathsf{O}^{\mathsf{M}}_{\underline{\mathsf{C}}} \longrightarrow \mathsf{O}^{\mathsf{M}}_{\underline{\mathsf{D}}}$$

is a homotopy Morita equivalence; i.e., the change-of-operad adjunction is a Quillen equivalence.

Proof. For assertion (1), in both model categories $\mathsf{Alg}_{\mathsf{M}}\big(\mathsf{O}^{\mathsf{M}}_{\underline{\mathsf{C}}}\big)$ and $\mathsf{Alg}_{\mathsf{M}}\big(\mathsf{O}^{\mathsf{M}}_{\underline{\mathsf{D}}}\big)$, fibrations and weak equivalences are defined entrywise in M. So by Definition 5.1.5 the right adjoint $(\mathsf{O}^{\mathsf{M}}_F)^*$ preserves fibrations and acyclic fibrations.

For assertion (2), by [Hovey (1999)] Corollary 1.3.16, it is enough to show that for each cofibrant object $X \in \mathsf{Alg}_{\mathsf{M}}\big(\mathsf{O}^{\mathsf{M}}_{\underline{\mathsf{C}}}\big)$, the derived unit

$$X \longrightarrow (\mathsf{O}^{\mathsf{M}}_F)^* R(\mathsf{O}^{\mathsf{M}}_F)_! X$$

is a weak equivalence, where R is the functorial fibrant replacement in $\mathsf{Alg}_{\mathsf{M}}\big(\mathsf{O}^{\mathsf{M}}_{\underline{\mathsf{D}}}\big)$. The derived unit is the composition

$$X \xrightarrow[\cong]{\eta_X} (\mathsf{O}^{\mathsf{M}}_F)^* (\mathsf{O}^{\mathsf{M}}_F)_! X \xrightarrow{(\mathsf{O}^{\mathsf{M}}_F)^* r} (\mathsf{O}^{\mathsf{M}}_F)^* R(\mathsf{O}^{\mathsf{M}}_F)_! X$$

with

- η_X the unit of the change-of-operad adjunction and
- $r : (\mathsf{O}^{\mathsf{M}}_F)_! X \longrightarrow R(\mathsf{O}^{\mathsf{M}}_F)_! X$ the fibrant replacement in $\mathsf{Alg}_{\mathsf{M}}\big(\mathsf{O}^{\mathsf{M}}_{\underline{\mathsf{D}}}\big)$.

Since the change-of-operad adjunction is an adjoint equivalence by Theorem 8.3.6(6), the unit and the counit are both natural isomorphisms. So it remains to see that the morphism $(\mathsf{O}^{\mathsf{M}}_F)^* r$ is a weak equivalence in $\mathsf{Alg}_{\mathsf{M}}\big(\mathsf{O}^{\mathsf{M}}_{\underline{\mathsf{C}}}\big)$, i.e., an entrywise weak equivalence in M. Since r is an entrywise weak equivalence, by the definition of the right adjoint $(\mathsf{O}^{\mathsf{M}}_F)^*$ in Definition 5.1.5, $(\mathsf{O}^{\mathsf{M}}_F)^* r$ is an entrywise weak equivalence in M.   □

Interpretation 8.5.2. If two orthogonal categories are orthogonally equivalent (i.e., there is an orthogonal equivalence between them), then their categories of algebraic quantum field theories have equivalent homotopy theories. In particular, these two categories of algebraic quantum field theories are equivalent both before and after inverting the weak equivalences.   ◇

Example 8.5.3 (Chiral conformal, Euclidean, and locally covariant QFT). In the context of Example 2.2.15 and Example 8.4.7, recall that Man^d is a small category equivalent to the entire category of d-dimensional oriented manifolds with orientation-preserving open embeddings as morphisms. Two different choices yield two equivalent orthogonal categories. So by Theorem 8.3.6(6) and Theorem 8.5.1(2) the change-of-operad adjunction between their categories of algebraic quantum field theories is both an adjoint equivalence and a Quillen equivalence. The same can be said for Euclidean quantum field theories in Example 8.4.10 and locally covariant quantum field theories in Example 8.4.11.   ◇

Chapter 9

Homotopy Algebraic Quantum Field Theories

In this chapter, we define homotopy algebraic quantum field theories on an orthogonal category $\overline{\mathsf{C}}$. We observe that each of them has a homotopy coherent C-diagram structure and a compatible objectwise A_∞-algebra structure, and satisfies a homotopy coherent version of the causality axiom. If a set of morphisms in C is chosen, then each homotopy algebraic quantum field theory also satisfies a homotopy coherent version of the time-slice axiom.

9.1 Overview

In Section 9.2 we define homotopy algebraic quantum field theories over an orthogonal category $\overline{\mathsf{C}}$ as algebras over the Boardman-Vogt construction $\mathsf{WO}_{\overline{\mathsf{C}}}^{\mathsf{M}}$ of the colored operad $\mathsf{O}_{\overline{\mathsf{C}}}^{\mathsf{M}}$, which is the image in M of the colored operad $\mathsf{O}_{\overline{\mathsf{C}}}$ in Definition 8.3.2. Then we record some of their categorical properties. This definition makes sense because, as we saw in (8.3.7), the category of algebraic quantum field theories on an orthogonal category $\overline{\mathsf{C}}$, as in Definition 8.2.3, is canonically isomorphic to the category of $\mathsf{O}_{\overline{\mathsf{C}}}^{\mathsf{M}}$-algebras. Each colored operad O is equipped with an augmentation $\eta : \mathsf{WO} \longrightarrow \mathsf{O}$ from its Boardman-Vogt construction. In Section 6.5 we saw that in favorable cases the Boardman-Vogt construction has the correct homotopy type in the sense that the augmentation $\eta : \mathsf{WO} \longrightarrow \mathsf{O}$ is a weak equivalence and that the induced change-of-operad adjunction is a Quillen equivalence. Furthermore, in Section 7.3 to Section 7.6 we observed that the Boardman-Vogt construction WO of a colored operad O encodes O-algebras up to coherent higher homotopies.

In Section 9.3 we present a long list of examples of homotopy algebraic quantum field theories, using mostly the orthogonal categories in Section 8.4. Among the examples are homotopy chiral conformal quantum field theories, homotopy Euclidean quantum field theories, homotopy locally covariant quantum field theories, and homotopy quantum field theories on spacetimes with additional geometric structure or timelike boundary.

Our main tool for understanding the structure in homotopy algebraic quantum field theories is the Coherence Theorem in Section 9.4. This coherence result describes a homotopy algebraic quantum field theory in terms of explicit structure morphisms indexed by trees and four generating axioms. In the remaining sections in this chapter, we describe structure that exists on every homotopy algebraic quantum field theory using the Coherence Theorem.

In Section 9.5 we observe that each homotopy algebraic quantum field theory satisfies a homotopy coherent version of the causality axiom. The causality axiom for an algebraic quantum field theory $\mathfrak{A} \in \mathsf{QFT}(\overline{\mathsf{C}})$ says that, for an orthogonal pair $(g_1 : a \to c, g_2 : b \to c)$ in $\overline{\mathsf{C}}$, the images of $\mathfrak{A}(a)$ and $\mathfrak{A}(b)$ in $\mathfrak{A}(c)$ commute. The homotopy coherent version says that the diagram defining the causality axiom is homotopy commutative via specified homotopies that are also structure morphisms.

In Section 9.6 we observe that every homotopy algebraic quantum field theory on $\overline{\mathsf{C}}$ has an underlying homotopy coherent C-diagram structure. This is the homotopy coherent version of the fact that each algebraic quantum field theory $\mathfrak{A} : \overline{\mathsf{C}} \longrightarrow \mathsf{Mon}(\mathsf{M})$ can be composed with the forgetful functor to M to yield a C-diagram in M. In Section 9.7 we observe that this homotopy coherent C-diagram structure satisfies a homotopy coherent version of the time-slice axiom. The time-slice axiom for an algebraic quantum field theory says that certain structure morphisms are isomorphisms. The homotopy coherent version of the time-slice axiom says that certain structure morphisms admit two-sided homotopy inverses via specified homotopies, where the homotopy inverses and the homotopies are also structure morphisms.

In Section 9.8 we observe that each homotopy algebraic quantum field theory has an objectwise A_∞-algebra structure. This is the homotopy coherent version of the fact that, for each algebraic quantum field theory $\mathfrak{A} : \overline{\mathsf{C}} \longrightarrow \mathsf{Mon}(\mathsf{M})$, each object $\mathfrak{A}(c)$ is a monoid in M for $c \in \mathsf{C}$. We saw in Section 7.5 that an A_∞-algebra is a homotopy coherent version of a monoid. Furthermore, in Section 9.9 we show that this objectwise A_∞-algebra structure is compatible with the homotopy coherent C-diagram structure via specified homotopies that are also structure morphisms. This is the homotopy coherent version of the fact that an algebraic quantum field theory is, in particular, a diagram of monoids.

An important point to keep in mind is that all of the above homotopy coherent structures, including the homotopies, are already encoded in the Boardman-Vogt construction $\mathsf{WO}^{\mathsf{M}}_{\overline{\mathsf{C}}}$. This is of course the entire reason for using the Boardman-Vogt construction to define homotopy algebraic quantum field theories. Our coend definition of the Boardman-Vogt construction plays a critical role here. In fact, the Coherence Theorem 7.2.1 for algebras over the Boardman-Vogt construction crucially depends on our coend definition of WO. A special case of this theorem is the Coherence Theorem for homotopy algebraic quantum field theories in Section 9.4, from which the results in Sections 9.5 to 9.9 follow.

Throughout this chapter $(\mathsf{M}, \otimes, \mathbb{1})$ is a cocomplete symmetric monoidal closed category with an initial object $\varnothing$ and a commutative segment $(J, \mu, 0, 1, \epsilon)$ as in Definition 6.2.1.

9.2 Homotopy AQFT as Operad Algebras

In this section, we define homotopy algebraic quantum field theories using the Boardman-Vogt construction in Chapter 6 and record their basic categorical properties.

Recollection 9.2.1. For a $\mathfrak{C}$-colored operad O in M, recall that the Boardman-Vogt construction of O is a $\mathfrak{C}$-colored operad WO, which is entrywise defined as a coend

$$\mathsf{WO}\binom{d}{\underline{c}} = \int^{T \in \underline{\mathsf{Tree}}^{\mathfrak{C}}\binom{d}{\underline{c}}} J[T] \otimes \mathsf{O}[T] \in \mathsf{M},$$

where $\underline{\mathsf{Tree}}^{\mathfrak{C}}\binom{d}{\underline{c}}$ is the substitution category of $\mathfrak{C}$-colored trees with profile $\binom{d}{\underline{c}}$ in Definition 3.2.11. The functors

$$J : \underline{\mathsf{Tree}}^{\mathfrak{C}}\binom{d}{\underline{c}}^{\mathsf{op}} \longrightarrow \mathsf{M} \quad \text{and} \quad \mathsf{O} : \underline{\mathsf{Tree}}^{\mathfrak{C}}\binom{d}{\underline{c}} \longrightarrow \mathsf{M}$$

are induced by J and O and are defined in Definition 6.2.5 and Corollary 4.4.15, respectively. Geometrically $J[T] \otimes \mathsf{O}[T]$ is the $\mathfrak{C}$-colored tree T whose internal edges are decorated by J and whose vertices are decorated by O. Via the coend, the substitution category parametrizes the relations among such decorated trees.

The operad structure of the Boardman-Vogt construction, defined in Definition 6.3.7, is induced by tree substitution. It is equipped with a natural augmentation $\eta : \mathsf{WO} \longrightarrow \mathsf{O}$ of $\mathfrak{C}$-colored operads, defined in Theorem 6.4.4. Intuitively the augmentation forgets the lengths of the internal edges (i.e., the J-component) and composes in the colored operad O.

Since the colored operad $\mathsf{O}_{\overline{\mathsf{C}}}$, defined in Definition 8.3.2, of an orthogonal category $\overline{\mathsf{C}}$ is defined over Set, we will have to first transfer it to M. Recall from Example 5.3.3 that the strong symmetric monoidal functor $\mathsf{Set} \longrightarrow \mathsf{M}$, sending a set S to the S-indexed coproduct $\coprod_S \mathbb{1}$, yields the change-of-category functor

$$(-)^{\mathsf{M}} : \mathsf{Operad}^{\mathfrak{C}}(\mathsf{Set}) \longrightarrow \mathsf{Operad}^{\mathfrak{C}}(\mathsf{M}).$$

The image of $\mathsf{O}_{\overline{\mathsf{C}}}$ in $\mathsf{Operad}^{\mathfrak{C}}(\mathsf{M})$ will be denoted by $\mathsf{O}_{\overline{\mathsf{C}}}^{\mathsf{M}}$. Also recall from Definition 4.5.5 the category of algebras over a colored operad.

Definition 9.2.2. Suppose $\overline{\mathsf{C}} = (\mathsf{C}, \perp)$ is an orthogonal category with object set $\mathfrak{C}$, and $\mathsf{WO}_{\overline{\mathsf{C}}}^{\mathsf{M}} \in \mathsf{Operad}^{\mathfrak{C}}(\mathsf{M})$ is the Boardman-Vogt construction of $\mathsf{O}_{\overline{\mathsf{C}}}^{\mathsf{M}} \in \mathsf{Operad}^{\mathfrak{C}}(\mathsf{M})$. We define the category

$$\mathsf{HQFT}(\overline{\mathsf{C}}) = \mathsf{Alg}_{\mathsf{M}}(\mathsf{WO}_{\overline{\mathsf{C}}}^{\mathsf{M}}),$$

whose objects are called *homotopy algebraic quantum field theories* on $\overline{\mathsf{C}}$.

Remark 9.2.3. In Definition 9.2.2 we first transfer $\mathsf{O}_{\overline{\mathsf{C}}} \in \mathsf{Operad}^{\mathfrak{C}}(\mathsf{Set})$ to $\mathsf{O}_{\overline{\mathsf{C}}}^{\mathsf{M}} \in \mathsf{Operad}^{\mathfrak{C}}(\mathsf{M})$, and then we apply W to $\mathsf{O}_{\overline{\mathsf{C}}}^{\mathsf{M}}$. In particular, the Boardman-Vogt construction is *not* apply to $\mathsf{O}_{\overline{\mathsf{C}}}$ because it depends on a choice of a commutative segment in M. ◇

Interpretation 9.2.4. One should think of the $\mathcal{C}$-colored operad $\mathsf{WO}^{\mathsf{M}}_{\overline{\mathcal{C}}}$ as made up of $\mathcal{C}$-colored trees whose internal edges are decorated by the commutative segment J and whose vertices are decorated by elements in $\mathsf{O}_{\overline{\mathcal{C}}}$ with the correct profile. A homotopy algebraic quantum field theory has structure morphisms indexed by these decorated $\mathcal{C}$-colored trees. The precise statement is the Coherence Theorem 9.4.1 below. ◇

The following observation compares algebraic quantum field theories and homotopy algebraic quantum field theories. It is a special case of Theorem 5.2.7(1), Corollary 6.4.7, Corollary 6.5.10, and (8.3.7).

Corollary 9.2.5. *Suppose* $\overline{\mathsf{C}} = (\mathsf{C}, \perp)$ *is an orthogonal category.*

(1) *The augmentation* $\eta : \mathsf{WO}^{\mathsf{M}}_{\overline{\mathcal{C}}} \longrightarrow \mathsf{O}^{\mathsf{M}}_{\overline{\mathcal{C}}}$ *induces a change-of-operad adjunction*

$$\mathsf{HQFT}(\overline{\mathsf{C}}) = \mathsf{Alg}_{\mathsf{M}}\left(\mathsf{WO}^{\mathsf{M}}_{\overline{\mathcal{C}}}\right) \underset{\eta^*}{\overset{\eta_!}{\rightleftarrows}} \mathsf{Alg}_{\mathsf{M}}\left(\mathsf{O}^{\mathsf{M}}_{\overline{\mathcal{C}}}\right) \cong \mathsf{QFT}(\overline{\mathsf{C}}) \ .$$

(2) *If* M *is a monoidal model category in which the colored operads* $\mathsf{O}^{\mathsf{M}}_{\overline{\mathcal{C}}}$ *and* $\mathsf{WO}^{\mathsf{M}}_{\overline{\mathcal{C}}}$ *are admissible, then the change-of-operad adjunction is a Quillen adjunction.*

(3) *If* $\mathsf{M} = \mathsf{Chain}_{\mathbb{K}}$ *with* $\mathbb{K}$ *a field of characteristic zero, then the change-of-operad adjunction is a Quillen equivalence.*

Interpretation 9.2.6. The right adjoint η^* allows us to consider an algebraic quantum field theory on $\overline{\mathsf{C}}$ as a homotopy algebraic quantum field theory on $\overline{\mathsf{C}}$. The left adjoint $\eta_!$ rectifies a homotopy algebraic quantum field theory to an algebraic quantum field theory. Furthermore, if M is $\mathsf{Chain}_{\mathbb{K}}$, then the augmentation η is a homotopy Morita equivalence. In particular, the homotopy theory of homotopy algebraic quantum field theories is equivalent to the homotopy theory of algebraic quantum field theories over the same orthogonal category. So there is no loss of homotopical information by considering homotopy algebraic quantum field theories compared to algebraic quantum field theories. ◇

The next observation is about changing the orthogonal categories. It is a consequence of Theorem 5.2.7(1), Theorem 6.4.4, Theorem 8.5.1, and Corollary 9.2.5. The second assertion below uses the fact that Quillen equivalences have the 2-out-of-3 property.

Corollary 9.2.7. *Suppose* $F : \overline{\mathsf{C}} \longrightarrow \overline{\mathsf{D}}$ *is an orthogonal functor.*

(1) *There is an induced diagram of change-of-operad adjunctions*

$$
\begin{array}{ccc}
\mathsf{HQFT}(\overline{\mathsf{C}}) = \mathsf{Alg}_{\mathsf{M}}\left(\mathsf{WO}^{\mathsf{M}}_{\overline{\mathcal{C}}}\right) & \underset{(\mathsf{WO}^{\mathsf{M}}_F)^*}{\overset{(\mathsf{WO}^{\mathsf{M}}_F)_!}{\rightleftarrows}} & \mathsf{Alg}_{\mathsf{M}}\left(\mathsf{WO}^{\mathsf{M}}_{\overline{\mathcal{D}}}\right) = \mathsf{HQFT}(\overline{\mathsf{D}}) \\[2mm]
\eta_! \downarrow \uparrow \eta^* & & \eta_! \downarrow \uparrow \eta^* \\[2mm]
\mathsf{QFT}(\overline{\mathsf{C}}) \cong \mathsf{Alg}_{\mathsf{M}}\left(\mathsf{O}^{\mathsf{M}}_{\overline{\mathcal{C}}}\right) & \underset{(\mathsf{O}^{\mathsf{M}}_F)^*}{\overset{(\mathsf{O}^{\mathsf{M}}_F)_!}{\rightleftarrows}} & \mathsf{Alg}_{\mathsf{M}}\left(\mathsf{O}^{\mathsf{M}}_{\overline{\mathcal{D}}}\right) \cong \mathsf{QFT}(\overline{\mathsf{D}})
\end{array}
$$

such that

$$(O_F^M)_! \eta_! = \eta_! (WO_F^M)_! \quad and \quad \eta^* (O_F^M)^* = (WO_F^M)^* \eta^*.$$

(2) *If* M *is a monoidal model category in which the colored operads* $O_{\overline{C}}^M$, $O_{\overline{D}}^M$, $WO_{\overline{C}}^M$, *and* $WO_{\overline{D}}^M$ *are admissible, then all four change-of-operad adjunctions are Quillen adjunctions.*

(3) *If* F *is an orthogonal equivalence and if* M = $Chain_{\mathbb{K}}$ *with* $\mathbb{K}$ *a field of characteristic zero, then all four change-of-operad adjunctions are Quillen equivalences.*

Interpretation 9.2.8. The right adjoint $(WO_F^M)^*$ sends each homotopy algebraic quantum field theory on $\overline{D}$ to one on $\overline{C}$. The left adjoint $(WO_F^M)_!$ sends each homotopy algebraic quantum field theory on $\overline{C}$ to one on $\overline{D}$. The equality

$$(O_F^M)_! \eta_! = \eta_! (WO_F^M)_!$$

means that the left adjoint diagram is commutative. The equality

$$\eta^* (O_F^M)^* = (WO_F^M)^* \eta^*$$

means that the right adjoint diagram is commutative. Moreover, if F is an orthogonal equivalence and if M is $Chain_{\mathbb{K}}$, then all four operad morphisms in the commutative diagram

$$
\begin{array}{ccc}
WO_{\overline{C}}^M & \xrightarrow{WO_F^M} & WO_{\overline{D}}^M \\
\eta \downarrow & & \downarrow \eta \\
O_{\overline{C}}^M & \xrightarrow{O_F^M} & O_{\overline{D}}^M
\end{array}
$$

are homotopy Morita equivalences. In particular, the homotopy theory of homotopy algebraic quantum field theories on $\overline{C}$ is equivalent to the homotopy theory of homotopy algebraic quantum field theories on $\overline{D}$. ◇

9.3 Examples of Homotopy AQFT

In this section, we apply Corollary 9.2.5 and Corollary 9.2.7 to the orthogonal categories and orthogonal functors in Section 8.4 to obtain examples of homotopy algebraic quantum field theories.

Example 9.3.1 (Homotopy coherent diagrams of A_∞-algebras). For each orthogonal category $\overline{C} = (C, \perp)$, there are two orthogonal functors

$$\overline{C_{min}} = (C, \varnothing) \xrightarrow{\;i_0\;} \overline{C} \xrightarrow{\;i_1\;} \overline{C_{max}} = (C, \perp_{max})$$

as in Example 8.4.3. By Corollary 9.2.7 there is an induced diagram whose middle squares consist of change-of-operad adjunctions

$$
\begin{array}{ccccc}
\mathsf{HQFT}\big(\overline{\mathsf{C}}_{\min}\big) & & \mathsf{HQFT}\big(\overline{\mathsf{C}}\big) & & \mathsf{HQFT}\big(\overline{\mathsf{C}}_{\max}\big) \\
\| & & \| & & \| \\
\mathsf{Alg}_{\mathsf{M}}\big(\mathsf{WO}^{\mathsf{M}}_{\overline{\mathsf{C}}_{\min}}\big) \underset{(\mathsf{WO}^{\mathsf{M}}_{i_0})^*}{\overset{(\mathsf{WO}^{\mathsf{M}}_{i_0})_!}{\rightleftarrows}} & \mathsf{Alg}_{\mathsf{M}}\big(\mathsf{WO}^{\mathsf{M}}_{\overline{\mathsf{C}}}\big) & \underset{(\mathsf{WO}^{\mathsf{M}}_{i_1})^*}{\overset{(\mathsf{WO}^{\mathsf{M}}_{i_1})_!}{\rightleftarrows}} & \mathsf{Alg}_{\mathsf{M}}\big(\mathsf{WO}^{\mathsf{M}}_{\overline{\mathsf{C}}_{\max}}\big) \\
\eta_! \uparrow\!\!\downarrow \eta^* & & \eta_! \uparrow\!\!\downarrow \eta^* & & \eta_! \uparrow\!\!\downarrow \eta^* \\
\mathsf{Alg}_{\mathsf{M}}\big(\mathsf{O}^{\mathsf{M}}_{\overline{\mathsf{C}}_{\min}}\big) \underset{(\mathsf{O}^{\mathsf{M}}_{i_0})^*}{\overset{(\mathsf{O}^{\mathsf{M}}_{i_0})_!}{\rightleftarrows}} & \mathsf{Alg}_{\mathsf{M}}\big(\mathsf{O}^{\mathsf{M}}_{\overline{\mathsf{C}}}\big) & \underset{(\mathsf{O}^{\mathsf{M}}_{i_1})^*}{\overset{(\mathsf{O}^{\mathsf{M}}_{i_1})_!}{\rightleftarrows}} & \mathsf{Alg}_{\mathsf{M}}\big(\mathsf{O}^{\mathsf{M}}_{\overline{\mathsf{C}}_{\max}}\big) \\
\downarrow \cong & & \downarrow \cong & & \downarrow \cong \\
\mathsf{Mon}(\mathsf{M})^{\mathsf{C}} = \mathsf{QFT}\big(\overline{\mathsf{C}}_{\min}\big) & & \mathsf{QFT}\big(\overline{\mathsf{C}}\big) & & \mathsf{QFT}\big(\overline{\mathsf{C}}_{\max}\big) = \mathsf{Com}(\mathsf{M})^{\mathsf{C}}
\end{array}
$$

with commutative left/right adjoint diagrams. Since $\mathsf{Mon}(\mathsf{M})^{\mathsf{C}}$ is the category of C-diagrams of monoids in M, in view of Theorems 7.3.5 and 7.5.5, we interpret

$$
\mathsf{HQFT}\big(\overline{\mathsf{C}}_{\min}\big) = \mathsf{Alg}_{\mathsf{M}}\big(\mathsf{WO}^{\mathsf{M}}_{\overline{\mathsf{C}}_{\min}}\big)
$$

as the category of homotopy coherent C-diagrams of A_∞-algebras. The right adjoint $(\mathsf{WO}^{\mathsf{M}}_{i_0})^*$ sends each homotopy algebraic quantum field theory on $\overline{\mathsf{C}}$ to its underlying homotopy coherent C-diagram of A_∞-algebras. We will explain this structure in more details in Section 9.9.   ◇

Example 9.3.2 (Homotopy chiral conformal, Euclidean, and locally covariant QFT). Applied to the orthogonal category $\overline{\mathsf{Man}^d} = (\mathsf{Man}^d, \perp)$ in Example 8.4.7, Corollary 9.2.5 gives a change-of-operad adjunction

$$
\mathsf{HQFT}\big(\overline{\mathsf{Man}^d}\big) = \mathsf{Alg}_{\mathsf{M}}\big(\mathsf{WO}^{\mathsf{M}}_{\overline{\mathsf{Man}^d}}\big) \underset{\eta^*}{\overset{\eta_!}{\rightleftarrows}} \mathsf{Alg}_{\mathsf{M}}\big(\mathsf{O}^{\mathsf{M}}_{\overline{\mathsf{Man}^d}}\big) \cong \mathsf{QFT}\big(\overline{\mathsf{Man}^d}\big)
$$

between chiral conformal quantum field theories and homotopy chiral conformal quantum field theories. Moreover, this adjunction is a Quillen equivalence when $\mathsf{M} = \mathsf{Chain}_{\mathbb{K}}$ with $\mathbb{K}$ a field of characteristic zero. There are similar adjunctions for:

- (homotopy) Euclidean quantum field theories associated to the orthogonal category $\overline{\mathsf{Riem}^d} = (\mathsf{Riem}^d, \perp)$ in Example 8.4.10;
- (homotopy) locally covariant quantum field theories associated to the orthogonal category $\overline{\mathsf{Loc}^d} = (\mathsf{Loc}^d, \perp)$ in Example 8.4.11;
- (homotopy) algebraic quantum field theories on spacetime with timelike boundary associated to the orthogonal category $\overline{\mathsf{Reg}(X)} = (\mathsf{Reg}(X), \perp)$ in Example 8.4.17.   ◇

Example 9.3.3 (Homotopy chiral conformal QFT on discs). By Corollary 9.2.7, the orthogonal functor

$$
j : \overline{\mathsf{Disc}^d} \longrightarrow \overline{\mathsf{Man}^d}
$$

in Example 8.4.8 induces a diagram of change-of-operad adjunctions

$$\mathsf{HQFT}\big(\overline{\mathsf{Disc}^d}\big) = \mathsf{Alg}_\mathsf{M}\Big(\mathsf{WO}^\mathsf{M}_{\underline{\mathsf{Disc}^d}}\Big) \underset{(\mathsf{WO}^\mathsf{M}_j)^*}{\overset{(\mathsf{WO}^\mathsf{M}_j)_!}{\rightleftarrows}} \mathsf{Alg}_\mathsf{M}\Big(\mathsf{WO}^\mathsf{M}_{\underline{\mathsf{Man}^d}}\Big) = \mathsf{HQFT}\big(\overline{\mathsf{Man}^d}\big)$$

$$\eta_! \Big\Updownarrow \eta^* \qquad\qquad\qquad \eta_! \Big\Updownarrow \eta^*$$

$$\mathsf{QFT}\big(\overline{\mathsf{Disc}^d}\big) \cong \mathsf{Alg}_\mathsf{M}\Big(\mathsf{O}^\mathsf{M}_{\underline{\mathsf{Disc}^d}}\Big) \underset{(\mathsf{O}^\mathsf{M}_j)^*}{\overset{(\mathsf{O}^\mathsf{M}_j)_!}{\rightleftarrows}} \mathsf{Alg}_\mathsf{M}\Big(\mathsf{O}^\mathsf{M}_{\underline{\mathsf{Man}^d}}\Big) \cong \mathsf{QFT}\big(\overline{\mathsf{Man}^d}\big)$$

with commutative left/right adjoint diagrams. When $d = 1$ the vertical adjunction on the left goes between chiral conformal quantum field theories defined on intervals and their homotopy analogues. There are similar diagrams of adjunctions associated to the orthogonal functors in Examples 8.4.9 and 8.4.12–8.4.16.   ◇

9.4   Coherence Theorem

For the rest of this chapter, we will study the structure of homotopy algebraic quantum field theories. In Definition 9.2.2 we defined a homotopy algebraic quantum field theory on an orthogonal category $\overline{\mathsf{C}} = (\mathsf{C}, \perp)$ as an algebra over the colored operad $\mathsf{WO}^\mathsf{M}_{\underline{\mathsf{C}}} \in \mathsf{Operad}^{\mathfrak{C}}(\mathsf{M})$, which is the Boardman-Vogt construction of the colored operad $\mathsf{O}^\mathsf{M}_{\underline{\mathsf{C}}} \in \mathsf{Operad}^{\mathfrak{C}}(\mathsf{M})$. Recall that $\mathsf{O}^\mathsf{M}_{\underline{\mathsf{C}}}$ is the image under the change-of-category functor

$$\mathsf{Operad}^{\mathfrak{C}}(\mathsf{Set}) \longrightarrow \mathsf{Operad}^{\mathfrak{C}}(\mathsf{M})$$

of the colored operad $\mathsf{O}_{\overline{\mathsf{C}}}$ in Definition 8.3.2.

The following coherence result describes homotopy algebraic quantum field theories in terms of generating structure morphisms and generating relations. Recall from Notation 4.5.1 the shorthand

$$X_{\underline{c}} = X_{c_1} \otimes \cdots \otimes X_{c_m}$$

for each $\mathfrak{C}$-colored object X and $\underline{c} = (c_1, \ldots, c_m) \in \mathsf{Prof}(\mathfrak{C})$. Also recall from Notation 4.4.6 that for $A \in \mathsf{M}^{\mathsf{Prof}(\mathfrak{C}) \times \mathfrak{C}}$ and a vertex v in a $\mathfrak{C}$-colored tree, $A(v)$ is the shorthand for the entry $A\binom{\mathsf{out}(v)}{\mathsf{in}(v)}$.

Theorem 9.4.1. *Suppose $\overline{\mathsf{C}} = (\mathsf{C}, \perp)$ is an orthogonal category with object set $\mathfrak{C}$. Then a homotopy algebraic quantum field theory on $\overline{\mathsf{C}}$ is exactly a pair (X, λ) consisting of*

- *a $\mathfrak{C}$-colored object $X = \{X_c\}_{c \in \mathfrak{C}}$ in M and*
- *a structure morphism*

$$\mathsf{J}[T] \otimes X_{\underline{c}} \xrightarrow{\ \lambda_T^{\{f^v\}}\ } X_d \in \mathsf{M} \tag{9.4.2}$$

   for

   – *each $T \in \underline{\mathsf{Tree}}^{\mathfrak{C}}\binom{d}{\underline{c}}$ with $\binom{d}{\underline{c}} \in \mathsf{Prof}(\mathfrak{C}) \times \mathfrak{C}$ and*

$-$ *each* $\{f^v\} \in \prod_{v \in \mathsf{Vt}(T)} \mathsf{O}_{\overline{C}}(v)$

that satisfies the following four conditions.

Associativity *For* $\big(\underline{c} = (c_1, \ldots, c_n); d\big) \in \mathsf{Prof}(\mathfrak{C}) \times \mathfrak{C}$ *with* $n \geq 1$, $T \in \underline{\mathsf{Tree}}^{\mathfrak{C}}(\underline{d}_c)$, $T_j \in \underline{\mathsf{Tree}}^{\mathfrak{C}}(\underline{c_j}{\underline{b_j}})$ *for* $1 \leq j \leq n$, $\underline{b} = (\underline{b}_1, \ldots, \underline{b}_n)$,

$$G = \mathsf{Graft}(T; T_1, \ldots, T_n) \in \underline{\mathsf{Tree}}^{\mathfrak{C}}(\underline{d}_b)$$

the grafting (3.3.1), $\{f^v\} \in \prod_{v \in \mathsf{Vt}(T)} \mathsf{O}_{\overline{C}}(v)$, *and* $\{f_j^u\} \in \prod_{u \in \mathsf{Vt}(T_j)} \mathsf{O}_{\overline{C}}(u)$ *for* $1 \leq j \leq n$, *the diagram*

$$
\begin{array}{ccc}
\mathsf{J}[T] \otimes \Big(\overset{n}{\underset{j=1}{\bigotimes}} \mathsf{J}[T_j] \Big) \otimes X_{\underline{b}} & \xrightarrow{\ \pi\ } & \mathsf{J}[G] \otimes X_{\underline{b}} \\
\end{array}
\tag{9.4.3}
$$

permute $\Big\downarrow \cong$

$$\mathsf{J}[T] \otimes \overset{n}{\underset{j=1}{\bigotimes}} \Big(\mathsf{J}[T_j] \otimes X_{\underline{b}_j} \Big) \qquad\qquad \lambda_G^{\{f^v\},\{f_j^u\}_{j=1}^n}$$

$\Big(\mathsf{Id}, \otimes_j \lambda_{T_j}^{\{f_j^u\}} \Big) \Big\downarrow$

$$
\begin{array}{ccc}
\mathsf{J}[T] \otimes X_{\underline{c}} & \xrightarrow{\ \lambda_T^{\{f^v\}}\ } & X_d
\end{array}
$$

is commutative. Here $\pi = \otimes_S 1$ *is the morphism in Lemma 6.2.7 for the grafting* G.

Unity *For each* $c \in \mathfrak{C}$, *the composition*

$$
X_c \xrightarrow{\ \cong\ } \mathsf{J}[\uparrow_c] \otimes X_c \xrightarrow{\ \lambda_{\uparrow_c}^{\varnothing}\ } X_c
\tag{9.4.4}
$$

is the identity morphism of X_c.

Equivariance *For each* $T \in \underline{\mathsf{Tree}}^{\mathfrak{C}}(\underline{d}_c)$, *permutation* $\sigma \in \Sigma_{|\underline{c}|}$, *and* $\{f^v\} \in \prod_{v \in \mathsf{Vt}(T)} \mathsf{O}_{\overline{C}}(v)$, *the diagram*

$$
\begin{array}{ccc}
\mathsf{J}[T] \otimes X_{\underline{c}} & \xrightarrow{\ \lambda_T^{\{f^v\}}\ } & X_d \\
(\mathsf{Id}, \sigma^{-1}) \Big\downarrow & & \Big\| \\
\mathsf{J}[T\sigma] \otimes X_{\underline{c}\sigma} & \xrightarrow{\ \lambda_{T\sigma}^{\{f^v\}}\ } & X_d
\end{array}
\tag{9.4.5}
$$

is commutative, in which $T\sigma \in \underline{\mathsf{Tree}}^{\mathfrak{C}}(\underline{d}_{c\sigma})$ *is the same as* T *except that its ordering is* $\zeta_T \sigma$ *with* ζ_T *the ordering of* T. *The permutation* $\sigma^{-1} : X_{\underline{c}} \xrightarrow{\ \cong\ } X_{\underline{c}\sigma}$ *permutes the factors in* $X_{\underline{c}}$.

Wedge Condition *For* $T \in \underline{\mathsf{Tree}}^{\mathfrak{C}}(\underline{d}_c)$, $H_v \in \underline{\mathsf{Tree}}^{\mathfrak{C}}(v)$ *for each* $v \in \mathsf{Vt}(T)$, $K = T(H_v)_{v \in T}$ *the tree substitution, and* $\{f_v^u\} \in \prod_{u \in \mathsf{Vt}(H_v)} \mathsf{O}_{\overline{C}}(u)$ *for each* $v \in \mathsf{Vt}(T)$, *the diagram*

$$
\begin{array}{ccc}
\mathsf{J}[T] \otimes X_{\underline{c}} & \xrightarrow{\ \lambda_T^{\{h^v\}}\ } & X_d \\
(\mathsf{J}, \mathsf{Id}) \Big\downarrow & & \Big\| \\
\mathsf{J}[K] \otimes X_{\underline{c}} & \xrightarrow{\ \lambda_K^{\{f_v^u\}_{u \in K}}\ } & X_d
\end{array}
\tag{9.4.6}
$$

is commutative, in which

$$h^v = \gamma_{H_v}^{O_{\overline{C}}}\big(\{f_v^u\}_{u \in H_v}\big) \in O_{\overline{C}}(v)$$

for each $v \in \mathsf{Vt}(T)$ with

$$\gamma_{H_v}^{O_{\overline{C}}} : O_{\overline{C}}[H_v] \longrightarrow O_{\overline{C}}(v) \in \mathsf{Set}$$

the operadic structure morphism (4.4.10) *of $O_{\overline{C}}$ for H_v.*

A morphism $f : (X, \lambda^X) \longrightarrow (Y, \lambda^Y)$ of homotopy algebraic quantum field theories on $\overline{C}$ is a morphism of the underlying $\mathfrak{C}$-colored objects that respects the structure morphisms in (9.4.2) *in the obvious sense.*

Proof. This is the special case of the Coherence Theorem 7.2.1 for the $\mathfrak{C}$-colored operad $O_{\overline{C}}^{\mathsf{M}}$. Indeed, recall that the $\mathfrak{C}$-colored operad $O_{\overline{C}}^{\mathsf{M}}$ has entries

$$O_{\overline{C}}^{\mathsf{M}}\binom{d}{\underline{c}} = \coprod_{O_{\overline{C}}\binom{d}{\underline{c}}} \mathbb{1}$$

for $(\underline{c}; d) \in \mathsf{Prof}(\mathfrak{C}) \times \mathfrak{C}$. For each $\mathfrak{C}$-colored tree T, there is a natural isomorphism

$$O_{\overline{C}}^{\mathsf{M}}[T] = \bigotimes_{v \in T} O_{\overline{C}}^{\mathsf{M}}\binom{\mathrm{out}(v)}{\mathrm{in}(v)} = \bigotimes_{v \in T}\Big[\coprod_{O_{\overline{C}}(v)} \mathbb{1}\Big] \cong \coprod_{\prod_{v \in T} O_{\overline{C}}(v)} \mathbb{1}.$$

It follows that there is a natural isomorphism

$$J[T] \otimes O_{\overline{C}}^{\mathsf{M}}[T] \otimes X_{\underline{c}} \cong \coprod_{\prod_{v \in T} O_{\overline{C}}(v)} J[T] \otimes X_{\underline{c}}.$$

Therefore, the structure morphism λ_T in (7.2.2) is uniquely determined by the restrictions $\lambda_T^{\{f^v\}}$ as stated in (9.4.2). The associativity, unity, equivariance, and wedge conditions (9.4.3)-(9.4.6) are exactly those in the Coherence Theorem 7.2.1. $\qquad\square$

9.5 Homotopy Causality Axiom

In this section, we explain that every homotopy algebraic quantum field theory satisfies a homotopy coherent version of the causality axiom.

Motivation 9.5.1. An algebraic quantum field theory $\mathfrak{A} \in \mathsf{QFT}(\overline{C})$ on an orthogonal category $\overline{C} = (C, \perp)$ satisfies the causality axiom (8.2.4). It says that for each orthogonal pair $(g_1 : a \to c, g_2 : b \to c) \in \perp$, the diagram

$$\mathfrak{A}(a) \otimes \mathfrak{A}(b) \xrightarrow{\big(\mathfrak{A}(g_1), \mathfrak{A}(g_2)\big)} \mathfrak{A}(c) \otimes \mathfrak{A}(c) \xrightarrow[\mu_c]{\mu_c(1\,2)} \mathfrak{A}(c)$$

is commutative, where $(1\,2)$ is the symmetry permutation on $\mathfrak{A}(c)^{\otimes 2}$. For a homotopy algebraic quantum field theory, we should expect this diagram to commute up to specified homotopies. $\qquad\diamond$

To explain the homotopy version of the causality axiom, we need the following notations. As before, using the canonical bijection in Example 8.3.4, for a morphism $f \in C(c, d)$, we will abbreviate an element $[\mathrm{id}_1, f] \in O_{\overline{C}}\binom{d}{c}$ to just f.

Assumption 9.5.2. *Suppose $\overline{C} = (C, \perp)$ is an orthogonal category with object set $\mathfrak{C}$, and $(g_1 : a \to c, g_2 : b \to c)$ is an orthogonal pair in $\overline{C}$.*

- Suppose $C = \mathrm{Cor}_{(c,c;c)} \in \underline{\mathsf{Tree}}^{\mathfrak{C}}\binom{c}{c,c}$, and $C_{ab} = \mathrm{Cor}_{(a,b;c)} \in \underline{\mathsf{Tree}}^{\mathfrak{C}}\binom{c}{a,b}$.
- Suppose $L_1 = \mathrm{Lin}_{(a,c)} \in \underline{\mathsf{Tree}}^{\mathfrak{C}}\binom{c}{a}$, and $L_2 = \mathrm{Lin}_{(b,c)} \in \underline{\mathsf{Tree}}^{\mathfrak{C}}\binom{c}{b}$.
- Define the grafting $T = \mathsf{Graft}(C; L_1, L_2) \in \underline{\mathsf{Tree}}^{\mathfrak{C}}\binom{c}{a,b}$, which we may visualize as follows.

Note that T is the 2-level tree $T\big((a), (b); (c, c); c\big)$ in Example 3.1.23, and it has two internal edges.

- Denote by Id_c^2 the element $\big[\mathrm{id}_2, \{\mathrm{Id}_c, \mathrm{Id}_c\}\big] \in O_{\overline{C}}\binom{c}{c,c}$.
- Denote by τ the element $\big[(1\ 2), \{\mathrm{Id}_c, \mathrm{Id}_c\}\big] \in O_{\overline{C}}\binom{c}{c,c}$, where $(1\ 2)$ is the non-identity permutation in Σ_2.
- Denote by $\underline{g}$ the element $\big[\mathrm{id}_2, \{g_1, g_2\}\big] \in O_{\overline{C}}\binom{c}{a,b}$.

The following result is the homotopy coherent version of the causality axiom. To simplify the notation, we will omit writing some of the identity morphisms below.

Theorem 9.5.3. *In the context of Assumption 9.5.2, suppose (X, λ) is a homotopy algebraic quantum field theory on $\overline{C}$. Then the diagram*

$$
\begin{array}{ccccc}
\mathbb{1}^{\otimes 2} \otimes X_a \otimes X_b & \xleftarrow{\ \cong\ } & J[C] \otimes \big(J[L_1] \otimes X_a\big) \otimes \big(J[L_2] \otimes X_b\big) & \xrightarrow{\big(\lambda_{L_1}^{g_1}, \lambda_{L_2}^{g_2}\big)} & J[C] \otimes X_c^{\otimes 2} \\
\downarrow{\scriptstyle 1^{\otimes 2}} & & \downarrow{\scriptstyle \lambda_T^{\{\mathrm{Id}_c^2, g_1, g_2\}}} & & \downarrow{\scriptstyle \lambda_C^{\mathrm{Id}_c^2}} \\
J^{\otimes 2} \otimes X_a \otimes X_b & \xrightarrow{\ \cong\ } & J[T] \otimes X_a \otimes X_b & \longrightarrow & X_c \\
\uparrow{\scriptstyle 0^{\otimes 2}} & & (1) & & \| \\
\mathbb{1}^{\otimes 2} \otimes X_a \otimes X_b & \xleftarrow{\ \cong\ } & J[C_{ab}] \otimes X_a \otimes X_b & \xrightarrow{\ \lambda_{C_{ab}}^{\underline{g}}\ } & X_c \\
\downarrow{\scriptstyle 0^{\otimes 2}} & & (2) & & \| \\
J^{\otimes 2} \otimes X_a \otimes X_b & \xrightarrow{\ \cong\ } & J[T] \otimes X_a \otimes X_b & \xrightarrow{\ \lambda_T^{\{\tau, g_1, g_2\}}\ } & X_c \\
\uparrow{\scriptstyle 1^{\otimes 2}} & & & & \uparrow{\scriptstyle \lambda_C^{\tau}} \\
\mathbb{1}^{\otimes 2} \otimes X_a \otimes X_b & \xleftarrow{\ \cong\ } & J[C] \otimes \big(J[L_1] \otimes X_a\big) \otimes \big(J[L_2] \otimes X_b\big) & \xrightarrow{\big(\lambda_{L_1}^{g_1}, \lambda_{L_2}^{g_2}\big)} & J[C] \otimes X_c^{\otimes 2}
\end{array}
$$

in M *is commutative, where* $0, 1 : \mathbb{1} \longrightarrow J$ *are part of the commutative segment* J.

Proof. This is a consequence of the Coherence Theorem 9.4.1 for homotopy algebraic quantum field theories. Indeed, in the above diagram:

(1) The top and bottom rectangles are commutative by the associativity condition (9.4.3) and the grafting definition of T.
(2) The rectangle (1) is commutative by the wedge condition (9.4.6) applied to the tree substitution

$$T = C_{ab}(T) \in \underline{\mathsf{Tree}}^{\mathfrak{C}}\binom{c}{a,b}$$

and by the equalities

$$\gamma_T^{O_{\overline{C}}}\big(\mathrm{Id}_c^2; g_1, g_2\big) = \big[\mathrm{id}_2, \{g_1, g_2\}\big] = \underline{g} \in O_{\overline{C}}\binom{c}{a,b}.$$

(3) The rectangle (2) is commutative by the same wedge condition and the equalities

$$\underline{g} = \big[(1\ 2), \{g_1, g_2\}\big] = \gamma_T^{O_{\overline{C}}}\big(\tau; g_1, g_2\big) \in O_{\overline{C}}\binom{c}{a,b},$$

the first of which holds because g_1 and g_2 are orthogonal.

$\square$

Interpretation 9.5.4. Theorem 9.5.3 is a precise form of the statement that the diagram

$$\begin{array}{ccc} X_a \otimes X_b & \xrightarrow{(X_{g_1}, X_{g_2})} & X_c \otimes X_c \\ {\scriptstyle (X_{g_1}, X_{g_2})}\downarrow & & \downarrow{\scriptstyle \mu_2^c} \\ X_c \otimes X_c & \xrightarrow{\mu_2^c(1\ 2)} & X_c \end{array}$$

is commutative up to homotopy. Here μ_2^c is the binary multiplication in the A_∞-algebra X_c as in Example 7.5.11. So Theorem 9.5.3 says that each homotopy algebraic quantum field theory satisfies the causality axiom up to specified homotopies that are also structure morphisms. ◇

Example 9.5.5 (Homotopy causality in homotopy chiral conformal QFT). Applied to the orthogonal category $\overline{\mathsf{Man}^d}$ in Example 8.4.7, Theorem 9.5.3 says that every $\mathsf{WO}^{\mathsf{M}}_{\overline{\mathsf{Man}^d}}$-algebra satisfies the causality axiom up to specified homotopies. There are similar statements for all the other orthogonal categories in Section 8.4. ◇

9.6 Homotopy Coherent Diagrams

Using the Coherence Theorem 9.4.1, for the next few sections we will explain the structure that exists in homotopy algebraic quantum field theories, i.e., in $\mathsf{WO}^{\mathsf{M}}_{\overline{\mathsf{C}}}$-algebras. In this section, we explain the homotopy coherent diagram structure that exists on each homotopy algebraic quantum field theory.

Motivation 9.6.1. Recall from Example 4.5.21 that for each small category C with object set $\mathfrak{C}$, there is a $\mathfrak{C}$-colored operad $\mathsf{C}^{\mathrm{diag}}$ in M whose algebras are C-diagrams in M. The algebras over the Boardman-Vogt construction $\mathsf{WC}^{\mathrm{diag}}$ are homotopy coherent C-diagrams in M, as we explained in Section 7.3. An algebraic quantum field theory on an orthogonal category $\overline{\mathsf{C}}$ is, first of all, a functor $\mathsf{C} \longrightarrow \mathsf{Mon}(\mathsf{M})$. Composing with the forgetful functor to M, each algebraic quantum field theory yields an underlying C-diagram in M. So we should expect a homotopy algebraic quantum field theory on $\overline{\mathsf{C}}$ to have the structure of a homotopy coherent C-diagram in M. ◇

For any colors $c, d \in \mathfrak{C}$, recall from Example 8.3.4 that there is a canonical bijection

$$\mathsf{C}(c,d) \cong \mathsf{O}_{\overline{\mathsf{C}}}\binom{d}{c},$$

sending $f \in \mathsf{C}(c,d)$ to $[\mathrm{id}_1, f]$. This induces a canonical isomorphism

$$\mathsf{O}_{\overline{\mathsf{C}}}^{\mathsf{M}}\binom{d}{c} = \coprod_{\mathsf{O}_{\overline{\mathsf{C}}}\binom{d}{c}} \mathbb{1} \cong \coprod_{\mathsf{C}(c,d)} \mathbb{1},$$

which we will use below.

Theorem 9.6.2. *Suppose* $\overline{\mathsf{C}} = (\mathsf{C}, \perp)$ *is an orthogonal category with object set* $\mathfrak{C}$.

(1) There is a morphism of $\mathfrak{C}$-*colored operads*

$$p : \mathsf{C}^{\mathrm{diag}} \longrightarrow \mathsf{O}_{\overline{\mathsf{C}}}^{\mathsf{M}}$$

in M defined entrywise as follows.

- *For $c, d \in \mathfrak{C}$, the morphism*

$$\mathsf{C}^{\mathrm{diag}}\binom{d}{c} = \coprod_{\mathsf{C}(c,d)} \mathbb{1} \xrightarrow[\cong]{p} \mathsf{O}_{\overline{\mathsf{C}}}^{\mathsf{M}}\binom{d}{c} \in \mathsf{M}$$

  sends the copy of $\mathbb{1}$ in $\mathsf{C}^{\mathrm{diag}}\binom{d}{c}$ corresponding to $f \in \mathsf{C}(c,d)$ to the copy of $\mathbb{1}$ in $\mathsf{O}_{\overline{\mathsf{C}}}^{\mathsf{M}}\binom{d}{c}$ corresponding to $[\mathrm{id}_1, f] \in \mathsf{O}_{\overline{\mathsf{C}}}\binom{d}{c}$.
- *If $|\underline{c}| \neq 1$, then*

$$\mathsf{C}^{\mathrm{diag}}\binom{d}{\underline{c}} = \varnothing \xrightarrow{p} \mathsf{O}_{\overline{\mathsf{C}}}^{\mathsf{M}}\binom{d}{\underline{c}} \in \mathsf{M}$$

  is the unique morphism from the initial object $\varnothing$.

(2) The morphism p induces a change-of-operad adjunction

$$\mathsf{Alg}_{\mathsf{M}}\big(\mathsf{WC}^{\mathrm{diag}}\big) \xrightleftharpoons[(\mathsf{W}p)^*]{(\mathsf{W}p)_!} \mathsf{Alg}_{\mathsf{M}}\big(\mathsf{WO}_{\overline{\mathsf{C}}}^{\mathsf{M}}\big) = \mathsf{HQFT}(\overline{\mathsf{C}}).$$

between the category of homotopy coherent C-diagrams in M and the category of homotopy algebraic quantum field theories on $\overline{\mathsf{C}}$.

Proof. For assertion (1), note that the equivariant structure on C^{diag} is trivial, since it is concentrated in unary entries, and its operadic composition is given by the categorical composition in C. A direct inspection of Definition 8.3.2 of $O_{\overline{C}}$ shows that p is entrywise well-defined and respects the colored units and operadic composition. Therefore, p is a morphism of $\mathfrak{C}$-colored operads.

Assertion (2) follows from assertion (1), the naturality of the Boardman-Vogt construction in Proposition 6.4.3, and Theorem 5.1.8. □

Interpretation 9.6.3. The right adjoint $(Wp)^*$ sends each homotopy algebraic quantum field theory on $\overline{C}$ to its underlying WC^{diag}-algebra, i.e., homotopy coherent C-diagram in M. More explicitly, suppose $(X, \lambda) \in \mathsf{HQFT}(\overline{C})$, and recall the Coherence Theorem 7.3.5 for homotopy coherent C-diagrams in M. Suppose given a profile $\underline{c} = (c_0, \dots, c_n) \in \mathsf{Prof}(\mathfrak{C})$ and a sequence of composable morphisms

$$\underline{f} = \{f_j\} \in \prod_{j=1}^{n} \mathsf{C}(c_{j-1}, c_j) \cong \prod_{j=1}^{n} \mathsf{O}_{\overline{C}}\binom{c_j}{c_{j-1}}.$$

Then the structure morphism

$$J[\mathsf{Lin}_{\underline{c}}] \otimes \big[(Wp)^* X\big]_{c_0} = J[\mathsf{Lin}_{\underline{c}}] \otimes X_{c_0} \xrightarrow{\lambda_{\underline{c}}^{\underline{f}}} X_{c_n} = \big[(Wp)^* X\big]_{c_n} \in \mathsf{M} \qquad (9.6.4)$$

in (7.3.6) of the homotopy coherent C-diagram $(Wp)^* X$ is given by the structure morphism $\lambda_{\mathsf{Lin}_{\underline{c}}}^{\{[\mathrm{id}_1, f_j]\}}$ in (9.4.2). ◇

Example 9.6.5 (Homotopy coherent diagram in homotopy chiral conformal QFT). Applied to the orthogonal category $\overline{\mathsf{Man}^d}$ in Example 8.4.7, we obtain the adjunction

$$\mathsf{Alg}_{\mathsf{M}}\big(W(\mathsf{Man}^d)^{\text{diag}}\big) \underset{(Wp)^*}{\overset{(Wp)_!}{\rightleftarrows}} \mathsf{Alg}_{\mathsf{M}}\big(W\mathsf{O}_{\mathsf{Man}^d}^{\mathsf{M}}\big) = \mathsf{HQFT}(\overline{\mathsf{Man}^d})$$

between the category of homotopy coherent Man^d-diagrams in M and the category of homotopy chiral conformal quantum field theories. There are similar adjunctions for all the other orthogonal categories in Section 8.4. ◇

9.7 Homotopy Time-Slice Axiom

In this section, we explain that for an orthogonal category $\overline{C} = (C, \perp)$ and a set S of morphisms in C, every homotopy algebraic quantum field theory on the orthogonal category $\overline{C[S^{-1}]} = \big(C[S^{-1}], \ell_*(\perp)\big)$ satisfies a homotopy coherent version of the time-slice axiom.

Motivation 9.7.1. For a chosen set S of morphisms in C, recall from Definition 8.2.3 that an algebraic quantum field theory $\mathfrak{A}$ on $\overline{C}$ satisfies the time-slice axiom with respect to S if the structure morphism $\mathfrak{A}(s) \in \mathsf{Mon}(\mathsf{M})$ is an isomorphism for each morphism $s \in S$. By Lemma 8.2.7 we know that algebraic quantum field theories on $\overline{C}$ that satisfy the time-slice axiom with respect to S are exactly the algebraic quantum field theories on $\overline{C[S^{-1}]} = \big(C[S^{-1}], \ell_*(\perp)\big)$, with

- $C[S^{-1}]$ the S-localization of the category C and
- $\ell_*(\perp)$ the pushforward of the orthogonality relation $\perp$ in C along the S-localization $\ell : C \longrightarrow C[S^{-1}]$.

Therefore, we should expect each homotopy algebraic quantum field theory on $\overline{C[S^{-1}]}$, i.e., $\mathsf{WO}^M_{\overline{C[S^{-1}]}}$-algebra, to have homotopy inverses for the structure morphisms in S.

In Theorem 9.6.2 we constructed a morphism of $\mathfrak{C}$-colored operads

$$p : C[S^{-1}]^{\mathrm{diag}} \longrightarrow \mathsf{O}^M_{\overline{C[S^{-1}]}}$$

together with an induced change-of-operad adjunction

$$\mathsf{Alg}_M\!\left(\mathsf{WC}[S^{-1}]^{\mathrm{diag}}\right) \underset{(Wp)^*}{\overset{(Wp)_!}{\rightleftarrows}} \mathsf{Alg}_M\!\left(\mathsf{WO}^M_{\overline{C[S^{-1}]}}\right) = \mathsf{HQFT}(\overline{C[S^{-1}]})$$

between the category of homotopy coherent $C[S^{-1}]$-diagrams in M and the category of homotopy algebraic quantum field theories on $\overline{C[S^{-1}]}$. The right adjoint $(Wp)^*$ leaves the underlying entries unchanged, so each $\mathsf{WO}^M_{\overline{C[S^{-1}]}}$-algebra has an underlying homotopy coherent $C[S^{-1}]$-diagram in M. ◇

For the following result on homotopy time-slice, recall from Example 3.1.19 the notation $\mathrm{Lin}_{\underline{c}}$ for a linear graph associated to a profile $\underline{c}$ and from Definition 6.2.1 the morphisms $0, 1 : \mathbb{1} \longrightarrow J$ as part of the commutative segment J. To simplify the notation, using the canonical bijection

$$C[S^{-1}](c, d) \overset{\cong}{\longrightarrow} \Sigma_1 \times C[S^{-1}](c, d) = \mathsf{O}_{\overline{C[S^{-1}]}}\binom{d}{c} , \qquad f \longmapsto [\mathrm{id}_1, f]$$

in Example 8.3.4, we will abbreviate an element $[\mathrm{id}_1, f] \in \mathsf{O}_{\overline{C[S^{-1}]}}\binom{d}{c}$ to f. We will use the notation $\lambda_T^{\{f^v\}}$ in (9.4.2) for a structure morphism of a homotopy algebraic quantum field theory.

Theorem 9.7.2. *Suppose $\overline{C} = (C, \perp)$ is an orthogonal category, and S is a set of morphisms in C. Suppose (X, λ) is a homotopy algebraic quantum field theory on $\overline{C[S^{-1}]} = (C[S^{-1}], \ell_*(\perp))$, i.e., a $\mathsf{WO}^M_{\overline{C[S^{-1}]}}$-algebra. Suppose $f : c \longrightarrow d$ is a morphism in S with inverse $f^{-1} : d \longrightarrow c \in C[S^{-1}]$. Then the structure morphism*

$$J[\mathrm{Lin}_{(d,c)}] \otimes X_d = \mathbb{1} \otimes X_d \xrightarrow{\lambda_{\mathrm{Lin}_{(d,c)}}^{\{f^{-1}\}}} X_c \in M$$

is a two-sided homotopy inverse of the structure morphism

$$J[\mathrm{Lin}_{(c,d)}] \otimes X_c = \mathbb{1} \otimes X_c \xrightarrow{\lambda_{\mathrm{Lin}_{(c,d)}}^{\{f\}}} X_d \in M$$

in the following sense.

(1) $\lambda_{\mathrm{Lin}_{(d,c)}}^{\{f^{-1}\}}$ *is a left homotopy inverse of* $\lambda_{\mathrm{Lin}_{(c,d)}}^{\{f\}}$ *in the sense that the diagram*

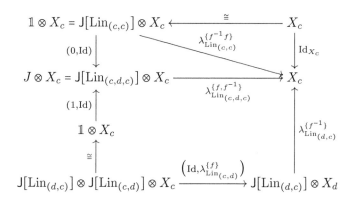

in M *is commutative.*

(2) $\lambda_{\mathrm{Lin}_{(d,c)}}^{\{f^{-1}\}}$ *is a right homotopy inverse of* $\lambda_{\mathrm{Lin}_{(c,d)}}^{\{f\}}$ *in the sense that the diagram*

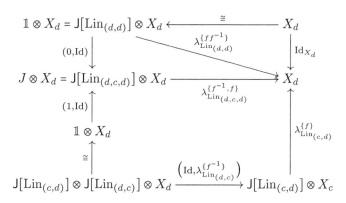

in M *is commutative.*

Proof. This is Corollary 7.4.2 applied to the underlying homotopy coherent $\mathsf{C}[S^{-1}]$-diagram in M of (X, λ). □

Interpretation 9.7.3. Every homotopy algebraic quantum field theory on $\overline{\mathsf{C}[S^{-1}]} = (\mathsf{C}[S^{-1}], \ell_*(\bot))$ satisfies a homotopy version of the time-slice axiom in the sense that each structure morphism $\lambda_{\mathrm{Lin}_{(c,d)}}^{\{f\}}$ for $f \in S$ has the structure morphism $\lambda_{\mathrm{Lin}_{(d,c)}}^{\{f^{-1}\}}$ as a two-sided homotopy inverse via specified homotopies. Furthermore, the homotopies $\lambda_{\mathrm{Lin}_{(c,d,c)}}^{\{f,f^{-1}\}}$ and $\lambda_{\mathrm{Lin}_{(d,c,d)}}^{\{f^{-1},f\}}$ are specified structure morphisms of the homotopy algebraic quantum field theory. In other words, each two-sided homotopy inverse and the corresponding homotopies are already encoded in the Boardman-Vogt construction $\mathrm{WO}_{\overline{\mathsf{C}[S^{-1}]}}^{\mathrm{M}}$ of $\mathsf{O}_{\overline{\mathsf{C}[S^{-1}]}}^{\mathrm{M}}$. ◇

Example 9.7.4 (Homotopy time-slice in homotopy locally covariant QFT). In the context of Example 8.4.11, every $\mathsf{WO}^{\mathsf{M}}_{\overline{\mathsf{Loc}^d[S^{-1}]}}$-algebra satisfies the homotopy time-slice axiom in the sense of Theorem 9.7.2. There are similar statements in the contexts of Examples 8.4.12–8.4.17.   ◇

9.8   Objectwise A_∞-Algebra

In this section, we explain that each homotopy algebraic quantum field theory on an orthogonal category is objectwise an A_∞-algebra.

Motivation 9.8.1. Recall from Example 4.5.17 that there is a 1-colored operad As whose algebras are precisely monoids in M. In Section 7.5 we observed that algebras over the Boardman-Vogt construction WAs, called A_∞-algebras, are in a precise sense monoids up to coherent higher homotopies. An algebraic quantum field theory $\mathfrak{A}$ on an orthogonal category $\overline{\mathsf{C}}$ is, first of all, a functor $\mathfrak{A} : \mathsf{C} \longrightarrow \mathsf{Mon}(\mathsf{M})$. So for each object $c \in \mathsf{C}$, $\mathfrak{A}(c)$ is a monoid in M. Therefore, it is reasonable to expect that a homotopy algebraic quantum field theory on $\overline{\mathsf{C}}$ is objectwise an A_∞-algebra in M.   ◇

In the following result, we assume the unique color for the 1-colored operad As is $*$. Recall from Definition 8.3.2 the set $\mathsf{O}_{\overline{\mathsf{C}}}^{(d)}(_c^c)$. We will write $\mathbb{1}_x$ for a copy of $\mathbb{1}$ indexed by an element x.

Theorem 9.8.2. *Suppose* $\overline{\mathsf{C}} = (\mathsf{C}, \perp)$ *is an orthogonal category, and* $c \in \mathsf{C}$.

(1) There is an operad morphism

$$j_c : \mathsf{As} \longrightarrow \mathsf{O}_{\overline{\mathsf{C}}}^{\mathsf{M}}$$

that sends $*$ *to* c *and is defined entrywise by the commutative diagrams*

$$
\begin{array}{ccc}
\mathbb{1}_\sigma & \xrightarrow{\;\;=\;\;} & \mathbb{1}_{[\sigma,\{\mathrm{Id}_c\}_{i=1}^n]} \\[4pt]
\scriptstyle\sigma\ \text{summand}\ \Big\downarrow & & \Big\downarrow\ \scriptstyle[\sigma,\{\mathrm{Id}_c\}_{i=1}^n]\ \text{summand} \\[4pt]
\mathsf{As}(n) = \coprod\limits_{\sigma \in \Sigma_n} \mathbb{1} & \xrightarrow{\;\;j_c\;\;} & \coprod\limits_{\mathsf{O}_{\overline{\mathsf{C}}}(_{c,\ldots,c}^c)} \mathbb{1} = \mathsf{O}_{\overline{\mathsf{C}}}^{\mathsf{M}}(_{c,\ldots,c}^c)
\end{array}
$$

for $n \geq 0$ *and* $\sigma \in \Sigma_n$, *where* $(c,\ldots,c)$ *has* n *copies of* c.

(2) The operad morphism j_c *induces a change-of-operad adjunction*

$$\mathsf{Alg}_{\mathsf{M}}(\mathsf{WAs}) \underset{(\mathsf{W}j_c)^*}{\overset{(\mathsf{W}j_c)_!}{\rightleftarrows}} \mathsf{Alg}_{\mathsf{M}}(\mathsf{WO}_{\overline{\mathsf{C}}}^{\mathsf{M}}) = \mathsf{HQFT}(\overline{\mathsf{C}})$$

between the category of A_∞-*algebras in* M *and the category of homotopy algebraic quantum field theories on* $\overline{\mathsf{C}}$.

Proof. It follows directly from Example 4.5.17 and Definition 8.3.2 that j_c is a well-defined operad morphism from the $\{*\}$-colored operad As to the $\mathsf{Ob}(\mathsf{C})$-colored operad $\mathsf{O}^{\mathsf{M}}_{\overline{\mathsf{C}}}$. Assertion (2) follows from assertion (1), the naturality of the Boardman-Vogt construction in Proposition 6.4.3, and Theorem 5.1.8. $\qquad\square$

Interpretation 9.8.3. For each $c \in \mathfrak{C} = \mathsf{Ob}(\mathsf{C})$, the right adjoint $(\mathsf{W}j_c)^*$ sends each homotopy algebraic quantum field theory (X, λ) on the orthogonal category $\overline{\mathsf{C}}$ to the A_∞-algebra X_c. Explicitly, for each $n \geq 0$, $T \in \underline{\mathsf{Tree}}^{\{*\}}(n)$, and $\{\sigma_v\} \in \prod_{v \in T} \Sigma_{|\mathsf{in}(v)|}$, the structure morphism

$$\mathsf{J}[T] \otimes \big[(\mathsf{W}j_c)^* X\big]^{\otimes n} = \mathsf{J}[T] \otimes X_c^{\otimes n} \xrightarrow{\;\lambda_T^{\{\sigma_v\}_{v \in T}}\;} X_c = (\mathsf{W}j_c)^* X$$

in (7.5.6) of the A_∞-algebra $(\mathsf{W}j_c)^* X$ is given by the structure morphism

$$\lambda_{T_c}^{\big\{[\sigma_v, \{\mathrm{Id}_c\}_{i=1}^{|\mathsf{in}(v)|}]\big\}_{v \in T_c}} \quad \text{with} \quad [\sigma_v, \{\mathrm{Id}_c\}_{i=1}^{|\mathsf{in}(v)|}] \in \mathsf{O}_{\overline{\mathsf{C}}}\binom{c}{c,\dots,c} \tag{9.8.4}$$

in (9.4.2). Here $T_c \in \underline{\mathsf{Tree}}^{\mathfrak{C}}\binom{c}{c,\dots,c}$ is the $\mathfrak{C}$-colored tree obtained from T by switching each of its edge color from $*$ to c. $\qquad\diamond$

Remark 9.8.5. We speculate that the objectwise A_∞-algebra structure in each homotopy algebraic quantum field theory may be related to non-associative quantum field theory [Dzhunushaliev (1994)] and non-associative gauge theory [Majid (2005); de Medeiros and Ramgoolam (2005); Okubo (1995); Ramgoolam (2004)]. $\diamond$

Example 9.8.6 (Objectwise A_∞-structure in homotopy chiral conformal QFT). In the context of Example 9.3.2, every $\mathsf{WO}^{\mathsf{M}}_{\mathsf{Man}^d}$-algebra has an A_∞-algebra structure in each color in the sense of Theorem 9.8.2. A similar statement holds for every other orthogonal category in Section 8.4. $\qquad\diamond$

9.9 Homotopy Coherent Diagrams of A_∞-Algebras

In Section 9.6 we explained that each homotopy algebraic quantum field theory on an orthogonal category $\overline{\mathsf{C}}$ has the structure of a homotopy coherent C-diagram in M. Moreover, in Section 9.8 we observed that each entry of a homotopy algebraic quantum field theory on $\overline{\mathsf{C}}$ has the structure of an A_∞-algebra. In this section, we explain how the homotopy coherent C-diagram structure and the objectwise A_∞-algebras in a homotopy algebraic quantum field theory are compatible with each other.

Motivation 9.9.1. For an orthogonal category $\overline{\mathsf{C}}$, an algebraic quantum field theory is, first of all, a functor $\mathfrak{A} : \mathsf{C} \longrightarrow \mathsf{Mon}(\mathsf{M})$, i.e., a C-diagram of monoids in M. For each morphism $g : c \longrightarrow d$ in C, its image $\mathfrak{A}(g) : \mathfrak{A}(c) \longrightarrow \mathfrak{A}(d)$ is a morphism of monoids in M. So it respects the multiplications and the units in the sense that

the diagrams

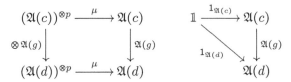

are commutative for all $p \geq 2$. In other words, the C-diagram structure and the monoid structure commute. When these structures are replaced by a homotopy coherent C-diagram and objectwise A_∞-algebras, we should expect the structures to commute up to specified homotopies. The homotopy coherent version is necessarily more involved because the structure morphisms in a homotopy coherent C-diagram (9.6.4) are indexed by linear graphs, while those in an A_∞-algebra (9.8.4) are indexed by trees. ◇

Recall the minimal orthogonal category $\overline{\mathsf{C}_{\min}} = (\mathsf{C}, \varnothing)$ in Example 8.4.1.

Corollary 9.9.2. *Suppose* C *is a small category with object set* $\mathfrak{C}$.

(1) *There is an equality*

$$\mathsf{O}^{\mathsf{M}}_{\overline{\mathsf{C}_{\min}}} = \mathsf{O}^{\mathsf{M}}_{\mathsf{C}} \in \mathsf{Operad}^{\mathfrak{C}}(\mathsf{M}),$$

where:

- $\mathsf{O}^{\mathsf{M}}_{\overline{\mathsf{C}_{\min}}}$ *is the image in* $\mathsf{Operad}^{\mathfrak{C}}(\mathsf{M})$ *of* $\mathsf{O}_{\overline{\mathsf{C}_{\min}}}$ *as in Definition 8.3.2.*
- $\mathsf{O}^{\mathsf{M}}_{\mathsf{C}}$ *is the* $\mathfrak{C}$-*colored operad in Example 4.5.22.*

(2) *Suppose* $\overline{\mathsf{C}} = (\mathsf{C}, \perp)$ *is an orthogonal category. Then the identity functor on* C *defines an orthogonal functor*

$$\overline{\mathsf{C}_{\min}} = (\mathsf{C}, \varnothing) \xrightarrow{\ i_0\ } \overline{\mathsf{C}}$$

and induces a diagram of change-of-operad adjunctions

$$
\begin{array}{ccc}
\mathsf{Alg}_{\mathsf{M}}(\mathsf{WO}^{\mathsf{M}}_{\mathsf{C}}) & \underset{(\mathsf{WO}^{\mathsf{M}}_{i_0})^*}{\overset{(\mathsf{WO}^{\mathsf{M}}_{i_0})_!}{\rightleftarrows}} & \mathsf{Alg}_{\mathsf{M}}(\mathsf{WO}^{\mathsf{M}}_{\overline{\mathsf{C}}}) = \mathsf{HQFT}(\overline{\mathsf{C}}) \\
{\scriptstyle \eta_!}\Big\uparrow\Big\downarrow{\scriptstyle \eta^*} & & {\scriptstyle \eta_!}\Big\uparrow\Big\downarrow{\scriptstyle \eta^*} \\
\mathsf{Mon}(\mathsf{M})^{\mathsf{C}} \cong \mathsf{Alg}_{\mathsf{M}}(\mathsf{O}^{\mathsf{M}}_{\mathsf{C}}) & \underset{(\mathsf{O}^{\mathsf{M}}_{i_0})^*}{\overset{(\mathsf{O}^{\mathsf{M}}_{i_0})_!}{\rightleftarrows}} & \mathsf{Alg}_{\mathsf{M}}(\mathsf{O}^{\mathsf{M}}_{\overline{\mathsf{C}}}) \cong \mathsf{QFT}(\overline{\mathsf{C}})
\end{array}
$$

with commuting left adjoint diagram and commuting right adjoint diagram.

Proof. The first assertion follows from the definition of $\mathsf{O}^{\mathsf{M}}_{\mathsf{C}}$ in Example 4.5.22 and Definition 8.3.2 of $\mathsf{O}^{\mathsf{M}}_{\overline{\mathsf{C}_{\min}}}$, where the equivalence relation $\sim$ is trivial for the empty orthogonality relation. The second assertion follows from the first assertion, Example 5.3.3, Corollary 6.4.9, and Theorem 8.3.6(2). ☐

Interpretation 9.9.3. Via the right adjoint $(\mathsf{WO}^{\mathsf{M}}_{i_0})^*$, every homotopy algebraic quantum field theory (X, λ) has an underlying $\mathsf{WO}^{\mathsf{M}}_{\mathfrak{C}}$-algebra, i.e., homotopy coherent C-diagram of A_∞-algebras as in Definition 7.7.2. Therefore, by Corollary 7.7.6, (X, λ) has an underlying homotopy coherent C-diagram in M and an objectwise A_∞-algebra structure. These are the structures in Theorem 9.6.2 and Theorem 9.8.2. $\diamond$

To explain the compatibility between these two structures precisely, we need the following notations. Recall our convention that, for a vertex v in a $\mathfrak{C}$-colored tree, we often abbreviate its profile $\mathsf{Prof}(v) \in \mathsf{Prof}(\mathfrak{C}) \times \mathfrak{C}$ to (v). Using the canonical bijection in Example 8.3.4, for a morphism $f \in C(c, d)$, we will abbreviate an element $[\mathsf{id}_1, f] \in O_{\overline{C}}\binom{d}{c}$ to just f.

Assumption 9.9.4. Suppose $\overline{\mathsf{C}} = (\mathsf{C}, \perp)$ is an orthogonal category with object set $\mathfrak{C}$. Suppose:

- $T_c \in \underline{\mathsf{Tree}}^{\{c\}}\binom{c}{c,\dots,c}$ is a c-colored tree for some color $c \in \mathfrak{C}$, where $(c, \dots, c) \in \mathsf{Prof}(\mathfrak{C})$ has $p \geq 0$ copies of c.
- $T_d \in \underline{\mathsf{Tree}}^{\{d\}}\binom{d}{d,\dots,d}$ is the d-colored tree obtained from T_c by replacing every edge color by d.
- $\underline{c} = \big(c = c_0, c_1, \dots, c_n = d\big) \in \mathsf{Prof}(\mathfrak{C})$ is a profile with $n \geq 1$.
- $L = \mathsf{Lin}_{\underline{c}} \in \underline{\mathsf{Tree}}^{\mathfrak{C}}\binom{d}{c}$ is the linear graph for the profile $\underline{c}$.

Define the graftings

$$T^1 = \mathsf{Graft}(L; T_c) \in \underline{\mathsf{Tree}}^{\mathfrak{C}}\binom{d}{c,\dots,c},$$

$$T^2 = \mathsf{Graft}\big(T_d; \underbrace{\mathsf{Lin}_{(c,d)}, \dots, \mathsf{Lin}_{(c,d)}}_{p \text{ copies}}\big) \in \underline{\mathsf{Tree}}^{\mathfrak{C}}\binom{d}{c,\dots,c}.$$

We may visualize T^1 (on the left) and T^2 (on the right) as follows.

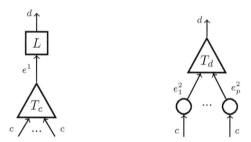

In T^1, the output flag of the c-colored tree T_c is grafted with the input leg of L. The c-colored internal edge in T_1 that extends from T_c to L, which is created by the grafting, is denoted by e^1. In T_2, the output flag of a copy of $\mathsf{Lin}_{(c,d)}$ is grafted with each of the p input legs of T_d. The d-colored internal edge that extends from the kth copy of $\mathsf{Lin}_{(c,d)}$ (from the left) to T_d is denoted by e^2_k. Note that if T_c is a corolla with $p = 0$, then

$$T_c = \overset{c\uparrow}{\bigcirc} \in \underline{\mathsf{Tree}}^{\{c\}}\binom{c}{\varnothing}, \qquad T^2 = T_d = \overset{d\uparrow}{\bigcirc} \in \underline{\mathsf{Tree}}^{\{d\}}\binom{d}{\varnothing}. \tag{9.9.5}$$

So in this case the grafting in T^2 is trivial because T_d has no input legs.

Suppose

$$\{g_j\} \in \prod_{j=1}^{n} \mathsf{C}(c_{j-1}, c_j) \cong \prod_{j=1}^{n} \mathsf{O}_{\overline{\mathsf{C}}}\binom{c_j}{c_{j-1}}$$

is a sequence of n composable morphisms in C. For each vertex v in T_c, suppose

$$f_v = \left[\sigma_v, \{f_v^i\}_{i=1}^{|\mathrm{in}(v)|}\right] \in \mathsf{O}_{\overline{\mathsf{C}}}(v) = \mathsf{O}_{\overline{\mathsf{C}}}\binom{c}{c,\dots,c} \tag{9.9.6}$$

in which $(c, \dots, c)$ has $|\mathrm{in}(v)|$ copies of c, $\sigma_v \in \Sigma_{|\mathrm{in}(v)|}$, and each $f_v^i \in \mathsf{C}(c, c)$. Define

$$q_v = \left[\sigma_v, \{\mathrm{Id}_d\}_{i=1}^{|\mathrm{in}(v)|}\right] \in \mathsf{O}_{\overline{\mathsf{C}}}\binom{d}{d,\dots,d},$$

in which $(d, \dots, d)$ has $|\mathrm{in}(v)|$ copies of d. Note that if

$$\gamma_{T_1}^{\mathsf{O}_{\overline{\mathsf{C}}}}\left(\{g_j\}_{j=1}^n, \{f_v\}_{v \in T_c}\right) = \left[\sigma, \{h_i\}_{i=1}^p\right] \in \mathsf{O}_{\overline{\mathsf{C}}}\binom{d}{c,\dots,c} \tag{9.9.7}$$

for some permutation $\sigma \in \Sigma_p$ and morphisms $h_i \in \mathsf{C}(c, d)$, then we also have

$$\gamma_{T_1}^{\mathsf{O}_{\overline{\mathsf{C}}}}\left(\{g_j\}_{j=1}^n, \{f_v\}_{v \in T_c}\right) = \gamma_{T_2}^{\mathsf{O}_{\overline{\mathsf{C}}}}\left(\{q_v\}_{v \in T_d}, \{h_i\}_{i=1}^p\right), \tag{9.9.8}$$

where by our notational convention $\{h_i\} \in \prod_{i=1}^p \mathsf{O}_{\overline{\mathsf{C}}}\binom{d}{c}$. This equality follows from the fact that each g_j is associated with the trivial permutation $\mathrm{id}_1 \in \Sigma_1$.

The following result is the homotopy coherent compatibility between the homotopy coherent C-diagram structure and the objectwise A_∞-algebra structure in a homotopy algebraic quantum field theory. A copy of the morphism $1 : \mathbb{1} \longrightarrow J$ corresponding to an internal edge e will be denoted by 1_e below. To simplify the notation, we will omit writing some of the identity morphisms below.

Theorem 9.9.9. *Suppose (X, λ) is a homotopy algebraic quantum field theory on $\overline{\mathsf{C}}$ in the setting of Assumption 9.9.4. Then the diagram* *is commutative, where the morphisms* $0, 1 : \mathbb{1} \longrightarrow J$ *are part of the commutative segment* J.

Proof. This is a consequence of the Coherence Theorem 9.4.1 for homotopy algebraic quantum field theories. Indeed, in the above diagram from top to bottom:

- The top rectangle is commutative by the associativity condition (9.4.3) and the grafting definition of T_1.
- The second rectangle is commutative by (9.9.7) and the wedge condition (9.4.6) applied to the tree substitution $T_1 = \mathrm{Cor}_{(c;d)}(T_1)$.
- The third rectangle is commutative by (9.9.8) and the wedge condition (9.4.6) applied to the tree substitution $T_2 = \mathrm{Cor}_{(c;d)}(T_2)$.
- The bottom rectangle is commutative by the associativity condition (9.4.3) and the grafting definition of T_2.

$\square$

Theorem 9.9.9 applies to all the homotopy algebraic quantum field theories in Section 9.3. In the following examples, we will explain some special cases of Theorem 9.9.9.

Example 9.9.10 (Homotopy compatibility). In (9.9.6) suppose each $f_v^i = \mathrm{Id}_c$. Then

$$f_v = \left[\sigma_v, \{\mathrm{Id}_c\}_{i=1}^{|\mathrm{in}(v)|} \right] \in \mathsf{O}_{\overline{\mathsf{C}}}(\genfrac{}{}{0pt}{}{c}{c,\ldots,c}) \quad \text{for} \quad v \in \mathsf{Vt}(T_c),$$
$$h_i = g_n \circ \cdots \circ g_1 \in \mathsf{C}(c,d) \cong \mathsf{O}_{\overline{\mathsf{C}}}(\genfrac{}{}{0pt}{}{d}{c}) \quad \text{for} \quad 1 \le i \le p.$$

In the commutative diagram in Theorem 9.9.9:

- The structure morphism

$$J[T_c] \otimes X_c^{\otimes p} \xrightarrow{\ \lambda_{T_c}^{\{f_v\}}\ } X_c$$

is part of the A_∞-algebra X_c as explained in (9.8.4).
- Similarly, the structure morphism

$$J[T_d] \otimes X_d^{\otimes p} \xrightarrow{\ \lambda_{T_d}^{\{q_v\}}\ } X_d$$

is part of the A_∞-algebra X_d.
- The structure morphisms

$$J[L] \otimes X_c \xrightarrow{\ \lambda_L^{\{g_j\}}\ } X_d \quad \text{and} \quad J[\mathrm{Lin}_{(c,d)}] \otimes X_c \xrightarrow{\ \lambda_{\mathrm{Lin}_{(c,d)}}^{\{h_i\}}\ } X_d$$

for $1 \le i \le p$ are part of the homotopy coherent C-diagram structure of (X, λ) as explained in (9.6.4).

Therefore, the entire commutative diagram in Theorem 9.9.9 is expressing an up-to-homotopy compatibility between the homotopy coherent C-diagram structure and the objectwise A_∞-algebra structure via specified homotopies in each homotopy algebraic quantum field theory. $\diamond$

Example 9.9.11 (Homotopy preservation of homotopy units). In the context of Example 9.9.10, suppose further that $T_c = \mathrm{Cor}_{(\varnothing;c)}$ with $p = 0$ as in (9.9.5) and that $n = 1$, so $L = \mathrm{Lin}_{(c,d)}$. Then the commutative diagram in Theorem 9.9.9 is a precise version of the statement that, for $g \in \mathsf{C}(c,d)$, the diagram

is commutative up to homotopy. Here X_g is the notation in Example 7.3.11 for a homotopy coherent C-diagram. Similarly, μ_0^c and μ_0^d are the two-sided homotopy units in the A_∞-algebras X_c and X_d, respectively, as explained in Examples 7.5.12 and 7.5.13. Therefore, in this case Theorem 9.9.9 says that, in each homotopy algebraic quantum field theory, the homotopy coherent C-diagram structure preserves the two-sided homotopy units in the objectwise A_∞-algebras up to specified homotopies. $\diamond$

Example 9.9.12 (Homotopy preservation of multiplication). In the context of Example 9.9.10, suppose further that

$$T_c = \mathrm{Cor}_{(c,\dots,c;c)}$$

is a c-colored corolla with p inputs and that $n = 1$, so $L = \mathrm{Lin}_{(c,d)}$. Then the commutative diagram in Theorem 9.9.9 is a precise version of the statement that, for $g \in \mathsf{C}(c,d)$, the diagram

$$
\begin{array}{ccc}
X_c^{\otimes p} & \xrightarrow{\ \mu_p^c\ } & X_c \\
{\scriptstyle \underset{i}{\otimes} X_g}\big\downarrow & & \big\downarrow{\scriptstyle X_g} \\
X_d^{\otimes p} & \xrightarrow{\ \mu_p^d\ } & X_d
\end{array}
$$

is commutative up to homotopy. Here μ_p^c and μ_p^d are the multiplications in the A_∞-algebras X_c and X_d, respectively, as in Example 7.5.11. Therefore, in this case Theorem 9.9.9 says that, in each homotopy algebraic quantum field theory, the homotopy coherent C-diagram structure preserves the multiplications in the objectwise A_∞-algebras up to specified homotopies. $\diamond$

Chapter 10

Prefactorization Algebras

In this chapter, we define prefactorization algebras on a configured category as algebras over a suitable colored operad and observe their basic structure. In Section 10.1 we briefly review prefactorization algebras on a topological space in the original sense of Costello-Gwilliam [Costello and Gwilliam (2017)]. Configured categories are abstractions of the category $\mathsf{Open}(X)$ for a topological space X and are defined in Section 10.2. The colored operad and prefactorization algebras associated to a configured category are defined in Section 10.3. The coherence theorems for prefactorization algebras, with or without the time-slice axiom, are also recorded in that section.

In Section 10.4 it is shown that every prefactorization algebra has an underlying pointed diagram. In Section 10.5 we observe that some entries of a prefactorization algebra are equipped with the structure of a commutative monoid. This applies, in particular, to the 0-entry of each prefactorization algebra on the configured category of a bounded lattice with least element 0. In Section 10.6, we show that, for each prefactorization algebra Y on the configured category of a bounded lattice, the commutative monoid Y_0 acts on every other entry, and the underlying diagram is a diagram of left Y_0-modules.

In Section 10.7 we show that every diagram of commutative monoids can be realized as a prefactorization algebra. In Sections 10.4 and 10.7 we also give evidence that prefactorization algebras and algebraic quantum field theories are closely related. A detailed study of their relationship is the subject of Chapter 12. In Section 10.8 we show that equivalences of configured categories yield equivalent and Quillen equivalent categories of prefactorization algebras.

Throughout this chapter, $(\mathsf{M}, \otimes, \mathbb{1})$ is a fixed cocomplete symmetric monoidal closed category, such as $\mathsf{Vect}_{\mathbb{K}}$ and $\mathsf{Chain}_{\mathbb{K}}$.

10.1   Costello-Gwilliam Prefactorization Algebras

Prefactorization algebras and their variants in the sense of Costello-Gwilliam [Costello and Gwilliam (2017)] provide a mathematical framework for quantum field theories that is analogous to deformation quantization in quantum mechanics.

In [Costello and Gwilliam (2017)] 3.1.1 a *prefactorization algebra* on a topological space X valued in M is defined as a functor

$$\mathcal{F} : \mathsf{Open}(X) \longrightarrow \mathsf{M}$$

that functorially assigns to each open subset $U \subset X$ an object $\mathcal{F}(U) \in \mathsf{M}$. If $U_1, \ldots, U_n \subset V \in \mathsf{Open}(X)$ are pairwise disjoint open subsets of V, then $\mathcal{F}$ is also equipped with a structure morphism

$$\mathcal{F}(U_1) \otimes \cdots \otimes \mathcal{F}(U_n) \xrightarrow{\ \mathcal{F}^V_{U_1,\ldots,U_n}\ } \mathcal{F}(V) \ \in \mathsf{M}.$$

These structure morphisms are required to satisfy some natural associativity, unity, and equivariance axioms.

In particular, if $\varnothing_X \subset X$ denotes the empty subset, then $\mathcal{F}(\varnothing_X)$ is equipped with an associative and commutative multiplication

$$\mathcal{F}(\varnothing_X) \otimes \mathcal{F}(\varnothing_X) \xrightarrow{\ \mathcal{F}^{\varnothing_X}_{\varnothing_X,\varnothing_X}\ } \mathcal{F}(\varnothing_X) \ .$$

If this multiplicative structure is equipped with a two-sided unit, making $\mathcal{F}(\varnothing_X)$ into a commutative monoid in M, then $\mathcal{F}$ is called a *unital prefactorization algebra on X*.

Physically X is the spacetime of interest. A prefactorization algebra $\mathcal{F}$ assigns to each open subset $U \subset X$ an object $\mathcal{F}(U)$ of quantum observables. For an inclusion $U \subset V$ of open subsets of X, the structure morphism

$$\mathcal{F}(U) \xrightarrow{\ \mathcal{F}^V_U\ } \mathcal{F}(V) \ \in \mathsf{M}$$

sends observables in U to observables in V. If the open subsets $U_1, \ldots, U_n \subset V$ are suitably disjoint, then we can combine the observables in the U_i's via the structure morphism $\mathcal{F}^V_{U_1,\ldots,U_n}$.

There is also a G-equivariant analogue of prefactorization algebras when the topological space X is equipped with an action by a group G. In this case, the category $\mathsf{Open}(X)$ is replaced by its G-equivariant analogue $\mathsf{Open}(X)_G$ in Example 2.2.14. As defined in [Costello and Gwilliam (2017)] 3.7.1.1, a *G-equivariant prefactorization algebra on X* is a prefactorization algebra on X defined by a functor

$$\mathcal{F} : \mathsf{Open}(X)_G \longrightarrow \mathsf{M},$$

so now there are structure isomorphisms

$$\mathcal{F}(U) \xrightarrow[\cong]{\ \mathcal{F}(g)\ } \mathcal{F}(gU) \ \in \mathsf{M}$$

for open subsets $U \subset X$ and elements $g \in G$. If $U_1, \ldots, U_n \subset V$ are pairwise disjoint open subsets of $V \in \mathsf{Open}(X)$ and if $g \in G$, then it is required that the diagram

$$
\begin{array}{ccc}
\mathcal{F}(U_1) \otimes \cdots \otimes \mathcal{F}(U_n) & \xrightarrow{\ (\mathcal{F}(g),\ldots,\mathcal{F}(g))\ } & \mathcal{F}(gU_1) \otimes \cdots \otimes \mathcal{F}(gU_n) \\
\ \downarrow{\scriptstyle \mathcal{F}^V_{U_1,\ldots,U_n}} & & \ \downarrow{\scriptstyle \mathcal{F}^{gV}_{gU_1,\ldots,gU_n}} \\
\mathcal{F}(V) & \xrightarrow{\ \mathcal{F}(g)\ } & \mathcal{F}(gV)
\end{array}
$$

in M be commutative.

Since prefactorization algebras on a topological space and algebraic quantum field theories on an orthogonal category are both mathematical frameworks for quantum field theories, one might wonder what the difference is. An algebraic quantum field theory on an orthogonal category is entrywise a monoid in M, so observables in the same object of observables can always be multiplied. On the other hand, a prefactorization algebra on X is entrywise an object in M. In particular, observables in a prefactorization algebra on X cannot be multiplied unless they come from pairwise disjoint open subsets. Despite this difference, in Sections 10.4, Section 10.7, and Chapter 12, we will see that these two mathematical approaches to quantum field theories–algebraic quantum field theories and prefactorization algebras–are actually closely related.

To facilitate the comparison between prefactorization algebras and algebraic quantum field theories, we will take a more categorical approach to the former. A Costello-Gwilliam prefactorization algebra on a topological space X is defined as a functor $\mathsf{Open}(X) \longrightarrow \mathsf{M}$ with some extra structure and properties. As we pointed out in Example 2.2.13, the category $\mathsf{Open}(X)$ is a bounded lattice, i.e., a lattice with both a least element and a greatest element. We take the abstraction one step further. In order to specify the structure morphism $\mathcal{F}^V_{U_1,\dots,U_n}$, we need two things:

- Each U_i is equipped with a morphism $U_i \longrightarrow V$.
- The U_i's are pairwise disjoint in a suitable sense.

We will achieve this below by a new concept called a configured category. Basically, we simply incorporate the finite families of morphisms $\{U_i \longrightarrow V\}_{i=1}^n$ into the data of our category and impose some natural axioms as suggested by $\mathsf{Open}(X)$. This is analogous to an orthogonal category in Definition 8.2.1, where a concept of disjointedness is built into the data of the category via a set $\perp$ of pairs of morphisms with a common codomain.

10.2 Configured Categories

In this section, we define configured categories, from which we will later define prefactorization algebras, and provide some key examples.

Definition 10.2.1. A *configured category* $\widehat{\mathsf{C}} = (\mathsf{C}, \Delta^\mathsf{C})$ is a pair consisting of

- a small category C and
- a set Δ^C in which each element, called a *configuration*, is a pair $(d; \{f_i\})$ with
  - $d \in \mathsf{C}$ and
  - $\{f_i\}$ a finite, possibly empty, sequence $\{f_i : c_i \longrightarrow d\}_{i=1}^n$ of morphisms in C with codomain d.

It is required that the following four axioms hold.

Symmetry If $\left(d; \{f_i\}_{i=1}^n\right) \in \Delta^{\mathsf{C}}$ and if $\sigma \in \Sigma_n$, then $\left(d; \{f_{\sigma(i)}\}_{i=1}^n\right) \in \Delta^{\mathsf{C}}$.

Subset If $(d; \{f_i\}) \in \Delta^{\mathsf{C}}$ and $\{f_j'\}$ is a possibly empty subsequence of $\{f_i\}$, then $(d; \{f_j'\}) \in \Delta^{\mathsf{C}}$.

Inclusivity If $f : c \longrightarrow d$ is a morphism in C, then $(d; \{f\}) \in \Delta^{\mathsf{C}}$.

Composition If $\left(d; \{f_i : c_i \longrightarrow d\}_{i=1}^n\right) \in \Delta^{\mathsf{C}}$ with $n \geq 1$ and if for each $1 \leq i \leq n$, $\left(c_i; \{g_{ij} : b_{ij} \longrightarrow c_i\}_{j=1}^{k_i}\right) \in \Delta^{\mathsf{C}}$, then the *composition*

$$\left(d; \{f_i g_{ij} : b_{ij} \longrightarrow d\}_{1 \leq i \leq n,\, 1 \leq j \leq k_i}\right) \tag{10.2.2}$$

is also a configuration.

Example 10.2.3. The composition (10.2.2) of $\left(d; \{f_1, f_2\}\right)$, $\left(c_1; \{g_{11}, g_{12}\}\right)$, and $\left(c_2; \{g_{21}, g_{22}, g_{23}\}\right)$ is

$$\left(d; \{f_1 g_{11}, f_1 g_{12}, f_2 g_{21}, f_2 g_{22}, f_2 g_{23}\}\right).$$

If we replace $\left(c_2; \{g_{21}, g_{22}, g_{23}\}\right)$ with $(c_2; \varnothing)$, then the composition becomes

$$\left(d; \{f_1 g_{11}, f_1 g_{12}\}\right).$$

$\diamond$

Interpretation 10.2.4. From a physical perspective, one should think of the objects in a configured category $(\mathsf{C}, \Delta^{\mathsf{C}})$ as the spacetime regions of interest, e.g., oriented manifolds of a fixed dimension. A morphism $f : c \longrightarrow d$ in C should be thought of as an inclusion of the spacetime region c into a bigger spacetime region d. A configuration $\left(d; \{f_i : c_i \longrightarrow d\}_{i=1}^n\right)$ is expressing the idea that the spacetime regions $c_1, \ldots, c_n$ are pairwise disjoint in d.

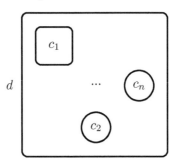

All four axioms in Definition 10.2.1 are physically motivated by this picture. Indeed, the symmetry axiom is just about relabeling the pairwise disjoint regions. The subset and inclusivity axioms are immediate. The composition (10.2.2) says that, if the spacetime regions c_i's are pairwise disjoint in d, and if the spacetime regions b_{ij}'s are pairwise disjoint in c_i for each i, then the entire collection $\{b_{ij}\}_{i,j}$ is pairwise disjoint in d.

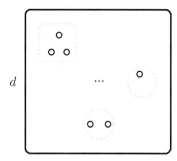

In this picture, the three small discs inside c_1 are the pairwise disjoint regions b_{11}, b_{12}, and b_{13}, and the two small discs inside c_2 are the disjoint regions b_{21} and b_{22}. The only small disc inside c_n is b_{n1}. ◇

Definition 10.2.5. Suppose $(\mathsf{C}, \triangle^\mathsf{C})$ is a configured category.

(1) For objects $c_1, \ldots, c_n, d \in \mathsf{C}$, the set of configurations $\left(d; \{c_i \longrightarrow d\}_{i=1}^n\right)$ is denoted by $\triangle^\mathsf{C}\binom{d}{c}$, where $\underline{c} = (c_1, \ldots, c_n)$.

(2) A *configured functor*

$$F : (\mathsf{C}, \triangle^\mathsf{C}) \longrightarrow (\mathsf{D}, \triangle^\mathsf{D})$$

between two configured categories is a functor $F : \mathsf{C} \longrightarrow \mathsf{D}$ that preserves the configurations, i.e.,

$$(Fd; \{Ff_i\}) \in \triangle^\mathsf{D} \quad \text{if} \quad (d; \{f_i\}) \in \triangle^\mathsf{C}.$$

(3) The category of configured categories and configured functors is denoted by ConfCat.

Lemma 10.2.6. *Suppose $(\mathsf{C}, \triangle^\mathsf{C})$ is a configured category.*

(1) For each object $d \in \mathsf{C}$, $(d; \varnothing)$ belongs to $\triangle^\mathsf{C}\binom{d}{\varnothing}$.

(2) The composition (10.2.2) belongs to $\triangle^\mathsf{C}\binom{d}{b}$, where $\underline{b}_i = (b_{i1}, \ldots, b_{ik_i})$ for $1 \leq i \leq n$ and $\underline{b} = (\underline{b}_1, \ldots, \underline{b}_n)$.

Proof. For the first assertion, first note that $(d; \{\mathrm{Id}_d\}) \in \triangle^\mathsf{C}$ by the inclusivity axiom. So by the subset axiom, we infer that $(d; \varnothing) \in \triangle^\mathsf{C}$. The second assertion follows directly from the definition. □

Notation 10.2.7. Suppose $(\mathsf{C}, \triangle^\mathsf{C})$ is a configured category.

(1) We call $(d; \varnothing) \in \triangle^\mathsf{C}\binom{d}{\varnothing}$ the *empty configuration at d*.

(2) To simplify the presentation, for a configuration $\left(d; \{f_i\}_{i=1}^n\right) \in \triangle^\mathsf{C}\binom{d}{c}$, we will often omit d, which is the common codomain of the morphisms f_i for $1 \leq i \leq n$, and simply write $\{f_i\}_{i=1}^n$ or $\{f_i\}$.

(3) If $\{f_i\}_{i=1}^n \in \triangle^\mathsf{C}$, then we call it an *$n$-ary configuration*.

Some examples of configured categories follow. Many more examples will be given in Section 12.1, where we will show that every orthogonal category yields a configured category.

Example 10.2.8 (Minimal configured category). Suppose C is a small category. Define Δ^C_{min} to be the set consisting of

- $(d; \varnothing)$ for all objects $d \in C$ and
- $(d; \{f\})$ for all morphisms $f \in C(c, d)$ with $c, d \in C$.

Then
$$\widehat{C_{min}} = (C, \Delta^C_{min})$$
is a configured category, called the *minimal configured category on* C. ◇

Example 10.2.9 (Maximal configured category). Suppose C is a small category. Define Δ^C_{max} to be the set of all pairs $(d; \{f_i\})$ with $d \in C$ and $\{f_i\}$ any possibly empty finite sequence of morphisms in C with codomain d. Then
$$\widehat{C_{max}} = (C, \Delta^C_{max})$$
is a configured category, called the *maximal configured category on* C. For objects $c_1, \dots, c_n, d \in C$ with $n \geq 0$ and $\underline{c} = (c_1, \dots, c_n)$, by our notational convention above, we have
$$\Delta^C_{max}\binom{d}{\underline{c}} = \begin{cases} \prod_{i=1}^n C(c_i, d) & \text{if } n \geq 1, \\ \{(d; \varnothing)\} & \text{if } n = 0. \end{cases}$$
If (C, Δ^C) is a configured category, then there are configured functors
$$\left(C, \Delta^C_{min}\right) \xrightarrow{\ i_0\ } \left(C, \Delta^C\right) \xrightarrow{\ i_1\ } \left(C, \Delta^C_{max}\right)$$
in which i_0 and i_1 are both the identity functors on C. ◇

Example 10.2.10 (Configured categories of bounded lattices). Suppose $(L, \leq)$ is a bounded lattice with least element 0, considered as a small category as in Example 2.2.12. For $c_1, \dots, c_n, d \in L$, suppose $\underline{c} = (c_1, \dots, c_n)$. Define the set
$$\Delta^L\binom{d}{\underline{c}} = \begin{cases} \prod_{i=1}^n L(c_i, d) & \text{if each } c_i \leq d \text{ and } c_p \wedge c_q = 0 \text{ for all } 1 \leq p \neq q \leq n, \\ \varnothing & \text{otherwise.} \end{cases}$$
In the first case, $\Delta^L\binom{d}{\underline{c}}$ is a one-element set because each morphism set $L(c_i, d)$ with $c_i \leq d$ is a one-element set if $n \geq 1$, while an empty product is also a one-element set if $n = 0$. This forms a configured category
$$\widehat{L} = \left(L, \Delta^L\right).$$
To check the composition axiom in (10.2.2), the key point is that if $a \wedge b = 0$ (i.e., if a and b have the least element 0 as their greatest lower bound), then 0 is the only lower bound of a and b. The other three axioms are immediate from the definition.

Note that if $(d, \{c_i \leq d\})$ is a configuration, then we can add any finite number of copies of $0 \leq d$ to the finite sequence $\{c_i \leq d\}$ to yield another configuration. In particular, for each $d \in L$, $\left(d; \{0 \leq d\}_{i=1}^n\right)$ is a configuration for each $n \geq 0$. ◇

Example 10.2.11 (Configured categories of topological spaces). Suppose X is a topological space. Recall from Example 2.2.13 the bounded lattice $(\mathsf{Open}(X), \subset)$. By Example 10.2.10 this yields a configured category

$$\widehat{\mathsf{Open}(X)} = \left(\mathsf{Open}(X), \triangle^X\right).$$

More explicitly, the category $\mathsf{Open}(X)$ has open subsets of X as objects and subset inclusions as morphisms. For open subsets $U_1, \ldots, U_n, V \subset X$, suppose $\underline{U} = (U_1, \ldots, U_n)$. Then

$$\triangle^X\left(\substack{V\\\underline{U}}\right) = \begin{cases} \prod_{i=1}^{n} \mathsf{Open}(X)(U_i, V) & \text{if the } U_i\text{'s are pairwise disjoint subsets of } V, \\ \varnothing & \text{otherwise.} \end{cases}$$

As in the previous example, in the first case, $\triangle^X\left(\substack{V\\\underline{U}}\right)$ is a one-element set. Moreover, for each open subset $V \subset X$, $\left(V; \{\varnothing_X \subset V\}_{i=1}^n\right)$ is a configuration for each $n \geq 0$, where $\varnothing_X$ is the empty subset of X. ◇

Example 10.2.12 (Configured categories of equivariant topological spaces). Suppose G is a group, and X is a topological space in which G acts on the left by homeomorphisms. Suppose $\mathsf{Open}(X)_G$ is the category in Example 2.2.14. For open subsets $U_1, \ldots, U_n, V \subset X$, suppose $\underline{U} = (U_1, \ldots, U_n)$. Define the set $\triangle_G^X\left(\substack{V\\\underline{U}}\right)$ as consisting of finite sequences

$$(g_1, \ldots, g_n) \in \prod_{i=1}^{n} \mathsf{Open}(X)_G(U_i, V)$$

such that:

- The $g_i U_i$'s are pairwise disjoint subsets of V.
- Each $g_i \in G$ is regarded as the composition

$$U_i \xrightarrow{\ \ g_i\ \ } g_i U_i \xrightarrow{\ \ \text{inclusion}\ \ } V$$

  in $\mathsf{Open}(X)_G$.

This defines a configured category

$$\widehat{\mathsf{Open}(X)}_G = \left(\mathsf{Open}(X)_G, \triangle_G^X\right).$$

If G is the trivial group, then we recover the configured category $\widehat{\mathsf{Open}(X)}$ in Example 10.2.11. ◇

10.3   Prefactorization Algebras as Operad Algebras

In this section, we define prefactorization algebras as algebras over some colored operads associated with configured categories. We record the coherence theorems for prefactorization algebras, with or without the time-slice axiom. We recover the prefactorization algebras of Costello-Gwilliam [Costello and Gwilliam (2017)] when the configured category is $\widehat{\mathsf{Open}(X)}$ for a topological space X. We also recover their equivariant prefactorization algebras when the configured category is $\widehat{\mathsf{Open}(X)}_G$ for a topological space X equipped with an action by a group G.

Motivation 10.3.1. Given a configured category $\widehat{C} = (C, \triangle^C)$, for the moment let us think of its objects as the spacetime regions of interest as in Interpretation 10.2.4. From the prefactorization algebra perspective, a quantum field theory $\mathcal{F}$ on $\widehat{C}$ is an assignment that associates to each spacetime region $c \in C$ an object $\mathcal{F}(c)$, say a chain complex, of quantum observables on c. If $c_1, \ldots, c_n$ are suitably disjoint spacetime regions in d, then we should be able to combine the observables in the form of a map

$$\mathcal{F}(c_1) \otimes \cdots \otimes \mathcal{F}(c_n) \longrightarrow \mathcal{F}(d).$$

These multiplication maps should satisfy some natural conditions with respect to sub-regions. The following colored operad is designed to model this structure.   ◇

Definition 10.3.2. Suppose $\widehat{C} = (C, \triangle)$ is a configured category with object set $\mathfrak{C}$. Define the following sets and functions.

Entries Define the object $O_{\widehat{C}} \in \mathsf{Set}^{\mathsf{Prof}(\mathfrak{C}) \times \mathfrak{C}}$ entrywise as

$$O_{\widehat{C}}\binom{d}{\underline{c}} = \triangle\binom{d}{\underline{c}} \quad \text{for} \quad \binom{d}{\underline{c}} \in \mathsf{Prof}(\mathfrak{C}) \times \mathfrak{C}.$$

Equivariance For $\sigma \in \Sigma_{|\underline{c}|}$, define the map

$$O_{\widehat{C}}\binom{d}{\underline{c}} \xrightarrow{\;\sigma\;} O_{\widehat{C}}\binom{d}{\underline{c}\sigma} \quad \text{by} \quad \{f_i\}\sigma = \{f_{\sigma(i)}\}$$

for $\{f_i\} \in \triangle\binom{d}{\underline{c}}$.

Colored Units For each $c \in \mathfrak{C}$, the c-colored unit in $O_{\widehat{C}}\binom{c}{c}$ is $\{\mathrm{Id}_c\}$.

Operadic Composition For $(\underline{c}; d) \in \mathsf{Prof}(\mathfrak{C}) \times \mathfrak{C}$ with $|\underline{c}| = n \geq 1$, $\underline{b}_i = (b_{i1}, \ldots, b_{ik_i}) \in \mathsf{Prof}(\mathfrak{C})$ for $1 \leq i \leq n$, and $\underline{b} = (\underline{b}_1, \ldots, \underline{b}_n)$, define the map

$$O_{\widehat{C}}\binom{d}{\underline{c}} \times \prod_{i=1}^{n} O_{\widehat{C}}\binom{c_i}{\underline{b}_i} \xrightarrow{\;\gamma\;} O_{\widehat{C}}\binom{d}{\underline{b}}$$

as the composition

$$\gamma\left(\{f_i\}_{i=1}^{n}; \{g_{1j}\}_{j=1}^{k_1}, \ldots, \{g_{nj}\}_{j=1}^{k_n}\right) = \{f_i g_{ij}\}_{1 \leq i \leq n,\, 1 \leq j \leq k_i}$$

in (10.2.2) for $\{f_i\} \in O_{\widehat{C}}\binom{d}{\underline{c}}$ and $\{g_{ij}\}_{j=1}^{k_i} \in O_{\widehat{C}}\binom{c_i}{\underline{b}_i}$ with $1 \leq i \leq n$.

Lemma 10.3.3. *In the setting of Definition 10.3.2:*

(1) $O_{\widehat{C}}$ is a $\mathfrak{C}$-colored operad in Set.

(2) This construction defines a functor

$$O_{(-)} : \mathsf{ConfCat} \longrightarrow \mathsf{Operad}(\mathsf{Set})$$

from the category of configured categories to the category of colored operads in Set.

Proof. One checks directly that $O_{\widehat{C}}$ satisfies the axioms in Definition 4.2.1, so it is a $\mathfrak{C}$-colored operad in Set. The naturality of this construction is also checked by a direct inspection.   □

Recall from Example 5.3.3 the strong symmetric monoidal functor $\mathsf{Set} \longrightarrow \mathsf{M}$, which sends each set S to the coproduct $\coprod_S \mathbb{1}$, and the induced change-of-category functor

$$(-)^{\mathsf{M}} : \mathsf{Operad}^{\mathfrak{C}}(\mathsf{Set}) \longrightarrow \mathsf{Operad}^{\mathfrak{C}}(\mathsf{M})$$

between the categories of $\mathfrak{C}$-colored operads. We will consider the image in M of $\mathsf{O}_{\widehat{\mathsf{C}}}$, denoted by $\mathsf{O}_{\widehat{\mathsf{C}}}^{\mathsf{M}}$. Also recall from Definition 5.4.3 the S-localization $\mathsf{O}[S^{-1}]$ of a $\mathfrak{C}$-colored operad O in Set for a set S of unary elements in O.

Definition 10.3.4. Suppose $\widehat{\mathsf{C}} = (\mathsf{C}, \triangle)$ is a configured category.

(1) Define the category

$$\mathsf{PFA}(\widehat{\mathsf{C}}) = \mathsf{Alg}_{\mathsf{M}}\big(\mathsf{O}_{\widehat{\mathsf{C}}}^{\mathsf{M}}\big),$$

whose objects are called *prefactorization algebras on* $\widehat{\mathsf{C}}$.

(2) Suppose S is a set of morphisms in C, regarded as a subset of $\triangle$ by the inclusivity axiom. Define the category

$$\mathsf{PFA}(\widehat{\mathsf{C}}, S) = \mathsf{Alg}_{\mathsf{M}}\big(\mathsf{O}_{\widehat{\mathsf{C}}}[S^{-1}]^{\mathsf{M}}\big),$$

whose objects are called *prefactorization algebras on* $\widehat{\mathsf{C}}$ *satisfying the time-slice axiom with respect to* S.

Remark 10.3.5. In the previous definition, a morphism $s : c \longrightarrow d \in S$ is regarded as a configuration $\{s\} \in \triangle\binom{d}{c} = \mathsf{O}_{\widehat{\mathsf{C}}}\binom{d}{c}$, hence also a unary element in $\mathsf{O}_{\widehat{\mathsf{C}}}$. So the S-localization $\mathsf{O}_{\widehat{\mathsf{C}}}[S^{-1}]$ exists by Theorem 5.4.6. $\diamond$

The following coherence theorem is a special case of Theorem 5.5.1. It explains precisely what a prefactorization algebra is.

Theorem 10.3.6. *Suppose* $\widehat{\mathsf{C}} = (\mathsf{C}, \triangle)$ *is a configured category with object set* $\mathfrak{C}$. *Then an* $\mathsf{O}_{\widehat{\mathsf{C}}}^{\mathsf{M}}$-*algebra is precisely a pair* (X, λ) *consisting of*

- *a* $\mathfrak{C}$-*colored object* $X = \{X_c\}_{c \in \mathfrak{C}}$ *in* M *and*
- *a structure morphism*

$$\bigotimes_{i=1}^{n} X_{c_i} \xrightarrow{\ \lambda\{f_i\}_{i=1}^{n}\ } X_d \in \mathsf{M} \qquad (10.3.7)$$

for

- *each* $\binom{d}{\underline{c}} = \binom{d}{c_1,\ldots,c_n} \in \mathsf{Prof}(\mathfrak{C}) \times \mathfrak{C}$ *and*
- *each configuration* $\{f_i\}_{i=1}^{n} \in \triangle\binom{d}{\underline{c}}$

that satisfies the following associativity, unity, and equivariance axioms.

Associativity *For* $(\underline{c}; d) \in \mathsf{Prof}(\mathfrak{C}) \times \mathfrak{C}$ *with* $|\underline{c}| = n \geq 1$, $\underline{b}_i = (b_{i1}, \ldots, b_{ik_i}) \in \mathsf{Prof}(\mathfrak{C})$
for $1 \leq i \leq n$, $\underline{b} = (\underline{b}_1, \ldots, \underline{b}_n)$, *configurations* $\{f_i\} \in \Delta\binom{d}{\underline{c}}$, *and* $\{g_{ij}\} \in \Delta\binom{c_i}{\underline{b}_i}$ *for*
$1 \leq i \leq n$, *the associativity diagram*

$$
\begin{array}{ccc}
\overset{n}{\underset{i=1}{\otimes}} \overset{k_i}{\underset{j=1}{\otimes}} X_{b_{ij}} & \xrightarrow{\ \lambda\{f_i g_{ij}\}_{i,j}\ } & X_d \\
{\scriptstyle \overset{n}{\underset{i=1}{\otimes}} \lambda\{g_{ij}\}_{j=1}^{k_i}} \downarrow & & \| \\
\overset{n}{\underset{i=1}{\otimes}} X_{c_i} & \xrightarrow{\ \lambda\{f_i\}\ } & X_d
\end{array}
\tag{10.3.8}
$$

in M *is commutative.*

Unity *For each* $c \in \mathfrak{C}$, $\lambda\{\mathrm{Id}_c\}$ *is equal to* Id_{X_c}.

Equivariance *For each configuration* $\{f_i\} \in \Delta\binom{d}{\underline{c}}$ *with* $|\underline{c}| = n$ *and* $\sigma \in \Sigma_n$, *the equivariance diagram*

$$
\begin{array}{ccc}
\overset{n}{\underset{i=1}{\otimes}} X_{c_i} & \xrightarrow{\ \sigma^{-1}\ } & \overset{n}{\underset{i=1}{\otimes}} X_{c_{\sigma(i)}} \\
& {\scriptstyle \lambda\{f_i\}} \searrow \quad \swarrow {\scriptstyle \lambda\{f_{\sigma(i)}\}} & \\
& X_d &
\end{array}
\tag{10.3.9}
$$

in M *is commutative.*

A *morphism of* $\mathsf{O}^{\mathsf{M}}_{\overline{\mathfrak{C}}}$-*algebras* $\varphi : (X, \lambda^X) \longrightarrow (Y, \lambda^Y)$ *is a morphism* $\varphi : X \longrightarrow Y$ *of* $\mathfrak{C}$-*colored objects in* M *that respects the structure morphisms in* (10.3.7) *in the sense that the diagram*

$$
\begin{array}{ccc}
\overset{n}{\underset{i=1}{\otimes}} X_{c_i} & \xrightarrow{\ \overset{n}{\underset{i=1}{\otimes}} \varphi_{c_i}\ } & \overset{n}{\underset{i=1}{\otimes}} Y_{c_i} \\
{\scriptstyle \lambda^X\{f_i\}} \downarrow & & \downarrow {\scriptstyle \lambda^Y\{f_i\}} \\
X_d & \xrightarrow{\ \varphi_d\ } & Y_d
\end{array}
\tag{10.3.10}
$$

in M *is commutative for each configuration* $\{f_i\} \in \Delta\binom{d}{\underline{c}}$ *with* $|\underline{c}| = n$.

Example 10.3.11 (Costello-Gwilliam prefactorization algebras). Consider the configured category $\widetilde{\mathsf{Open}(X)}$ in Example 10.2.11 for a topological space X. A prefactorization algebra on $\widetilde{\mathsf{Open}(X)}$, i.e., an $\mathsf{O}^{\mathsf{M}}_{\overline{\mathsf{Open}(X)}}$-algebra in Theorem 10.3.6, is precisely a *unital prefactorization algebra on* X in the sense of [Costello and Gwilliam (2017)] 3.1.1.1 and 3.1.2.3. $\diamond$

Example 10.3.12 (Costello-Gwilliam equivariant prefactorization algebras). Consider the configured category $\widetilde{\mathsf{Open}(X)}_G$ in Example 10.2.12 for a topological space X equipped with a left action by a group G. A prefactorization algebra on $\widetilde{\mathsf{Open}(X)}_G$, i.e., an $\mathsf{O}^{\mathsf{M}}_{\overline{\mathsf{Open}(X)}_G}$-algebra in Theorem 10.3.6, is precisely a *unital G-equivariant prefactorization algebra on* X in the sense of [Costello and Gwilliam (2017)] 3.7.1.1. $\diamond$

The following coherence theorem is a special case of Theorems 5.5.5 and 5.5.6. It explains precisely what a prefactorization algebra satisfying the time-slice axiom is.

Theorem 10.3.13. *Suppose $\widehat{\mathsf{C}} = (\mathsf{C}, \triangle)$ is a configured category, and S is a set of morphisms in C.*

(1) The S-localization morphism $\ell : \mathsf{O}_{\widehat{\mathsf{C}}} \longrightarrow \mathsf{O}_{\widehat{\mathsf{C}}}[S^{-1}]$ induces the change-of-operad adjunction

$$\mathsf{PFA}(\widehat{\mathsf{C}}) = \mathsf{Alg}_\mathsf{M}\left(\mathsf{O}_{\widehat{\mathsf{C}}}^\mathsf{M}\right) \underset{(\ell^\mathsf{M})^*}{\overset{\ell_!^\mathsf{M}}{\rightleftarrows}} \mathsf{Alg}_\mathsf{M}\left(\mathsf{O}_{\widehat{\mathsf{C}}}[S^{-1}]^\mathsf{M}\right) = \mathsf{PFA}(\widehat{\mathsf{C}}, S)$$

whose right adjoint $(\ell^\mathsf{M})^*$ is full and faithful and whose counit

$$\epsilon : \ell_!^\mathsf{M}(\ell^\mathsf{M})^* \overset{\cong}{\longrightarrow} \mathrm{Id}_{\mathsf{Alg}_\mathsf{M}\left(\mathsf{O}_{\widehat{\mathsf{C}}}[S^{-1}]^\mathsf{M}\right)}$$

is a natural isomorphism.

(2) Via the right adjoint $(\ell^\mathsf{M})^*$, $\mathsf{O}_{\widehat{\mathsf{C}}}[S^{-1}]^\mathsf{M}$-algebras are equivalent to $\mathsf{O}_{\widehat{\mathsf{C}}}^\mathsf{M}$-algebras whose structure morphisms $\lambda\{s\}$ are isomorphisms for all $s \in S$.

Interpretation 10.3.14. A prefactorization algebra on a configured category $\widehat{\mathsf{C}}$ satisfies the time-slice axiom with respect to S precisely when the structure morphisms $\lambda\{s\}$ are invertible for all $s \in S$. This is the exact analogue of the time-slice axiom for algebraic quantum field theories in Definition 8.2.3, which may also be implemented by replacing the orthogonal category with its S-localization as in Lemma 8.2.7. For prefactorization algebras, the time-slice axiom may be implemented by replacing the colored operad $\mathsf{O}_{\widehat{\mathsf{C}}}$ with its S-localization $\mathsf{O}_{\widehat{\mathsf{C}}}[S^{-1}]$. ◇

The following result compares prefactorization algebras, with or without the time-slice axiom, on different configured categories.

Corollary 10.3.15. *Suppose $F : \widehat{\mathsf{C}} = (\mathsf{C}, \triangle^\mathsf{C}) \longrightarrow (\mathsf{D}, \triangle^\mathsf{D}) = \widehat{\mathsf{D}}$ is a configured functor, and S is a set of morphisms in D. Define*

$$S_0 = F^{-1}(S) = \left\{g \in \mathrm{Mor}(\mathsf{C}) : Fg \in S\right\}$$

to be the F-pre-image of S. Then there is an induced diagram of change-of-operad adjunctions

$$\mathsf{PFA}(\widehat{\mathsf{C}}) = \mathsf{Alg}_\mathsf{M}\left(\mathsf{O}_{\widehat{\mathsf{C}}}^\mathsf{M}\right) \underset{(\mathsf{O}_F^\mathsf{M})^*}{\overset{(\mathsf{O}_F^\mathsf{M})_!}{\rightleftarrows}} \mathsf{Alg}_\mathsf{M}\left(\mathsf{O}_{\widehat{\mathsf{D}}}^\mathsf{M}\right) = \mathsf{PFA}(\widehat{\mathsf{D}})$$

$$\mathsf{PFA}(\widehat{\mathsf{C}}, S_0) = \mathsf{Alg}_\mathsf{M}\left(\mathsf{O}_{\widehat{\mathsf{C}}}[S_0^{-1}]^\mathsf{M}\right) \underset{(\mathsf{O}_{F'}^\mathsf{M})^*}{\overset{(\mathsf{O}_{F'}^\mathsf{M})_!}{\rightleftarrows}} \mathsf{Alg}_\mathsf{M}\left(\mathsf{O}_{\widehat{\mathsf{D}}}[S^{-1}]^\mathsf{M}\right) = \mathsf{PFA}(\widehat{\mathsf{D}}, S)$$

in which

$$(\mathsf{O}_{F'}^\mathsf{M})_! \ell_!^\mathsf{M} = \ell_!^\mathsf{M}(\mathsf{O}_F^\mathsf{M})_! \quad and \quad (\ell^\mathsf{M})^*(\mathsf{O}_{F'}^\mathsf{M})^* = (\mathsf{O}_F^\mathsf{M})^*(\ell^\mathsf{M})^*.$$

Proof. Consider the solid-arrow diagram

$$
\begin{array}{ccc}
\mathsf{O}_{\widehat{\mathfrak{C}}} & \xrightarrow{\;\;\mathsf{O}_F\;\;} & \mathsf{O}_{\widehat{\mathfrak{D}}} \\
\ell \downarrow & & \downarrow \ell \\
\mathsf{O}_{\widehat{\mathfrak{C}}}[S_0^{-1}] & \xdashrightarrow{\;\;\mathsf{O}_{F'}\;\;} & \mathsf{O}_{\widehat{\mathfrak{D}}}[S^{-1}]
\end{array}
$$

of colored operads in Set, where

$$\ell : \mathsf{O}_{\widehat{\mathfrak{C}}} \longrightarrow \mathsf{O}_{\widehat{\mathfrak{C}}}[S_0^{-1}] \quad \text{and} \quad \ell : \mathsf{O}_{\widehat{\mathfrak{D}}} \longrightarrow \mathsf{O}_{\widehat{\mathfrak{D}}}[S^{-1}]$$

are the S_0-localization of $\mathsf{O}_{\widehat{\mathfrak{C}}}$ and the S-localization of $\mathsf{O}_{\widehat{\mathfrak{D}}}$, respectively. Since every unary element in

$$\ell \mathsf{O}_F(S_0) \subset \ell(S)$$

is invertible in $\mathsf{O}_{\widehat{\mathfrak{D}}}[S^{-1}]$, by the universal property of S_0-localization, there is a unique operad morphism $\mathsf{O}_{F'}$ that makes the entire diagram commutative. This diagram becomes a commutative diagram of colored operads in M once we apply the change-of-category functor $(-)^\mathsf{M}$. The desired diagram of change-of-operad adjunctions is obtained by applying Theorem 5.1.8.   □

Example 10.3.16 (Costello-Gwilliam locally constant prefactorization algebras). In the configured category $\mathsf{Open}(\mathbb{R})$, suppose S is the set of inclusions of open intervals. By Theorem 10.3.13 a prefactorization algebra on $\widehat{\mathsf{Open}(\mathbb{R})}$ satisfying the time-slice axiom with respect to S is equivalent to a prefactorization algebra on $\widehat{\mathsf{Open}(\mathbb{R})}$ whose structure morphisms $\lambda\{s\}$ are isomorphisms for all $s \in S$. These are precisely the *locally constant* unital prefactorization algebras on $\mathbb{R}$ in [Costello and Gwilliam (2017)] 3.2.0.1.   ◇

10.4   Pointed Diagram Structure

In the following few sections, we will provide more examples of prefactorization algebras. Along the way, we provide evidence that prefactorization algebras are closely related to algebraic quantum field theories, a relationship that will be made precise in Chapter 12. In this section, we observe that every prefactorization algebra has an underlying pointed diagram, which itself can be realized as a prefactorization algebra on the minimal configured category. There is a free-forgetful adjunction between the category of prefactorization algebras on the minimal configured category and the category of algebraic quantum field theories on the minimal orthogonal category.

First we need the following definition.

Definition 10.4.1. Suppose C is a small category.

(1) A C-diagram $\mathcal{F} : \mathsf{C} \longrightarrow \mathsf{M}$ is *pointed* if it is equipped with a c-colored unit

$$1_c : \mathbb{1} \longrightarrow \mathcal{F}(c) \in \mathsf{M}$$

for each object $c \in \mathsf{C}$ such that the diagram

$$
\begin{array}{ccc}
\mathbb{1} & =\!=\!=\!= & \mathbb{1} \\
{\scriptstyle 1_c}\big\downarrow & & \big\downarrow{\scriptstyle 1_d} \\
\mathcal{F}(c) & \xrightarrow{\ \mathcal{F}(f)\ } & \mathcal{F}(d)
\end{array}
$$

is commutative for each morphism $f : c \longrightarrow d \in \mathsf{C}$.

(2) A natural transformation between two pointed C-diagrams is *pointed* if it preserves the colored units.

(3) The category of pointed C-diagrams in M and pointed natural transformations is denoted by $\mathsf{M}^{\mathsf{C}}_*$.

Example 10.4.2. By forgetting the multiplicative structure, every C-diagram of monoids in M has an underlying pointed C-diagram, where the colored units are the units of the monoids. ◇

Proposition 10.4.3. *Suppose* C *is a small category. Then there is a canonical isomorphism*

$$
\mathsf{Alg}_{\mathsf{M}}\big(\mathsf{O}^{\mathsf{M}}_{\widehat{\mathsf{C}_{\min}}}\big) = \mathsf{PFA}(\widehat{\mathsf{C}_{\min}}) \cong \mathsf{M}^{\mathsf{C}}_*
$$

between the category of prefactorization algebras on the minimal configured category $\widehat{\mathsf{C}_{\min}} = (\mathsf{C}, \triangle^{\mathsf{C}}_{\min})$ *in Example 10.2.8 and the category of pointed C-diagrams in* M.

Proof. Both a pointed C-diagram in M and a prefactorization algebra on $\widehat{\mathsf{C}_{\min}}$ assign to each object $c \in \mathsf{C}$ an object $\mathcal{F}(c) \in \mathsf{M}$. To see that pointed C-diagrams in M are precisely the prefactorization algebras on $\widehat{\mathsf{C}_{\min}}$, we use the Coherence Theorem 10.3.6. There are only two kinds of configurations in $\triangle^{\mathsf{C}}_{\min}$:

- $(c; \varnothing)$ for all objects $c \in \mathsf{C}$ and
- $(d; \{f\})$ for all morphisms $f \in \mathsf{C}(c, d)$ with $c, d \in \mathsf{C}$.

If $\mathcal{F}$ is a prefactorization algebra on $\widehat{\mathsf{C}_{\min}}$, then its only structure morphisms (10.3.7) are

$$
\lambda\big\{(c; \varnothing)\big\} = 1_c : \mathbb{1} \longrightarrow \mathcal{F}(c) \in \mathsf{M}
$$

for objects $c \in \mathsf{C}$ and

$$
\lambda\big\{(d; \{f\})\big\} = \mathcal{F}(f) : \mathcal{F}(c) \longrightarrow \mathcal{F}(d) \in \mathsf{M}
$$

for morphisms $f \in \mathsf{C}(c, d)$.

The equivariance condition (10.3.9) is trivial, and the unity condition says that

$$
\mathcal{F}(\mathrm{Id}_c) = \mathrm{Id}_{\mathcal{F}(c)} \quad \text{for} \quad c \in \mathsf{C}.
$$

The associativity condition (10.3.8) must have $n = 1$. If $k_1 = 0$, then the associativity condition is the diagram in Definition 10.4.1 that defines pointed C-diagrams. If

$k_1 = 1$, then the associativity condition is the commutative diagram

$$
\begin{array}{ccc}
\mathcal{F}(b) & \xrightarrow{\ \mathcal{F}(fg)\ } & \mathcal{F}(d) \\
{\scriptstyle \mathcal{F}(g)}\big\downarrow & & \big\| \\
\mathcal{F}(c) & \xrightarrow{\ \mathcal{F}(f)\ } & \mathcal{F}(d)
\end{array}
$$

for all objects $b, c, d \in \mathsf{C}$ and composable morphisms $(f, g) \in \mathsf{C}(c, d) \times \mathsf{C}(b, c)$. There-fore, a prefactorization algebra on $\widehat{\mathsf{C}}_{\mathsf{min}}$ is precisely a pointed C-diagram in M.

Similarly, to see the correspondence between morphisms, we use (10.3.10) in the Coherence Theorem 10.3.6. If $n = 0$ then (10.3.10) is the preservation of colored units. If $n = 1$ then (10.3.10) is the commutative square that defines a natural transformation between two C-diagrams in M. ☐

Recall from Definition 10.3.2 the colored operad $\mathsf{O}_{\widehat{\mathsf{C}}}$ for a configured category $\widehat{\mathsf{C}}$. The following observation is a consequence of Corollary 10.3.15 and Proposition 10.4.3.

Corollary 10.4.4. *Suppose* $\widehat{\mathsf{C}} = (\mathsf{C}, \triangle)$ *is a configured category. Then the configured functor*

$$
\widehat{\mathsf{C}}_{\mathsf{min}} = (\mathsf{C}, \triangle^{\mathsf{C}}_{\mathsf{min}}) \xrightarrow{\ i_0\ } (\mathsf{C}, \triangle) = \widehat{\mathsf{C}} \,,
$$

whose underlying functor is the identity functor on C, *induces a change-of-operad adjunction*

$$
\mathsf{M}^{\mathsf{C}}_* \cong \mathsf{PFA}(\widehat{\mathsf{C}}_{\mathsf{min}}) = \mathsf{Alg}_{\mathsf{M}}\left(\mathsf{O}^{\mathsf{M}}_{\widehat{\mathsf{C}}_{\mathsf{min}}}\right) \underset{(\mathsf{O}^{\mathsf{M}}_{i_0})^*}{\overset{(\mathsf{O}^{\mathsf{M}}_{i_0})_!}{\rightleftarrows}} \mathsf{Alg}_{\mathsf{M}}\left(\mathsf{O}^{\mathsf{M}}_{\widehat{\mathsf{C}}}\right) = \mathsf{PFA}(\widehat{\mathsf{C}})
$$

between the category of pointed C-*diagrams in* M *and the category of prefactorization algebras on* $\widehat{\mathsf{C}}$.

Interpretation 10.4.5. Each prefactorization algebra (X, λ) on a configured cat-egory $\widehat{\mathsf{C}} = (\mathsf{C}, \triangle)$ has an underlying pointed C-diagram in M. For a morphism $f : c \longrightarrow d$ in C, the corresponding morphism is the structure morphism

$$
X_c \xrightarrow{\ \lambda\{f\}\ } X_d \in \mathsf{M}
$$

with $\{f\} \in \triangle\binom{d}{c}$. For each object $c \in \mathsf{C}$, the c-colored unit is the structure morphism

$$
\mathbb{1} \xrightarrow{\ \lambda\{(c;\varnothing)\}\ } X_c \in \mathsf{M}
$$

with $(c; \varnothing) \in \triangle\binom{c}{\varnothing}$. ◇

Recall from Definition 8.3.2 the colored operad $\mathsf{O}_{\overline{\mathsf{C}}}$ for an orthogonal category $\overline{\mathsf{C}}$.

Example 10.4.6. In Example 8.4.1 we noted that the category $\mathsf{QFT}(\overline{\mathsf{C}_{\min}})$ of algebraic quantum field theories on $\overline{\mathsf{C}_{\min}} = (\mathsf{C}, \varnothing)$, where $\varnothing$ is the empty orthogonality relation, is the category $\mathsf{Mon}(\mathsf{M})^{\mathsf{C}}$ of C-diagrams in $\mathsf{Mon}(\mathsf{M})$. There is a forgetful functor

$$\mathsf{Alg}_{\mathsf{M}}\!\left(\mathsf{O}^{\mathsf{M}}_{\overline{\mathsf{C}_{\min}}}\right) \cong \mathsf{QFT}(\overline{\mathsf{C}_{\min}}) = \mathsf{Mon}(\mathsf{M})^{\mathsf{C}} \longrightarrow \mathsf{M}^{\mathsf{C}}_{*} \cong \mathsf{PFA}(\overline{\mathsf{C}_{\min}}) = \mathsf{Alg}_{\mathsf{M}}\!\left(\mathsf{O}^{\mathsf{M}}_{\overline{\mathsf{C}_{\min}}}\right)$$

from the category of C-diagrams of monoids in M to the category of pointed C-diagrams in M that forgets about the multiplicative structure. This relationship between algebraic quantum field theories and prefactorization algebras is conceptual rather than random, as we now explain. ◇

Proposition 10.4.7. *Suppose* C *is a small category with object set* $\mathfrak{C}$. *Then there is a morphism*

$$\mathsf{O}_{\overline{\mathsf{C}_{\min}}} \xrightarrow{\ \delta_{\min}\ } \mathsf{O}_{\overline{\mathsf{C}_{\min}}} \ \in \mathsf{Operad}^{\mathfrak{C}}(\mathsf{Set})$$

that is entrywise defined as follows.

- $\delta_{\min}$ *is a canonical bijection on each 0-ary entry* $\binom{c}{\varnothing}$ *for* $c \in \mathfrak{C}$ *and each unary entry* $\binom{d}{c}$ *for* $c, d \in \mathfrak{C}$.
- $\delta_{\min}$ *is the unique morphism from the empty set to* $\mathsf{O}_{\overline{\mathsf{C}_{\min}}}\binom{d}{\underline{c}}$ *if* $|\underline{c}| \geq 2$.

Proof. Recall that $\triangle^{\mathsf{C}}_{\min}$ only has 0-ary configurations $(c; \varnothing) \in \triangle^{\mathsf{C}}_{\min}\binom{c}{\varnothing}$ for $c \in \mathfrak{C}$ and unary configurations $(d; \{f\}) \in \triangle^{\mathsf{C}}_{\min}\binom{d}{c}$ for $f \in \mathsf{C}(c,d)$. So there are canonical bijections on the 0-ary entries

$$\mathsf{O}_{\overline{\mathsf{C}_{\min}}}\binom{c}{\varnothing} = \triangle^{\mathsf{C}}_{\min}\binom{c}{\varnothing} = \{(c;\varnothing)\} \xrightarrow[\cong]{\ \delta_{\min}\ } \Sigma_0 \times * = \mathsf{O}_{\overline{\mathsf{C}_{\min}}}\binom{c}{\varnothing}$$

for $c \in \mathfrak{C}$ and on the unary entries

$$\mathsf{O}_{\overline{\mathsf{C}_{\min}}}\binom{d}{c} = \triangle^{\mathsf{C}}_{\min}\binom{d}{c} = \mathsf{C}(c,d) \xrightarrow[\cong]{\ \delta_{\min}\ } \Sigma_1 \times \mathsf{C}(c,d) = \mathsf{O}_{\overline{\mathsf{C}_{\min}}}\binom{d}{c}$$

for $c, d \in \mathfrak{C}$. For $\underline{c} = (c_1, \ldots, c_n)$ with $n \geq 2$, the morphism $\delta_{\min}$ is defined as the unique morphism

$$\mathsf{O}_{\overline{\mathsf{C}_{\min}}}\binom{d}{\underline{c}} = \triangle^{\mathsf{C}}_{\min}\binom{d}{\underline{c}} = \varnothing \xrightarrow{\ \delta_{\min}\ } \Sigma_n \times \prod_{j=1}^{n} \mathsf{C}(c_j, d) = \mathsf{O}_{\overline{\mathsf{C}_{\min}}}\binom{d}{\underline{c}}\,.$$

A direct inspection shows that $\delta_{\min}$ is a well-defined morphism of C-colored operads in Set. Indeed, there is no equivariance relation to check because $\mathsf{O}_{\overline{\mathsf{C}_{\min}}}$ is concentrated in 0-ary and unary entries. The preservation by $\delta_{\min}$ of colored units and operadic composition follows from the fact that on both sides these structures are given by identity morphisms and composition in C. □

Recall the change-of-category functor

$$(-)^{\mathsf{M}} : \mathsf{Operad}^{\mathfrak{C}}(\mathsf{Set}) \longrightarrow \mathsf{Operad}^{\mathfrak{C}}(\mathsf{M})$$

in Example 5.3.3. The following result is a consequence of Theorem 5.1.8 and Proposition 10.4.7.

Corollary 10.4.8. *Suppose* C *is a small category with object set* $\mathfrak{C}$. *Then the morphism*

$$\mathsf{O}^{\mathsf{M}}_{\overline{\mathsf{C}_{\min}}} \xrightarrow{\;\delta^{\mathsf{M}}_{\min}\;} \mathsf{O}^{\mathsf{M}}_{\overline{\mathsf{C}_{\min}}}$$

of $\mathfrak{C}$-*colored operads induces a change-of-operad adjunction*

$$\mathsf{M}^{\mathsf{C}}_* \cong \mathsf{PFA}(\widehat{\mathsf{C}_{\min}}) = \mathsf{Alg}_{\mathsf{M}}\!\left(\mathsf{O}^{\mathsf{M}}_{\overline{\mathsf{C}_{\min}}}\right) \underset{(\delta^{\mathsf{M}}_{\min})^*}{\overset{(\delta^{\mathsf{M}}_{\min})_!}{\rightleftarrows}} \mathsf{Alg}_{\mathsf{M}}\!\left(\mathsf{O}^{\mathsf{M}}_{\overline{\mathsf{C}_{\min}}}\right) \cong \mathsf{QFT}(\overline{\mathsf{C}_{\min}}) = \mathsf{Mon}(\mathsf{M})^{\mathsf{C}}$$

whose right adjoint $(\delta^{\mathsf{M}}_{\min})^*$ *is the forgetful functor in Example 10.4.6.*

Interpretation 10.4.9. In the minimal case, there is a morphism $\delta^{\mathsf{M}}_{\min}$ from the $\mathfrak{C}$-colored operad $\mathsf{O}^{\mathsf{M}}_{\overline{\mathsf{C}_{\min}}}$ defining prefactorization algebras (= pointed C-diagrams in M) to the $\mathfrak{C}$-colored operad $\mathsf{O}^{\mathsf{M}}_{\overline{\mathsf{C}_{\min}}}$ for algebraic quantum field theories (= C-diagrams of monoids in M). This morphism of $\mathfrak{C}$-colored operads induces a free-forgetful adjunction between the algebra categories.   ◇

10.5   Commutative Monoid Structure

In this section, we observe that some entries of a prefactorization algebra are equipped with the structure of a commutative monoid. In particular, this applies to the empty subset for prefactorization algebras on a topological space. Recall from Example 4.5.19 the commutative operad Com, which is a 1-colored operad whose algebras are commutative monoids in M. We will denote its unique color by $*$. For the following result, the example to keep in mind is the empty subset $\varnothing_X \subset X$ in a topological space X.

Proposition 10.5.1. *Suppose* $\widehat{\mathsf{C}} = (\mathsf{C}, \triangle)$ *is a configured category with object set* $\mathfrak{C}$, *and* $c \in \mathfrak{C}$ *such that*

$$\{\mathrm{Id}_c\}_{i=1}^n \in \triangle\!\begin{pmatrix} c \\ c,\ldots,c \end{pmatrix}$$

for all n.

(1) Then there is an operad morphism

$$\mathsf{Com} \xrightarrow{\;\iota_c\;} \mathsf{O}^{\mathsf{M}}_{\widehat{\mathsf{C}}}$$

that sends $*$ *to* $c \in \mathfrak{C}$ *and that is entrywise defined by the summand inclusion*

$$\mathsf{Com}(n) = \mathbb{1} \xrightarrow[\text{summand}]{\{\mathrm{Id}_c\}_{i=1}^n} \coprod_{\triangle\left(\begin{smallmatrix} c \\ c,\ldots,c \end{smallmatrix}\right)} \mathbb{1} = \mathsf{O}^{\mathsf{M}}_{\widehat{\mathsf{C}}}\!\begin{pmatrix} c \\ c,\ldots,c \end{pmatrix} \quad \text{for} \quad n \geq 0.$$

(2) There is an induced change-of-operad adjunction

$$\mathsf{Com}(\mathsf{M}) = \mathsf{Alg}_{\mathsf{M}}(\mathsf{Com}) \underset{\iota_c^*}{\overset{(\iota_c)_!}{\rightleftarrows}} \mathsf{Alg}_{\mathsf{M}}\!\left(\mathsf{O}^{\mathsf{M}}_{\widehat{\mathsf{C}}}\right) = \mathsf{PFA}(\widehat{\mathsf{C}}) .$$

Proof. For the first assertion, one checks directly that this is a well-defined operad morphism. The second assertion follows from the first assertion and Theorem 5.1.8. □

Interpretation 10.5.2. With $c \in \mathfrak{C}$ as in Proposition 10.5.1, if (Y, λ) is a prefactorization algebra on $\widehat{\mathsf{C}}$, then Y_c is equipped with the structure of a commutative monoid. More explicitly, in the context of the Coherence Theorem 10.3.6, the monoid multiplication in Y_c is the structure morphism

$$Y_c \otimes Y_c \xrightarrow{\lambda\{\mathrm{Id}_c, \mathrm{Id}_c\}} Y_c \in \mathsf{M}$$

with $\{\mathrm{Id}_c, \mathrm{Id}_c\} \in \Delta\binom{c}{c,c}$, and its unit is the structure morphism

$$\mathbb{1} \xrightarrow{\lambda\{(c;\varnothing)\}} Y_c \in \mathsf{M}$$

with $(c; \varnothing) \in \Delta\binom{c}{\varnothing}$. ◇

Example 10.5.3 (Commutative monoid structure in prefactorization algebras on bounded lattices). Consider the configured category $\widehat{L} = (L, \Delta^L)$ for a bounded lattice $(L, \leq)$ in Example 10.2.10. The least element $0 \in L$ has the property that

$$\{\mathrm{Id}_0\}_{i=1}^n \in \Delta^L\binom{0}{0,\ldots,0} \quad \text{for} \quad n \geq 0$$

because $0 \leq d$ for all $d \in L$ and $0 \wedge 0 = 0$. Therefore, by Proposition 10.5.1, if Y is a prefactorization algebra on $\widehat{L}$, then Y_0 is equipped with a commutative monoid structure whose multiplication is the structure morphism

$$Y_0 \otimes Y_0 \xrightarrow{\lambda\{\mathrm{Id}_0, \mathrm{Id}_0\}} Y_0 \in \mathsf{M}$$

with $\{\mathrm{Id}_0, \mathrm{Id}_0\} \in \Delta^L\binom{0}{0,0}$ and whose unit is the structure morphism

$$\mathbb{1} \xrightarrow{\lambda\{(0;\varnothing)\}} Y_0 \in \mathsf{M}$$

with $(0; \varnothing) \in \Delta^L\binom{0}{\varnothing}$. ◇

Example 10.5.4 (Commutative monoid structure in Costello-Gwilliam prefactorization algebras). Consider the empty subset $\varnothing_X \subset X$ in the configured category $\widehat{\mathsf{Open}(X)}$ in Example 10.2.11 for a topological space X. This is a special case of Example 10.5.3 with $L = \mathsf{Open}(X)$ and least element $0 = \varnothing_X$. Therefore, by Proposition 10.5.1, if Y is a prefactorization algebra on $\widehat{\mathsf{Open}(X)}$ (i.e., a Costello-Gwilliam unital prefactorization algebra on X), then $Y_{\varnothing_X}$ is equipped with a commutative monoid structure. ◇

Example 10.5.5 (Commutative monoid structure in Costello-Gwilliam equivariant prefactorization algebras). Consider the configured category $\widehat{\mathsf{Open}(X)}_G$ for a topological space X with a left action by a group G in Example 10.2.12. The empty subset $\varnothing_X \subset X$ has the property that

$$\{\mathrm{Id}_{\varnothing_X}\}_{i=1}^n \in \Delta^X_G\binom{\varnothing_X}{\varnothing_X,\ldots,\varnothing_X} \quad \text{for} \quad n \geq 0.$$

Therefore, by Proposition 10.5.1, if Y is a prefactorization algebra on $\widehat{\mathsf{Open}(X)}_G$ (i.e., a Costello-Gwilliam unital G-equivariant prefactorization algebra on X), then $Y_{\varnothing_X}$ is equipped with a commutative monoid structure. ◇

10.6   Diagrams of Modules over a Commutative Monoid

In Example 10.5.3 we observed that, for each prefactorization algebra Y on the configured category of a bounded lattice, the entry Y_0 is equipped with the structure of a commutative monoid. In this section, we first observe that every other entry of Y is equipped with the structure of a left Y_0-module in the sense of Definition 2.6.6. Then we show that these left Y_0-modules are compatible with the diagram structure in Corollary 10.4.4. As in Example 2.2.12, we will regard a lattice $(L, \leq)$ also as a category, where a morphism $c \longrightarrow d$ exists if and only if $c \leq d$.

Proposition 10.6.1. *Suppose $\widehat{L} = (L, \Delta^L)$ is the configured category of a bounded lattice $(L, \leq)$ with least element $0 \in L$ as in Example 10.2.10, and (Y, λ) is a prefactorization algebra on $\widehat{L}$. Then for each element $d \in L$, the entry Y_d is equipped with the structure of a left Y_0-module via the structure morphism*

$$Y_0 \otimes Y_d \xrightarrow{\ \lambda\{0_d, \mathrm{Id}_d\}\ } Y_d \in \mathsf{M}$$

with

- $0_d : 0 \longrightarrow d \in L$ *the unique morphism and*
- $\{0_d, \mathrm{Id}_d\} \in \Delta^L \binom{d}{0,d}$.

Proof. This is a consequence of the Coherence Theorem 10.3.6. To see that the required associativity diagram

$$
\begin{array}{ccc}
Y_0 \otimes Y_0 \otimes Y_d & \xrightarrow{\left(\mathrm{Id}, \lambda\{0_d, \mathrm{Id}_d\}\right)} & Y_0 \otimes Y_d \\
{\scriptstyle\left(\lambda\{\mathrm{Id}_0, \mathrm{Id}_0\}, \mathrm{Id}\right)}\Big\downarrow & & \Big\downarrow{\scriptstyle\lambda\{0_d, \mathrm{Id}_d\}} \\
Y_0 \otimes Y_d & \xrightarrow{\ \lambda\{0_d, \mathrm{Id}_d\}\ } & Y_d
\end{array}
$$

of a left Y_0-module is commutative, we apply the associativity condition (10.3.8) to the equalities

$$\gamma\Big(\{0_d, \mathrm{Id}_d\}; \{\mathrm{Id}_0, \mathrm{Id}_0\}, \{\mathrm{Id}_d\}\Big)$$

$$= \{0_d, 0_d, \mathrm{Id}_d\}$$

$$= \gamma\Big(\{0_d, \mathrm{Id}_d\}; \{\mathrm{Id}_0\}, \{0_d, \mathrm{Id}_d\}\Big) \in \Delta^L\binom{d}{0,0,d}.$$

Therefore, both composites in the previous diagram are equal to the structure morphism

$$Y_0 \otimes Y_0 \otimes Y_d \xrightarrow{\ \lambda\{0_d, 0_d, \mathrm{Id}_d\}\ } Y_d \in \mathsf{M}.$$

Similarly, to see that the required unity diagram

$$
\begin{array}{ccc}
\mathbb{1} \otimes Y_d & \xrightarrow{\left(\lambda\{(0;\varnothing)\}, \mathrm{Id}\right)} & Y_0 \otimes Y_d \\
{\scriptstyle\cong}\Big\downarrow & & \Big\downarrow{\scriptstyle\lambda\{0_d, \mathrm{Id}_d\}} \\
Y_d & =\!=\!=\!=\!=\!= & Y_d
\end{array}
$$

of a left Y_0-module is commutative, first note that

$$\mathrm{Id}_{Y_d} = \lambda\{\mathrm{Id}_d\}$$

by the unity condition in Theorem 10.3.6. Therefore, the associativity condition (10.3.8) applied to the equality

$$\gamma\Big(\{0_d, \mathrm{Id}_d\}; \{(0; \varnothing)\}, \{\mathrm{Id}_d\}\Big) = \{\mathrm{Id}_d\} \in \Delta^L\binom{d}{d}$$

yields the desired unity diagram. □

Motivation 10.6.2. In Proposition 10.6.1 we observed that, for a bounded lattice $(L, \leq)$ with least element 0 and for a prefactorization algebra (Y, λ) on the configured category $\widehat{L}$, the entry Y_0 is equipped with the structure of a commutative monoid, and every other entry Y_d is equipped with the structure of a left Y_0-module. These left Y_0-module structures should be compatible with the L-diagram structure. In the next result, we will consider the underlying L-diagram structure instead of pointed L-diagram. ◇

Corollary 10.6.3. *Suppose* $\widehat{L} = (L, \Delta^L)$ *is the configured category of a bounded lattice* $(L, \leq)$ *with least element* $0 \in L$ *as in Example 10.2.10, and* (Y, λ) *is a prefactorization algebra on* $\widehat{L}$. *Then the underlying L-diagram in* M *of* (Y, λ) *in Corollary 10.4.4 becomes an L-diagram of left Y_0-modules when equipped with the structure morphisms in Proposition 10.6.1.*

Proof. Suppose $g : c \longrightarrow d$ in L, i.e., $c \leq d$. We must show that the diagram

$$
\begin{array}{ccc}
Y_0 \otimes Y_c & \xrightarrow{\lambda\{0_c, \mathrm{Id}_c\}} & Y_c \\
{\scriptstyle (\mathrm{Id}, \lambda\{g\})} \downarrow & & \downarrow {\scriptstyle \lambda\{g\}} \\
Y_0 \otimes Y_d & \xrightarrow{\lambda\{0_d, \mathrm{Id}_d\}} & Y_d
\end{array}
$$

in M is commutative, where $0_c : 0 \longrightarrow c \in L$. There are equalities

$$\gamma\Big(\{0_d, \mathrm{Id}_d\}; \{\mathrm{Id}_0\}, \{g\}\Big)$$

$$= \{0_d, g\}$$

$$= \gamma\Big(\{g\}; \{0_c, \mathrm{Id}_c\}\Big) \in \Delta^L\binom{d}{0,c}.$$

Moreover, the unity condition in Theorem 10.3.6 implies that

$$\lambda\{\mathrm{Id}_0\} = \mathrm{Id}_{Y_0}.$$

So the associativity condition (10.3.8) applied to the above equalities implies that both composites in the above diagram are equal to the structure morphism

$$Y_0 \otimes Y_c \xrightarrow{\lambda\{0_d, g\}} Y_d$$

in M. □

Interpretation 10.6.4. For each prefactorization algebra Y on the configured category $\widehat{L}$ of a bounded lattice L:

- Y_0 is equipped with the structure of a commutative monoid.
- Every other entry Y_d is equipped with the structure of a left Y_0-module.
- The underlying L-diagram of Y is an L-diagram of left Y_0-modules.   ◇

Example 10.6.5 (Costello-Gwilliam prefactorization algebras). For the configured category $\widehat{\mathsf{Open}(X)}$ of a topological space X and for a prefactorization algebra (Y, λ) on $\widehat{\mathsf{Open}(X)}$, the entry $Y_{\varnothing_X}$ is a commutative monoid. Furthermore, for each open subset $U \subset X$, the entry Y_U is equipped with the structure of a left $Y_{\varnothing_X}$-module. Furthermore, the underlying $\mathsf{Open}(X)$-diagram in M of Y is an $\mathsf{Open}(X)$-diagram of left $Y_{\varnothing_X}$-modules.   ◇

Example 10.6.6 (Costello-Gwilliam equivariant prefactorization algebras). Suppose X is a topological space with a left action by a group G. Consider the configured category $\widehat{\mathsf{Open}(X)}_G$ in Example 10.2.12. There is a configured functor

$$\widehat{\mathsf{Open}(X)} \xrightarrow{\ \iota\ } \widehat{\mathsf{Open}(X)}_G$$

that is the identity assignment on objects (i.e., open subsets of X) and morphisms (i.e., inclusions of open subsets). By Lemma 10.3.3 and Example 5.3.3, it induces a morphism of $\mathfrak{C}$-colored operads

$$\mathsf{O}^{\mathsf{M}}_{\widehat{\mathsf{Open}(X)}} \xrightarrow{\ \mathsf{O}^{\mathsf{M}}_{\iota}\ } \mathsf{O}^{\mathsf{M}}_{\widehat{\mathsf{Open}(X)}_G}$$

where $\mathfrak{C} = \mathsf{Ob}\big(\mathsf{Open}(X)\big)$. By Theorem 5.1.8 there is a change-of-operad adjunction

$$\mathsf{PFA}\big(\widehat{\mathsf{Open}(X)}\big) = \mathsf{Alg}_{\mathsf{M}}\big(\mathsf{O}^{\mathsf{M}}_{\widehat{\mathsf{Open}(X)}}\big) \underset{(\mathsf{O}^{\mathsf{M}}_{\iota})^*}{\overset{(\mathsf{O}^{\mathsf{M}}_{\iota})_!}{\rightleftarrows}} \mathsf{Alg}_{\mathsf{M}}\big(\mathsf{O}^{\mathsf{M}}_{\widehat{\mathsf{Open}(X)}_G}\big) = \mathsf{PFA}\big(\widehat{\mathsf{Open}(X)}_G\big)\,,$$

in which the right adjoint $(\mathsf{O}^{\mathsf{M}}_{\iota})^*$ forgets about the structure isomorphisms

$$\lambda\{g\} : Y_U \xrightarrow{\ \cong\ } Y_{gU}$$

for $g \in G$ and $U \in \mathsf{Open}(X)$. Therefore, by Example 10.6.5 for each prefactorization algebra (Y, λ) on $\widehat{\mathsf{Open}(X)}_G$, $Y_{\varnothing_X}$ is a commutative monoid, and every other entry Y_U for $U \in \mathsf{Open}(X)$ is equipped with the structure of a left $Y_{\varnothing_X}$-module. Furthermore, the underlying $\mathsf{Open}(X)$-diagram in M of Y is an $\mathsf{Open}(X)$-diagram of left $Y_{\varnothing_X}$-modules.   ◇

10.7   Diagrams of Commutative Monoids

In this section, we observe that diagrams of commutative monoids can be realized as prefactorization algebras on the maximal configured category $\overline{\mathsf{C}_{\mathsf{max}}} = (\mathsf{C}, \triangle^{\mathsf{C}}_{\mathsf{max}})$ in Example 10.2.9. Furthermore, these diagrams of commutative monoids coincide with algebraic quantum field theories on the maximal orthogonal category

$\overline{C_{max}} = (C, \perp_{max})$ in Example 8.4.2. Taking C to be the category of (complex) n-manifolds, we recover prefactorization algebras on (complex) n-manifolds in the sense of Costello-Gwilliam.

Proposition 10.7.1. *Suppose C is a small category with object set $\mathfrak{C}$. Then there is a canonical isomorphism*

$$O_{\widehat{C_{max}}} \xrightarrow[\cong]{\delta_{max}} O_{\overline{C_{max}}}$$

of $\mathfrak{C}$-colored operads in Set.

Proof. By definition every pair of morphisms in C with the same codomain are orthogonal in $\overline{C_{max}}$. Therefore, two elements are equal

$$[\sigma, \underline{f}] = [\sigma', \underline{f}'] \in O_{\overline{C_{max}}}\binom{d}{\underline{c}}$$

if and only if

$$\underline{f} = \underline{f}' \in \prod_{i=1}^{|\underline{c}|} C(c_i, d).$$

- On 0-ary entries there is a canonical bijection

$$O_{\widehat{C_{max}}}\binom{c}{\varnothing} = \Delta^C_{max}\binom{c}{\varnothing} = \{(c; \varnothing)\} \xrightarrow{\cong} \Sigma_0 \times * = O_{\overline{C_{max}}}\binom{c}{\varnothing}$$

  for $c \in \mathfrak{C}$.
- For n-ary entries with $n \geq 1$, there is a canonical bijection

$$O_{\widehat{C_{max}}}\binom{d}{\underline{c}} = \Delta^C_{max}\binom{d}{\underline{c}} = \prod_{i=1}^{n} C(c_i, d) \xrightarrow{\cong} O_{\overline{C_{max}}}\binom{d}{\underline{c}}$$

  for $(\underline{c}; d) \in \mathsf{Prof}(\mathfrak{C}) \times \mathfrak{C}$ with $\underline{c} = (c_1, \dots, c_n)$.

The required morphism δ_{max} is defined as these canonical isomorphisms. Moreover, δ_{max} is a well-defined morphism of $\mathfrak{C}$-colored operads because on both sides the structures are defined using the identity morphisms and the categorical composition in C. $\qquad\square$

Example 10.7.2 (Prefactorization algebras are AQFT in the classical case). Applying the change-of-category functor

$$(-)^M : \mathsf{Operad}^{\mathfrak{C}}(\mathsf{Set}) \longrightarrow \mathsf{Operad}^{\mathfrak{C}}(M)$$

to δ_{max}, we obtain a canonical isomorphism

$$O^M_{\widehat{C_{max}}} \xrightarrow[\cong]{\delta^M_{max}} O^M_{\overline{C_{max}}}$$

of $\mathfrak{C}$-colored operads in M. Therefore, the induced functor on algebra categories

$$\mathsf{PFA}(\overline{C_{max}}) = \mathsf{Alg}_M\left(O^M_{\widehat{C_{max}}}\right) \xleftarrow[\cong]{(\delta^M_{max})^*} \mathsf{Alg}_M\left(O^M_{\overline{C_{max}}}\right) \cong \mathsf{QFT}(\overline{C_{max}}) = \mathsf{Com}(M)^{\mathfrak{C}}$$

is also an isomorphism, where the equality $\mathsf{QFT}(\overline{\mathsf{C}_{\mathsf{max}}}) = \mathsf{Com}(\mathsf{M})^{\mathsf{C}}$ is from Example 8.4.2. In other words, in the maximal case (i.e., with $\Delta_{\mathsf{max}}^{\mathsf{C}}$ and $\perp_{\mathsf{max}}$), prefactorization algebras coincide with algebraic quantum field theories, which in turn are precisely C-diagrams of commutative monoids in M. Physically we interpret this isomorphism as saying that the two mathematical approaches to quantum field theory both reduce to the classical case where observables form commutative monoids.

◇

Example 10.7.3 (Prefactorization algebras as symmetric monoidal functors). If the small category C has all small coproducts, then there is another nice description of C-diagrams of commutative monoids in M. Indeed, by Proposition 2.6.5 the category $\mathsf{Com}(\mathsf{M})^{\mathsf{C}}$ is canonically isomorphism to the category $\mathsf{SMFun}(\mathsf{C}, \mathsf{M})$ of symmetric monoidal functors, where C is regarded as a symmetric monoidal category under coproducts. So there are canonical isomorphisms

$$\mathsf{PFA}(\overline{\mathsf{C}_{\mathsf{max}}}) \cong \mathsf{QFT}(\overline{\mathsf{C}_{\mathsf{max}}}) = \mathsf{Com}(\mathsf{M})^{\mathsf{C}} \cong \mathsf{SMFun}(\mathsf{C}, \mathsf{M})$$

from the category of prefactorization algebras on $\overline{\mathsf{C}_{\mathsf{max}}}$ to the category of symmetric monoidal functors $\mathsf{C} \longrightarrow \mathsf{M}$.   ◇

Example 10.7.4 (Costello-Gwilliam prefactorization algebras on manifolds). Suppose Emb^{n} is a small category equivalent to the category of smooth n-manifolds with open embeddings as morphisms. Symmetric monoidal functors $\mathsf{Emb}^{n} \longrightarrow \mathsf{M}$ are called *prefactorization algebras on n-manifolds with values in* M in [Costello and Gwilliam (2017)] Definition 6.3.0.2. By Example 10.7.3 the category of such symmetric monoidal functors is isomorphic to the category of prefactorization algebras on the maximal configured category $\overline{\mathsf{Emb}_{\mathsf{max}}^{n}}$ and the category of algebraic quantum field theories on the maximal orthogonal category $\overline{\mathsf{Emb}_{\mathsf{max}}^{n}}$.   ◇

Example 10.7.5 (Costello-Gwilliam prefactorization algebras on complex manifolds). Suppose Hol^{n} is a small category equivalent to the category of complex n-manifolds with open holomorphic embeddings as morphisms. Symmetric monoidal functors $\mathsf{Hol}^{n} \longrightarrow \mathsf{M}$ are called *prefactorization algebras on complex n-manifolds with values in* M in [Costello and Gwilliam (2017)] Definition 6.3.2.2. By Example 10.7.3 the category of such symmetric monoidal functors is isomorphic to the category of prefactorization algebras on the maximal configured category $\overline{\mathsf{Hol}_{\mathsf{max}}^{n}}$ and the category of algebraic quantum field theories on the maximal orthogonal category $\overline{\mathsf{Hol}_{\mathsf{max}}^{n}}$.   ◇

10.8   Configured and Homotopy Morita Equivalences

In Theorem 8.3.6(6) we observed that the equivalence type of the category $\mathsf{QFT}(\overline{\mathsf{C}})$ of algebraic quantum field theories on $\overline{\mathsf{C}}$ is an invariant of the equivalence type of the orthogonal category $\overline{\mathsf{C}}$. In this section, we prove a prefactorization algebra analogue of this result as well as a homotopical version. First we need the following configured analogue of an orthogonal equivalence.

Definition 10.8.1. A *configured equivalence*

$$F : (\mathsf{C}, \triangle^{\mathsf{C}}) \longrightarrow (\mathsf{D}, \triangle^{\mathsf{D}})$$

between configured categories is an equivalence $F : \mathsf{C} \longrightarrow \mathsf{D}$ of categories such that

$$(d; \{f_i\}) \in \triangle^{\mathsf{C}} \quad \text{if and only if} \quad (Fd; \{Ff_i\}) \in \triangle^{\mathsf{D}}.$$

Example 10.8.2. In the context of Example 10.7.4, different choices of a small category equivalent to the category of smooth n-manifolds yield configured categories connected by configured equivalences. The same is true in the context of Example 10.7.5 for complex n-manifolds. ◇

Theorem 10.8.3. *Suppose* $F : \widehat{\mathsf{C}} \longrightarrow \widehat{\mathsf{D}}$ *is a configured equivalence. Then the change-of-operad adjunction*

$$\mathsf{PFA}(\widehat{\mathsf{C}}) = \mathsf{Alg}_\mathsf{M}\left(\mathsf{O}_{\widehat{\mathsf{C}}}^\mathsf{M}\right) \underset{(\mathsf{O}_F^\mathsf{M})^*}{\overset{(\mathsf{O}_F^\mathsf{M})_!}{\rightleftharpoons}} \mathsf{Alg}_\mathsf{M}\left(\mathsf{O}_{\widehat{\mathsf{D}}}^\mathsf{M}\right) = \mathsf{PFA}(\widehat{\mathsf{D}})$$

is an adjoint equivalence.

Proof. By Theorem 2.4.9 it is enough to show that the right adjoint $(\mathsf{O}_F^\mathsf{M})^*$ is an equivalence of categories, i.e., full, faithful, and essentially surjective. Since $F : \mathsf{C} \longrightarrow \mathsf{D}$ is an equivalence of categories, for each object $d \in \mathsf{D}$, we can choose

- an object $c_d \in \mathsf{C}$ and
- an isomorphism $h_d : Fc_d \overset{\cong}{\longrightarrow} d$ in D.

We can further insist that, if $d = Fc$ for some $c \in \mathsf{C}$, then

- c_{Fc} is chosen from within the F-pre-image of Fc, i.e., $Fc_{Fc} = Fc$, and
- h_{Fc} is Id_{Fc}.

By the inclusivity axiom, each $\{h_d\}$ is a configuration in $\widehat{\mathsf{D}}$. We now check the three required properties of $(\mathsf{O}_F^\mathsf{M})^*$. To simplify the presentation, we will write $(\mathsf{O}_F^\mathsf{M})^*(X, \lambda^X)$ as X^* for each $\mathsf{O}_{\widehat{\mathsf{D}}}^\mathsf{M}$-algebra (X, λ^X) and similarly for morphisms.

To see that $(\mathsf{O}_F^\mathsf{M})^*$ is faithful, suppose

$$\phi, \psi : (X, \lambda^X) \longrightarrow (Y, \lambda^Y)$$

are two morphisms of $\mathsf{O}_{\widehat{\mathsf{D}}}^\mathsf{M}$-algebras such that

$$\phi^* = \psi^* : X^* \longrightarrow Y^* \in \mathsf{Alg}_\mathsf{M}\left(\mathsf{O}_{\widehat{\mathsf{C}}}^\mathsf{M}\right).$$

We must show that $\phi = \psi$ in $\mathsf{Alg}_\mathsf{M}(\mathsf{O}_{\widehat{\mathsf{D}}}^\mathsf{M})$. It is sufficient to prove this equality color-wise, so suppose $d \in \mathsf{D}$. By the associativity and the unity conditions in the Coherence Theorem 10.3.6, the structure morphism

$$X_{Fc_d} \xrightarrow{\lambda^X\{h_d\}} X_d \in \mathsf{M}$$

is invertible with inverse $\lambda^X\{h_d^{-1}\}$. Since $\phi \in \mathsf{Alg}_{\mathsf{M}}(\mathsf{O}_{\mathsf{D}}^{\mathsf{M}})$, the diagram

$$
\begin{array}{ccc}
X_{c_d}^* = X_{Fc_d} & \xrightarrow[\cong]{\lambda^X\{h_d\}} & X_d \\
\phi_{c_d}^* = \Big\downarrow \phi_{Fc_d} & & \Big\downarrow \phi_d \\
Y_{c_d}^* = Y_{Fc_d} & \xrightarrow[\cong]{\lambda^Y\{h_d\}} & Y_d
\end{array}
$$

is a special case of (10.3.10), so it is commutative. By the invertibility of $\lambda^X\{h_d\}$, we infer the equality

$$\phi_d = \lambda^Y\{h_d\} \circ \phi_{c_d}^* \circ \lambda^X\{h_d^{-1}\}.$$

The same is true with $\psi \in \mathsf{Alg}_{\mathsf{M}}(\mathsf{O}_{\mathsf{D}}^{\mathsf{M}})$ in place of ϕ, so

$$\psi_d = \lambda^Y\{h_d\} \circ \psi_{c_d}^* \circ \lambda^X\{h_d^{-1}\}.$$

Since $\phi_{c_d}^* = \psi_{c_d}^*$ by assumption, we conclude that $\phi_d = \psi_d$. This proves that the right adjoint $(\mathsf{O}_F^{\mathsf{M}})^*$ is faithful.

To see that $(\mathsf{O}_F^{\mathsf{M}})^*$ is full, suppose

$$\varphi : X^* \longrightarrow Y^* \in \mathsf{Alg}_{\mathsf{M}}(\mathsf{O}_{\mathsf{C}}^{\mathsf{M}})$$

for some $\mathsf{O}_{\mathsf{D}}^{\mathsf{M}}$-algebras (X, λ^X) and (Y, λ^Y). We must show that

$$\varphi = \phi^* \quad \text{for some} \quad \phi : X \longrightarrow Y \in \mathsf{Alg}_{\mathsf{M}}(\mathsf{O}_{\mathsf{D}}^{\mathsf{M}}).$$

We define such a morphism ϕ entrywise as the composition

$$
\begin{array}{ccc}
X_{c_d}^* = X_{Fc_d} & \xleftarrow[\cong]{\lambda^X\{h_d^{-1}\}} & X_d \\
\varphi_{c_d} \Big\downarrow & & \Big\downarrow \phi_d \\
Y_{c_d}^* = Y_{Fc_d} & \xrightarrow[\cong]{\lambda^Y\{h_d\}} & Y_d
\end{array}
\tag{10.8.4}
$$

for $d \in \mathsf{D}$, where the object $c_d \in \mathsf{C}$ and the isomorphism $h_d : Fc_d \xrightarrow{\cong} d \in \mathsf{D}$ are as in the first paragraph.

To show that ϕ is a morphism of $\mathsf{O}_{\mathsf{D}}^{\mathsf{M}}$-algebras, suppose $\{f_i\} \in \Delta^{\mathsf{D}}\binom{d}{\underline{d}}$ with $\underline{d} = (d_1, \dots, d_n)$ and each $f_i \in \mathsf{D}(d_i, d)$. We must show that the diagram

$$
\begin{array}{ccc}
\bigotimes\limits_{i=1}^{n} X_{d_i} & \xrightarrow{\otimes_i \phi_{d_i}} & \bigotimes\limits_{i=1}^{n} Y_{d_i} \\
\lambda^X\{f_i\} \Big\downarrow & & \Big\downarrow \lambda^Y\{f_i\} \\
X_d & \xrightarrow{\phi_d} & Y_d
\end{array}
$$

in M is commutative. For each $1 \le i \le n$, the composition

$$
\begin{array}{ccc}
Fc_{d_i} & \xrightarrow{h_d^{-1} f_i h_{d_i}} & Fc_d \\
h_{d_i} \Big\downarrow \cong & & \cong \Big\uparrow h_d^{-1} \\
d_i & \xrightarrow{f_i} & d
\end{array}
$$

in D has a unique F-pre-image $g_i \in C(c_{d_i}, c_d)$ because F is full and faithful. Moreover, if f_i is the identity morphism of d, then g_i is the identity morphism of c_d. By the inclusivity axiom and the composition axiom in Definition 10.2.1, there is a configuration

$$\{Fg_i\}_{i=1}^n = \{h_d^{-1}f_i h_{d_i}\}_{i=1}^n \in \Delta^D\binom{Fc_d}{Fc_{d_1},\ldots,Fc_{d_n}}. \tag{10.8.5}$$

Since F is a configured equivalence, this implies that there is a configuration

$$\{g_i\}_{i=1}^n \in \Delta^C\binom{c_d}{c_{d_1},\ldots,c_{d_n}}. \tag{10.8.6}$$

The diagram

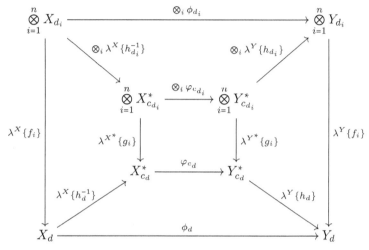

in M is commutative:

- The top and bottom trapezoids are commutative by the definition of ϕ in (10.8.4).
- The left and right trapezoids are commutative by the associativity condition (10.3.8) for λ^X and λ^Y and the equality in (10.8.5).
- The middle square is commutative by (10.3.10) and (10.8.6) because φ is a morphism of O_C^M-algebras.

So ϕ is a morphism of O_D^M-algebras.

To show that $\phi^* = \varphi$, first observe that for each object $d \in D$, there are equalities

$$\phi_{c_d}^* = \phi F c_d = \varphi_{c_{Fc_d}} = \varphi_{c_d}$$

by the definition of ϕ. Now suppose $c \in C$. We must show that $\phi_c^* = \varphi_c$. We just proved that

$$\phi_{c_{Fc}}^* = \varphi_{c_{Fc}}, \tag{10.8.7}$$

since $Fc \in D$. Note that $Fc_{Fc} = Fc$ by our choices of the objects $c_?$ in the first paragraph. Since F is full and faithful, there exists a unique isomorphism

$$r_c : c \xrightarrow{\cong} c_{Fc} \in C \quad \text{such that} \quad Fr_c = \mathrm{Id}_{Fc}. \tag{10.8.8}$$

Since φ is a morphism of O_C^M-algebras, the diagram

$$
\begin{array}{ccc}
X_c^* & \xrightarrow{\varphi_c} & Y_c^* \\
{\scriptstyle \lambda^{X^*}\{r_c\}}\Big\downarrow {\scriptstyle \cong} & & {\scriptstyle \cong}\Big\downarrow{\scriptstyle \lambda^{Y^*}\{r_c\}} \\
X_{cFc}^* & \xrightarrow{\varphi_{cFc}} & Y_{cFc}^*
\end{array}
$$

in M is commutative. So there is an equality

$$\varphi_c = \lambda^{Y^*}\{r_c^{-1}\} \circ \varphi_{cFc} \circ \lambda^{X^*}\{r_c\}. \tag{10.8.9}$$

Similarly, since ϕ^* is a morphism of O_C^M-algebras, there is an equality

$$\phi_c^* = \lambda^{Y^*}\{r_c^{-1}\} \circ \phi_{cFc}^* \circ \lambda^{X^*}\{r_c\}. \tag{10.8.10}$$

The desired equality $\phi_c^* = \varphi_c$ now follows from (10.8.7), (10.8.9), and (10.8.10). Therefore, the right adjoint $(O_F^M)^*$ is full.

Finally, to prove that $(O_F^M)^*$ is essentially surjective, suppose (Z, λ^Z) is an O_C^M-algebra. We must show that there exist

- $(W, \lambda^W) \in \mathsf{Alg}_M(O_D^M)$ and
- an isomorphism $Z \xrightarrow{\cong} W^*$ of O_C^M-algebras.

First we define the entries of W as

$$W_d = Z_{cd} \quad \text{for} \quad d \in \mathsf{D}.$$

For each $c \in \mathsf{C}$, there is a canonical isomorphism

$$Z_c \xrightarrow[\cong]{\lambda^Z\{r_c\}} Z_{cFc} = W_{Fc} = W_c^* \in \mathsf{C}$$

with $r_c \in \mathsf{C}(c, cFc)$ the isomorphism in (10.8.8).

To define the O_D^M-algebra structure morphism of W, suppose as above that $\{f_i\} \in \Delta^D\binom{d}{\underline{d}}$ with each $f_i \in \mathsf{D}(d_i, d)$. In (10.8.6) we observed that there exists a unique configuration

$$\{g_i\} \in \Delta^C \quad \text{such that} \quad \{Fg_i\} = \{h_d^{-1} f_i h_{d_i}\} \in \Delta^D.$$

We now define the structure morphism $\lambda^W\{f_i\}$ as in the commutative diagram

$$
\begin{array}{ccc}
\displaystyle\bigotimes_{i=1}^{n} W_{d_i} & \xrightarrow{\lambda^W\{f_i\}} & W_d \\
\Big\| & & \Big\| \\
\displaystyle\bigotimes_{i=1}^{n} Z_{cd_i} & \xrightarrow{\lambda^Z\{g_i\}} & Z_{cd}
\end{array}
\tag{10.8.11}
$$

in M. Since (Z, λ^Z) satisfies the associativity, unity, and equivariance conditions in the Coherence Theorem 10.3.6, it follows that (W, λ^W) is an O_D^M-algebra. It remains to show that $\{\lambda^Z\{r_c\}\}_{c \in \mathsf{C}}$ is a morphism of O_C^M-algebras.

Suppose

$$\{p_j\} \in \Delta^C\left(\begin{smallmatrix} c \\ c_1,\ldots,c_m \end{smallmatrix}\right)$$

is a configuration with each $p_j \in C(c_j, c)$, so

$$\{Fp_j\} \in \Delta^D\left(\begin{smallmatrix} Fc \\ Fc_1,\ldots,Fc_m \end{smallmatrix}\right).$$

We must show that the diagram

$$
\begin{array}{ccc}
\bigotimes\limits_{j=1}^{m} Z_{c_j} & \xrightarrow{\otimes_j \lambda^Z\{r_{c_j}\}} & \bigotimes\limits_{j=1}^{m} W^*_{c_j} = \bigotimes\limits_{j=1}^{m} Z_{cFc_j} \\
\lambda^Z\{p_j\} \downarrow & & \downarrow \lambda^{W^*}\{p_j\} = \lambda^W\{Fp_j\} \\
Z_c & \xrightarrow[\lambda^Z\{r_c\}]{} & W^*_c = Z_{cFc}
\end{array}
\tag{10.8.12}
$$

in M is commutative. Since $\{Fp_j\} \in \Delta^D$, as in (10.8.5) there exists a unique configuration

$$\{k_j\} \in \Delta^C\left(\begin{smallmatrix} cFc \\ cFc_1,\ldots,cFc_m \end{smallmatrix}\right) \quad \text{such that} \quad \{Fk_j\} = \{Fp_j\} \in \Delta^D\left(\begin{smallmatrix} Fc \\ Fc_1,\ldots,Fc_m \end{smallmatrix}\right) \tag{10.8.13}$$

since

$$h_{Fc}^{-1} = \mathrm{Id}_{Fc} \quad \text{and} \quad h_{Fc_j} = \mathrm{Id}_{Fc_j}.$$

By the definition (10.8.11) we have

$$\lambda^W\{Fp_j\} = \lambda^Z\{k_j\}.$$

Therefore, by the associativity condition (10.3.8) for (Z, λ^Z), to show the commutativity of the square (10.8.12), it is enough to prove that the diagram

$$
\begin{array}{ccc}
c_j & \xrightarrow{r_{c_j}} & cFc_j \\
p_j \downarrow & & \downarrow k_j \\
c & \xrightarrow[r_c]{} & cFc
\end{array}
$$

in C is commutative for each $1 \le j \le m$. Since F is faithful, it is enough to show that the F-image diagram

$$
\begin{array}{ccc}
Fc_j & \xrightarrow{Fr_{c_j}} & FcFc_j = Fc_j \\
Fp_j \downarrow & & \downarrow Fk_j \\
Fc & \xrightarrow[Fr_c]{} & FcFc = Fc
\end{array}
$$

in D is commutative. But since Fr_c and each Fr_{c_j} are the identity morphisms by (10.8.8) and since $Fp_j = Fk_j$ by (10.8.13), we conclude that the above square is commutative. Therefore, the right adjoint $(O_F^M)^*$ is essentially surjective. $\square$

Theorem 10.8.14. *Suppose* $F : \widehat{C} \longrightarrow \widehat{D}$ *is a configured functor, and* M *is a monoidal model category in which the colored operads* $O_{\widehat{C}}^M$ *and* $O_{\widehat{D}}^M$ *are admissible.*

(1) The change-of-operad adjunction

$$\mathsf{PFA}(\widehat{\mathsf{C}}) = \mathsf{Alg}_\mathsf{M}\big(\mathsf{O}^\mathsf{M}_{\widehat{\mathsf{C}}}\big) \underset{(\mathsf{O}^\mathsf{M}_F)^*}{\overset{(\mathsf{O}^\mathsf{M}_F)_!}{\rightleftarrows}} \mathsf{Alg}_\mathsf{M}\big(\mathsf{O}^\mathsf{M}_{\widehat{\mathsf{D}}}\big) = \mathsf{PFA}(\widehat{\mathsf{D}})$$

is a Quillen adjunction.

(2) If F is a configured equivalence, then the operad morphism

$$\mathsf{O}^\mathsf{M}_F : \mathsf{O}^\mathsf{M}_{\widehat{\mathsf{C}}} \longrightarrow \mathsf{O}^\mathsf{M}_{\widehat{\mathsf{D}}}$$

is a homotopy Morita equivalence; i.e., the change-of-operad adjunction is a Quillen equivalence.

Proof. The proof is the same as that of Theorem 8.5.1. Here we use Theorem 10.8.3 to infer that the unit of the change-of-operad adjunction is a natural isomorphism. ☐

Interpretation 10.8.15. If two configured categories are connected by a configured equivalence, then their categories of prefactorization algebras have equivalent homotopy theories. In particular, these two categories of prefactorization algebras are equivalent both before and after inverting the weak equivalences. ◇

Example 10.8.16 (Costello-Gwilliam prefactorization algebras on (complex) n-manifolds). In the context of Example 10.7.4, different choices of a small category equivalent to the category of smooth n-manifolds yield configured categories connected by configured equivalences. By Theorem 10.8.3 and Theorem 10.8.14, for any two different choices their categories of prefactorization algebras are connected by a change-of-operad adjunction that is both an adjoint equivalence and a Quillen equivalence. The same is true in the context of Example 10.7.5 for complex n-manifolds. ◇

Chapter 11

Homotopy Prefactorization Algebras

In this chapter, we define homotopy prefactorization algebras on a configured category and study their structure.

11.1 Overview

In Section 11.2 we define homotopy prefactorization algebras on a configured category $\widehat{\mathsf{C}}$ as algebras over the Boardman-Vogt construction $\mathsf{WO}_{\widehat{\mathsf{C}}}^{\mathsf{M}}$ of $\mathsf{O}_{\widehat{\mathsf{C}}}^{\mathsf{M}}$. This definition makes sense because prefactorization algebras on $\widehat{\mathsf{C}}$ are defined as algebras over the colored operad $\mathsf{O}_{\widehat{\mathsf{C}}}^{\mathsf{M}}$. The Boardman-Vogt construction comes with an augmentation $\eta : \mathsf{WO}_{\widehat{\mathsf{C}}}^{\mathsf{M}} \longrightarrow \mathsf{O}_{\widehat{\mathsf{C}}}^{\mathsf{M}}$, which induces a change-of-operad adjunction between the category of prefactorization algebras and the category of homotopy prefactorization algebras. In favorable cases, this change-of-operad adjunction is a Quillen equivalence.

Examples of homotopy prefactorization algebras are given in Section 11.3. They include homotopy coherent versions of prefactorization algebras satisfying the time-slice axiom, Costello-Gwilliam (equivariant) prefactorization algebras, and prefactorization algebras on (complex) manifolds. Moreover, homotopy coherent diagrams of E_∞-algebras are homotopy prefactorization algebras on the maximal configured category.

In Section 11.4 we record the coherence theorem for homotopy prefactorization algebras. This is a special case of the Coherence Theorem 7.2.1 for algebras over the Boardman-Vogt construction. This coherence theorem is our main tool for understanding the structure on homotopy prefactorization algebras.

In Section 11.5 we observe that every homotopy prefactorization algebra on a configured category $\widehat{\mathsf{C}}$ has an underlying homotopy coherent pointed C-diagram. This is the homotopy coherent version of the fact that every prefactorization algebra on $\widehat{\mathsf{C}}$ has an underlying pointed C-diagram. Compared to a homotopy coherent C-diagram, a homotopy coherent pointed C-diagram has additional structure morphisms parametrized by truncated linear graphs in Example 3.1.20.

In Section 11.6 we observe that, when a set of morphisms S in C is chosen, $\mathsf{WO}_{\widehat{\mathsf{C}}}[S^{-1}]^{\mathsf{M}}$-algebras are homotopy coherent versions of prefactorization algebras on $\widehat{C}$ satisfying the time-slice axiom with respect to S. In particular, the structure morphism corresponding to each morphism $f \in S$ is invertible up to specified homotopies that are also structure morphisms. This is the homotopical analogue of Theorem 10.3.13, which says that prefactorization algebras satisfying the time-slice axiom with respect to S are equivalent to prefactorization algebras whose structure morphisms corresponding to all $f \in S$ are invertible.

In Section 11.7 we show that some entries of a homotopy prefactorization algebra on a configured category $\widehat{C}$ are E_∞-algebras. For example, this is the case if $\widehat{C}$ comes from a bounded lattice L with least element 0. In this case, for each homotopy prefactorization algebra on $\widehat{L}$, the 0-entry is equipped with the structure of an E_∞-algebra. This is the homotopical analogue of the fact that, for each prefactorization algebra on $\widehat{L}$, the 0-entry is equipped with the structure of a commutative monoid. It follows that each homotopy Costello-Gwilliam (equivariant) prefactorization algebra has an E_∞-algebra at the entry corresponding to the empty subset.

In Section 11.8 we show that for a bounded lattice L with least element 0 and for a homotopy prefactorization algebra Y on $\widehat{L}$, every other entry of Y admits the structure of an E_∞-module over the E_∞-algebra Y_0. This is the homotopy coherent version of the fact that, for a prefactorization algebra Y on $\widehat{L}$, each entry Y_d is a left Y_0-module. In particular, this applies to homotopy Costello-Gwilliam (equivariant) prefactorization algebras.

In Section 11.9 we show that the objectwise E_∞-module structure in Section 11.8 is homotopically compatible with the homotopy coherent diagram structure in Section 11.5. This is the homotopical analogue of the fact that, for a prefactorization algebra Y on $\widehat{L}$, the underlying L-diagram of Y is actually an L-diagram of left Y_0-modules.

Finally, in Section 11.10 we observe that every homotopy coherent C-diagram of E_∞-algebras can be realized as a homotopy prefactorization algebra on the maximal configured category on C. This implies that homotopy Costello-Gwilliam prefactorization algebras on (complex) manifolds are homotopy coherent diagrams of E_∞-algebras. This is the homotopical analogue of the definition of Costello-Gwilliam prefactorization algebras on (complex) manifolds as symmetric monoidal functors.

Throughout this chapter $(M, \otimes, \mathbb{1})$ is a cocomplete symmetric monoidal closed category with an initial object $\varnothing$ and a commutative segment $(J, \mu, 0, 1, \epsilon)$ as in Definition 6.2.1. For a small category C, its object set will be denoted by $\mathfrak{C}$.

11.2 Homotopy Prefactorization Algebras as Operad Algebras

In this section, we define homotopy prefactorization algebras on a configured category using the Boardman-Vogt construction in Chapter 6 and record their basic categorical properties.

Recollection 11.2.1. For a $\mathfrak{C}$-colored operad O in M, its Boardman-Vogt construction WO is the $\mathfrak{C}$-colored operad with entries

$$\mathsf{WO}\binom{d}{c} = \int^{T \in \underline{\mathsf{Tree}}^{\mathfrak{C}}\binom{d}{c}} J[T] \otimes O[T] \in \mathsf{M},$$

where $\underline{\mathsf{Tree}}^{\mathfrak{C}}\binom{d}{c}$ is the substitution category of $\mathfrak{C}$-colored trees with profile $\binom{d}{c}$ in Definition 3.2.11. The functors

$$J : \underline{\mathsf{Tree}}^{\mathfrak{C}}\binom{d}{c}^{\mathrm{op}} \longrightarrow \mathsf{M} \quad \text{and} \quad O : \underline{\mathsf{Tree}}^{\mathfrak{C}}\binom{d}{c} \longrightarrow \mathsf{M}$$

are induced by J and O and are defined in Definition 6.2.5 and Corollary 4.4.15, respectively. The operad structure on WO, defined in Definition 6.3.7, is induced by tree substitution. It is equipped with a natural augmentation $\eta : \mathsf{WO} \longrightarrow \mathsf{O}$ of $\mathfrak{C}$-colored operads, defined in Theorem 6.4.4. One should think of the augmentation as forgetting the lengths of the internal edges (i.e., the J-component) and composing in the colored operad O.

The strong symmetric monoidal functor $\mathsf{Set} \longrightarrow \mathsf{M}$ in Example 5.3.3, sending a set S to the S-indexed coproduct $\coprod_S \mathbb{1}$, yields the change-of-category functor

$$(-)^{\mathsf{M}} : \mathsf{Operad}^{\mathfrak{C}}(\mathsf{Set}) \longrightarrow \mathsf{Operad}^{\mathfrak{C}}(\mathsf{M}).$$

For the colored operad $O_{\widehat{\mathsf{C}}}$ in Definition 10.3.2 for a configured category $\widehat{\mathsf{C}} = (\mathsf{C}, \triangle)$, its image in $\mathsf{Operad}^{\mathfrak{C}}(\mathsf{M})$ will be denoted by $\mathsf{O}_{\widehat{\mathsf{C}}}^{\mathsf{M}}$. Its entries are

$$\mathsf{O}_{\widehat{\mathsf{C}}}^{\mathsf{M}}\binom{d}{c} = \coprod_{\triangle\binom{d}{c}} \mathbb{1} \quad \text{for} \quad \binom{d}{c} \in \mathsf{Prof}(\mathfrak{C}) \times \mathfrak{C}.$$

Also recall from Definition 4.5.5 the category of algebras over a colored operad.

Definition 11.2.2. Suppose $\widehat{\mathsf{C}} = (\mathsf{C}, \triangle)$ is a configured category with object set $\mathfrak{C}$, and $\mathsf{WO}_{\widehat{\mathsf{C}}}^{\mathsf{M}} \in \mathsf{Operad}^{\mathfrak{C}}(\mathsf{M})$ is the Boardman-Vogt construction of $\mathsf{O}_{\widehat{\mathsf{C}}}^{\mathsf{M}} \in \mathsf{Operad}^{\mathfrak{C}}(\mathsf{M})$.

(1) We define the category

$$\mathsf{HPFA}(\widehat{\mathsf{C}}) = \mathsf{Alg}_{\mathsf{M}}\left(\mathsf{WO}_{\widehat{\mathsf{C}}}^{\mathsf{M}}\right),$$

whose objects are called *homotopy prefactorization algebras* on $\widehat{\mathsf{C}}$.

(2) Suppose S is a set of morphisms in C. We define the category

$$\mathsf{HPFA}(\widehat{\mathsf{C}}, S) = \mathsf{Alg}_{\mathsf{M}}\left(\mathsf{WO}_{\widehat{\mathsf{C}}}[S^{-1}]^{\mathsf{M}}\right),$$

whose objects are called *homotopy prefactorization algebras* on $\widehat{\mathsf{C}}$ satisfying the *homotopy time-slice axiom* with respect to S.

Interpretation 11.2.3. The $\mathfrak{C}$-colored operad $\mathsf{WO}_{\widehat{\mathsf{C}}}^{\mathsf{M}}$ is made up of $\mathfrak{C}$-colored trees whose internal edges are decorated by the commutative segment J and whose vertices are decorated by elements in the colored operad $O_{\widehat{\mathsf{C}}}$ (i.e., configurations in $\widehat{\mathsf{C}}$) with the correct profile. A homotopy prefactorization algebra has structure morphisms indexed by these decorated $\mathfrak{C}$-colored trees. The precise statement is the Coherence Theorem 11.4.1 below. ◇

The following observation compares prefactorization algebras and homotopy prefactorization algebras. It is a special case of Theorem 5.2.7(1), Corollary 6.4.7, and Corollary 6.5.10.

Corollary 11.2.4. *Suppose* $\widehat{\mathsf{C}} = (\mathsf{C}, \triangle)$ *is a configured category.*

(1) The augmentation $\eta : \mathsf{WO}_{\widehat{\mathsf{C}}}^{\mathsf{M}} \longrightarrow \mathsf{O}_{\widehat{\mathsf{C}}}^{\mathsf{M}}$ *induces a change-of-operad adjunction*

$$\mathsf{HPFA}(\widehat{\mathsf{C}}) = \mathsf{Alg}_{\mathsf{M}}\big(\mathsf{WO}_{\widehat{\mathsf{C}}}^{\mathsf{M}}\big) \underset{\eta^*}{\overset{\eta_!}{\rightleftarrows}} \mathsf{Alg}_{\mathsf{M}}\big(\mathsf{O}_{\widehat{\mathsf{C}}}^{\mathsf{M}}\big) = \mathsf{PFA}(\widehat{\mathsf{C}}) .$$

(2) If M *is a monoidal model category in which the colored operads* $\mathsf{O}_{\widehat{\mathsf{C}}}^{\mathsf{M}}$ *and* $\mathsf{WO}_{\widehat{\mathsf{C}}}^{\mathsf{M}}$ *are admissible, then the change-of-operad adjunction is a Quillen adjunction.*

(3) If $\mathsf{M} = \mathsf{Chain}_{\mathbb{K}}$ *with* $\mathbb{K}$ *a field of characteristic zero, then the change-of-operad adjunction is a Quillen equivalence.*

Interpretation 11.2.5. The right adjoint η^* allows us to consider a prefactorization algebra on $\widehat{\mathsf{C}}$ as a homotopy prefactorization algebra on $\widehat{\mathsf{C}}$. The left adjoint $\eta_!$ rectifies a homotopy prefactorization algebra to a prefactorization algebra. Furthermore, if M is $\mathsf{Chain}_{\mathbb{K}}$, then the augmentation η is a homotopy Morita equivalence. In particular, the homotopy theory of homotopy prefactorization algebras is equivalent to the homotopy theory of prefactorization algebras over the same configured category. So there is no loss of homotopical information by considering homotopy prefactorization algebras.   ⋄

The next observation is about changing the configured categories. It is a consequence of Theorem 5.2.7(1), Theorem 6.4.4, Theorem 10.8.14, and Corollary 11.2.4. The second assertion below uses the fact that Quillen equivalences have the 2-out-of-3 property.

Corollary 11.2.6. *Suppose* $F : \widehat{\mathsf{C}} \longrightarrow \widehat{\mathsf{D}}$ *is a configured functor.*

(1) There is an induced diagram of change-of-operad adjunctions

$$
\begin{array}{ccc}
\mathsf{HPFA}(\widehat{\mathsf{C}}) = \mathsf{Alg}_{\mathsf{M}}\big(\mathsf{WO}_{\widehat{\mathsf{C}}}^{\mathsf{M}}\big) & \underset{(\mathsf{WO}_F^{\mathsf{M}})^*}{\overset{(\mathsf{WO}_F^{\mathsf{M}})_!}{\rightleftarrows}} & \mathsf{Alg}_{\mathsf{M}}\big(\mathsf{WO}_{\widehat{\mathsf{D}}}^{\mathsf{M}}\big) = \mathsf{HPFA}(\widehat{\mathsf{D}}) \\[4pt]
\eta_! \Big\updownarrow \eta^* & & \eta_! \Big\updownarrow \eta^* \\[4pt]
\mathsf{PFA}(\widehat{\mathsf{C}}) = \mathsf{Alg}_{\mathsf{M}}\big(\mathsf{O}_{\widehat{\mathsf{C}}}^{\mathsf{M}}\big) & \underset{(\mathsf{O}_F^{\mathsf{M}})^*}{\overset{(\mathsf{O}_F^{\mathsf{M}})_!}{\rightleftarrows}} & \mathsf{Alg}_{\mathsf{M}}\big(\mathsf{O}_{\widehat{\mathsf{D}}}^{\mathsf{M}}\big) = \mathsf{PFA}(\widehat{\mathsf{D}})
\end{array}
$$

such that

$$(\mathsf{O}_F^{\mathsf{M}})_! \eta_! = \eta_! (\mathsf{WO}_F^{\mathsf{M}})_! \quad and \quad \eta^* (\mathsf{O}_F^{\mathsf{M}})^* = (\mathsf{WO}_F^{\mathsf{M}})^* \eta^* .$$

(2) If M *is a monoidal model category in which the colored operads* $\mathsf{O}_{\widehat{\mathsf{C}}}^{\mathsf{M}}$, $\mathsf{O}_{\widehat{\mathsf{D}}}^{\mathsf{M}}$, $\mathsf{WO}_{\widehat{\mathsf{C}}}^{\mathsf{M}}$, *and* $\mathsf{WO}_{\widehat{\mathsf{D}}}^{\mathsf{M}}$ *are admissible, then all four change-of-operad adjunctions are Quillen adjunctions.*

(3) If F is a configured equivalence and if M = Chain$_{\mathbb{K}}$ *with* $\mathbb{K}$ *a field of character-istic zero, then all four change-of-operad adjunctions are Quillen equivalences.*

Interpretation 11.2.7. The right adjoint $(\mathsf{WO}_F^{\mathsf{M}})^*$ sends each homotopy prefac-torization algebra on $\widehat{\mathsf{D}}$ to one on $\widehat{\mathsf{C}}$. The left adjoint $(\mathsf{WO}_F^{\mathsf{M}})_!$ sends each homotopy prefactorization algebra on $\widehat{\mathsf{C}}$ to one on $\widehat{\mathsf{D}}$. The equality

$$(\mathsf{O}_F^{\mathsf{M}})_! \eta_! = \eta_! (\mathsf{WO}_F^{\mathsf{M}})_!$$

means that the left adjoint diagram is commutative. The equality

$$\eta^* (\mathsf{O}_F^{\mathsf{M}})^* = (\mathsf{WO}_F^{\mathsf{M}})^* \eta^*$$

means that the right adjoint diagram is commutative. Moreover, if F is a configured equivalence and if M is Chain$_{\mathbb{K}}$, then all four operad morphisms in the commutative diagram

$$
\begin{array}{ccc}
\mathsf{WO}_{\widehat{\mathsf{C}}}^{\mathsf{M}} & \xrightarrow{\ \mathsf{WO}_F^{\mathsf{M}}\ } & \mathsf{WO}_{\widehat{\mathsf{D}}}^{\mathsf{M}} \\
{\scriptstyle\eta}\big\downarrow & & \big\downarrow{\scriptstyle\eta} \\
\mathsf{O}_{\widehat{\mathsf{C}}}^{\mathsf{M}} & \xrightarrow[\ \mathsf{O}_F^{\mathsf{M}}\ ]{} & \mathsf{O}_{\widehat{\mathsf{D}}}^{\mathsf{M}}
\end{array}
$$

are homotopy Morita equivalences. In particular, the homotopy theory of homotopy prefactorization algebras on $\widehat{\mathsf{C}}$ is equivalent to the homotopy theory of homotopy prefactorization algebras on $\widehat{\mathsf{D}}$. ◇

11.3 Examples

In this section, we list some examples of homotopy prefactorization algebras.

Example 11.3.1 (Homotopy coherent pointed diagrams). For a small category C, consider the minimal configured category $\widehat{\mathsf{C}_{\mathsf{min}}} = (\mathsf{C}, \triangle_{\mathsf{min}}^{\mathsf{C}})$ on C in Example 10.2.8. By Corollary 11.2.4 and Proposition 10.4.3, the augmentation

$$\eta : \mathsf{WO}_{\widehat{\mathsf{C}_{\mathsf{min}}}}^{\mathsf{M}} \longrightarrow \mathsf{O}_{\widehat{\mathsf{C}_{\mathsf{min}}}}^{\mathsf{M}}$$

induces a change-of-operad adjunction

$$\mathsf{HPFA}(\widehat{\mathsf{C}_{\mathsf{min}}}) = \mathsf{Alg}_{\mathsf{M}}\left(\mathsf{WO}_{\widehat{\mathsf{C}_{\mathsf{min}}}}^{\mathsf{M}}\right) \underset{\eta^*}{\overset{\eta_!}{\rightleftarrows}} \mathsf{Alg}_{\mathsf{M}}\left(\mathsf{O}_{\widehat{\mathsf{C}_{\mathsf{min}}}}^{\mathsf{M}}\right) = \mathsf{PFA}(\widehat{\mathsf{C}_{\mathsf{min}}}) \cong \mathsf{M}_*^{\mathsf{C}}$$

between the category of homotopy prefactorization algebras on $\widehat{\mathsf{C}_{\mathsf{min}}}$ and the cate-gory of pointed C-diagrams in M. Therefore, $\mathsf{WO}_{\widehat{\mathsf{C}_{\mathsf{min}}}}^{\mathsf{M}}$-algebras are homotopy coher-ent versions of pointed C-diagrams in M. We will study them in details in Section 11.5. Furthermore, if M = Chain$_{\mathbb{K}}$ with $\mathbb{K}$ a field of characteristic zero, then this adjunction is a Quillen equivalence. ◇

Example 11.3.2 (Underlying homotopy coherent pointed diagrams). Suppose $\widehat{\mathsf{C}} = (\mathsf{C}, \triangle)$ is a configured category. In Corollary 10.4.4 we saw that the identity functor on C induces a configured functor

$$\widehat{\mathsf{C}_{min}} = (\mathsf{C}, \triangle^{\mathsf{C}}_{min}) \xrightarrow{\ i_0\ } (\mathsf{C}, \triangle) = \widehat{\mathsf{C}}.$$

By Corollary 11.2.6 there is an induced diagram of change-of-operad adjunctions

$$
\begin{array}{ccc}
\mathsf{HPFA}(\widehat{\mathsf{C}_{min}}) = \mathsf{Alg}_\mathsf{M}\big(\mathsf{WO}^\mathsf{M}_{\widehat{\mathsf{C}_{min}}}\big) & \underset{(\mathsf{WO}^\mathsf{M}_{i_0})^*}{\overset{(\mathsf{WO}^\mathsf{M}_{i_0})_!}{\rightleftarrows}} & \mathsf{Alg}_\mathsf{M}\big(\mathsf{WO}^\mathsf{M}_{\widehat{\mathsf{C}}}\big) = \mathsf{HPFA}(\widehat{\mathsf{C}}) \\[4pt]
\eta_! \Big\uparrow\Big\downarrow \eta^* & & \eta_! \Big\uparrow\Big\downarrow \eta^* \\[4pt]
\mathsf{M}^{\mathsf{C}}_* \cong \mathsf{PFA}(\widehat{\mathsf{C}_{min}}) = \mathsf{Alg}_\mathsf{M}\big(\mathsf{O}^\mathsf{M}_{\widehat{\mathsf{C}_{min}}}\big) & \underset{(\mathsf{O}^\mathsf{M}_{i_0})^*}{\overset{(\mathsf{O}^\mathsf{M}_{i_0})_!}{\rightleftarrows}} & \mathsf{Alg}_\mathsf{M}\big(\mathsf{O}^\mathsf{M}_{\widehat{\mathsf{C}}}\big) = \mathsf{PFA}(\widehat{\mathsf{C}})
\end{array}
$$

with commuting left adjoint diagram and commuting right adjoint diagram. In particular, via the right adjoin $(\mathsf{WO}^\mathsf{M}_{i_0})^*$, every homotopy prefactorization algebra on $\widehat{\mathsf{C}}$ has an underlying $\mathsf{WO}^\mathsf{M}_{\widehat{\mathsf{C}_{min}}}$-algebra, i.e., homotopy coherent pointed C-diagram in M. ⋄

Example 11.3.3 (Homotopy prefactorization algebras with time-slice). Suppose $\widehat{\mathsf{C}} = (\mathsf{C}, \triangle)$ is a configured category, and S is a set of morphisms in C. By the naturality of the Boardman-Vogt construction in Theorem 6.4.4 and the change-of-category functor $(-)^\mathsf{M}$, the S-localization morphism $\ell : \mathsf{O}_{\widehat{\mathsf{C}}} \longrightarrow \mathsf{O}_{\widehat{\mathsf{C}}}[S^{-1}]$ yields a commutative diagram

$$
\begin{array}{ccc}
\mathsf{WO}^\mathsf{M}_{\widehat{\mathsf{C}}} & \xrightarrow{\ \mathsf{W}\ell^\mathsf{M}\ } & \mathsf{WO}_{\widehat{\mathsf{C}}}[S^{-1}]^\mathsf{M} \\[4pt]
\eta \Big\downarrow & & \Big\downarrow \eta \\[4pt]
\mathsf{O}^\mathsf{M}_{\widehat{\mathsf{C}}} & \xrightarrow{\ \ell^\mathsf{M}\ } & \mathsf{O}_{\widehat{\mathsf{C}}}[S^{-1}]^\mathsf{M}
\end{array}
$$

of $\mathfrak{C}$-colored operads in M, where $\mathfrak{C} = \mathsf{Ob}(\mathsf{C})$. By Corollary 11.2.6 there is an induced diagram of change-of-operad adjunctions

$$
\begin{array}{ccc}
\mathsf{HPFA}(\widehat{\mathsf{C}}) = \mathsf{Alg}_\mathsf{M}\big(\mathsf{WO}^\mathsf{M}_{\widehat{\mathsf{C}}}\big) & \underset{(\mathsf{W}\ell^\mathsf{M})^*}{\overset{\mathsf{W}\ell^\mathsf{M}_!}{\rightleftarrows}} & \mathsf{Alg}_\mathsf{M}\big(\mathsf{WO}_{\widehat{\mathsf{C}}}[S^{-1}]^\mathsf{M}\big) = \mathsf{HPFA}(\widehat{\mathsf{C}}, S) \\[4pt]
\eta_! \Big\uparrow\Big\downarrow \eta^* & & \eta_! \Big\uparrow\Big\downarrow \eta^* \\[4pt]
\mathsf{PFA}(\widehat{\mathsf{C}}) = \mathsf{Alg}_\mathsf{M}\big(\mathsf{O}^\mathsf{M}_{\widehat{\mathsf{C}}}\big) & \underset{(\ell^\mathsf{M})^*}{\overset{\ell^\mathsf{M}_!}{\rightleftarrows}} & \mathsf{Alg}_\mathsf{M}\big(\mathsf{O}_{\widehat{\mathsf{C}}}[S^{-1}]^\mathsf{M}\big) = \mathsf{PFA}(\widehat{\mathsf{C}}, S)
\end{array}
$$

with commuting left adjoint diagram and commuting right adjoint diagram. Objects in $\mathsf{Alg}_\mathsf{M}\big(\mathsf{WO}_{\widehat{\mathsf{C}}}[S^{-1}]^\mathsf{M}\big)$ are homotopy prefactorization algebras on $\widehat{\mathsf{C}}$ satisfying the homotopy time-slice axiom with respect to S in Definition 11.2.2.

In Theorem 10.3.13 we noted that $\mathsf{O}_{\widehat{\mathsf{C}}}[S^{-1}]^\mathsf{M}$-algebras are equivalent to prefactorization algebras on $\widehat{\mathsf{C}}$ whose structure morphisms $\lambda\{s\}$ are isomorphisms for all $s \in S$. Therefore, in each $\mathsf{WO}_{\widehat{\mathsf{C}}}[S^{-1}]^\mathsf{M}$-algebra, the structure morphisms $\lambda\{s\}$ should be invertible up to coherent homotopies. We will explain this in details in Section 11.6. ⋄

Example 11.3.4 (Homotopy prefactorization algebras on bounded lattices). For a bounded lattice $(L, \leq)$ with least element 0, consider the configured category $\widehat{L} = (L, \Delta^L)$ in Example 10.2.10. By Corollary 11.2.4 the augmentation

$$\eta : \mathsf{WO}^{\mathsf{M}}_{\widehat{L}} \longrightarrow \mathsf{O}^{\mathsf{M}}_{\widehat{L}}$$

induces a change-of-operad adjunction

$$\mathsf{HPFA}(\widehat{L}) = \mathsf{Alg}_{\mathsf{M}}\left(\mathsf{WO}^{\mathsf{M}}_{\widehat{L}}\right) \underset{\eta^*}{\overset{\eta_!}{\rightleftarrows}} \mathsf{Alg}_{\mathsf{M}}\left(\mathsf{O}^{\mathsf{M}}_{\widehat{L}}\right) = \mathsf{PFA}(\widehat{L})$$

between the category of homotopy prefactorization algebras on $\widehat{L}$ and the category of prefactorization algebras on $\widehat{L}$. Furthermore, if $\mathsf{M} = \mathsf{Chain}_{\mathbb{K}}$ with $\mathbb{K}$ a field of characteristic zero, then this adjunction is a Quillen equivalence.

For a prefactorization algebra (Y, λ) on $\widehat{L}$, we saw in Example 10.5.3 that Y_0 is equipped with the structure of a commutative monoid. Furthermore, in Corollary 10.6.3 we observed that the underlying L-diagram of Y is an L-diagram of left Y_0-modules. Therefore, for a homotopy prefactorization algebra (Y, λ) on $\widehat{L}$:

- Y_0 is equipped with the structure of an E_∞-algebra.
- Every other entry Y_d with $d \in L$ is an E_∞-module over Y_0.
- These E_∞-modules over Y_0 are compatible with the homotopy coherent L-diagram structure.

We will study these structures in details in Section 11.7 to Section 11.9. ◇

Example 11.3.5 (Homotopy Costello-Gwilliam prefactorization algebras). For a topological space X, consider the configured category $\widehat{\mathsf{Open}(X)}$ in Example 10.2.11. Since $\widehat{\mathsf{Open}(X)}$ is an example of $\widehat{L}$ for the bounded lattice $\mathsf{Open}(X)$, everything in Example 11.3.4 applies to $\widehat{\mathsf{Open}(X)}$. In particular, for a $\mathsf{WO}^{\mathsf{M}}_{\widehat{\mathsf{Open}(X)}}$-algebra (Y, λ):

- Y has an underlying homotopy coherent pointed $\mathsf{Open}(X)$-diagram in M by Example 11.3.1.
- $Y_{\varnothing_X}$ is equipped with the structure of an E_∞-algebra, where $\varnothing_X \subset X$ is the empty subset.
- Every other entry Y_U with $U \in \mathsf{Open}(X)$ is an E_∞-module over $Y_{\varnothing_X}$.
- These E_∞-modules over $Y_{\varnothing_X}$ are compatible with the homotopy coherent $\mathsf{Open}(X)$-diagram structure. ◇

Example 11.3.6 (Homotopy Costello-Gwilliam equivariant prefactorization algebras). For a topological space X equipped with a left action by a group G, consider the configured category $\widehat{\mathsf{Open}(X)}_G$ in Example 10.2.12. In Example 10.6.6 we saw that there is a configured functor

$$\widehat{\mathsf{Open}(X)} \overset{\iota}{\longrightarrow} \widehat{\mathsf{Open}(X)}_G .$$

By Corollary 11.2.6 there is an induced diagram of change-of-operad adjunctions

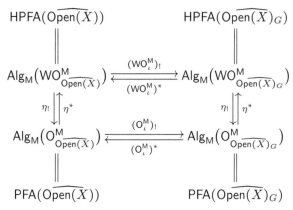

with commuting left adjoint diagram and commuting right adjoint diagram. In particular, via the right adjoin $(\mathsf{WO}_\iota^\mathsf{M})^*$, every homotopy prefactorization algebra on $\widehat{\mathsf{Open}(X)}_G$ also has the structure stated in Example 11.3.5.   ◇

Example 11.3.7 (Homotopy coherent diagrams of E_∞-algebras). For a small category C, consider the maximal configured category $\widehat{\mathsf{C}_\mathsf{max}} = (\mathsf{C}, \triangle_\mathsf{max}^\mathsf{C})$ on C in Example 10.2.9. By Corollary 11.2.4 and Example 10.7.2, the augmentation

$$\eta : \mathsf{WO}_{\widehat{\mathsf{C}_\mathsf{max}}}^\mathsf{M} \longrightarrow \mathsf{O}_{\widehat{\mathsf{C}_\mathsf{max}}}^\mathsf{M}$$

induces a change-of-operad adjunction

$$\mathsf{HPFA}(\widehat{\mathsf{C}_\mathsf{max}}) = \mathsf{Alg}_\mathsf{M}\left(\mathsf{WO}_{\widehat{\mathsf{C}_\mathsf{max}}}^\mathsf{M}\right) \underset{\eta^*}{\overset{\eta_!}{\rightleftarrows}} \mathsf{Alg}_\mathsf{M}\left(\mathsf{O}_{\widehat{\mathsf{C}_\mathsf{max}}}^\mathsf{M}\right) = \mathsf{PFA}(\widehat{\mathsf{C}_\mathsf{max}}) \cong \mathsf{Com}(\mathsf{M})^\mathsf{C}$$

between the category of homotopy prefactorization algebras on $\widehat{\mathsf{C}_\mathsf{max}}$ and the category of C-diagrams of commutative monoids in M. Therefore, $\mathsf{WO}_{\widehat{\mathsf{C}_\mathsf{max}}}^\mathsf{M}$-algebras should be homotopy coherent C-diagrams of E_∞-algebras. We will explain this in details in Section 11.10. Furthermore, if $\mathsf{M} = \mathsf{Chain}_\mathbb{K}$ with $\mathbb{K}$ a field of characteristic zero, then this adjunction is a Quillen equivalence.   ◇

Example 11.3.8 (Homotopy Costello-Gwilliam prefactorization algebras on manifolds). Suppose Emb^n is a small category equivalent to the category of smooth n-manifolds with open embeddings as morphisms. Recall from Example 10.7.4 that symmetric monoidal functors $\mathsf{Emb}^n \longrightarrow \mathsf{M}$ are called prefactorization algebras on n-manifolds with values in M. The category of such objects is isomorphic to the category $\mathsf{PFA}(\widehat{\mathsf{Emb}_\mathsf{max}^n})$ of prefactorization algebras on $\widehat{\mathsf{Emb}_\mathsf{max}^n}$. By Example 11.3.7 this category is related to the category of homotopy prefactorization algebras on $\widehat{\mathsf{Emb}_\mathsf{max}^n}$, i.e., $\mathsf{WO}_{\widehat{\mathsf{Emb}_\mathsf{max}^n}}^\mathsf{M}$-algebras, via the change-of-operad adjunction. In Section 11.10 we will see that homotopy prefactorization algebras on $\widehat{\mathsf{Emb}_\mathsf{max}^n}$ are homotopy coherent Emb^n-diagrams of E_∞-algebras. There are similar statements for Hol^n, which is a small category equivalent to the category of complex n-manifolds with open holomorphic embeddings as morphisms.   ◇

11.4 Coherence Theorem

In this section, we record the following coherence theorem for homotopy prefactorization algebras. In the remaining sections of this chapter, we will use this coherence theorem to study the structure of homotopy prefactorization algebras. For a vertex v in a colored tree, recall our convention of writing (v) for its profile. So if $\mathsf{Prof}(v) = \binom{d}{\underline{c}}$, then $\triangle(v) = \triangle\binom{d}{\underline{c}}$ for a configured category $(\mathsf{C}, \triangle)$ as in Definition 10.2.5.

Theorem 11.4.1. *Suppose* $\widehat{\mathsf{C}} = (\mathsf{C}, \triangle)$ *is a configured category with object set* $\mathfrak{C}$. *Then a* $\mathsf{WO}_{\widehat{\mathsf{C}}}^{\mathsf{M}}$*-algebra is precisely a pair* (X, λ) *consisting of*

- *a* $\mathfrak{C}$*-colored object* X *in* M *and*
- *a structure morphism*

$$\mathsf{J}[T] \otimes X_{\underline{c}} \xrightarrow{\quad \lambda_T \left\{ \underline{f}^v \right\}_{v \in T} \quad} X_d \ \in \mathsf{M} \tag{11.4.2}$$

for

- *each* $T \in \underline{\mathsf{Tree}}^{\mathfrak{C}}\binom{d}{\underline{c}}$ *with* $(\underline{c}; d) \in \mathsf{Prof}(\mathfrak{C}) \times \mathfrak{C}$ *and*
- *each* $\left\{ \underline{f}^v \right\}_{v \in T} \in \prod_{v \in T} \triangle(v)$

that satisfies the following four conditions.

Associativity *Suppose* $\left(\underline{c} = (c_1, \dots, c_n); d \right) \in \mathsf{Prof}(\mathfrak{C}) \times \mathfrak{C}$ *with* $n \geq 1$, $T \in \underline{\mathsf{Tree}}^{\mathfrak{C}}\binom{d}{\underline{c}}$, $T_j \in \underline{\mathsf{Tree}}^{\mathfrak{C}}\binom{c_j}{\underline{b}_j}$ *for* $1 \leq j \leq n$, $\underline{b} = (\underline{b}_1, \dots, \underline{b}_n)$,

$$G = \mathsf{Graft}(T; T_1, \dots, T_n) \in \underline{\mathsf{Tree}}^{\mathfrak{C}}\binom{d}{\underline{b}}$$

is the grafting (3.3.1), $\left\{ \underline{f}^v \right\}$ *is as above, and* $\left\{ \underline{f}^u \right\} \in \prod_{u \in T_j} \triangle(u)$ *for each* $1 \leq j \leq n$. *Then the diagram*

$$\mathsf{J}[T] \otimes \left(\bigotimes_{j=1}^{n} \mathsf{J}[T_j] \right) \otimes X_{\underline{b}} \xrightarrow{\quad (\pi, \mathrm{Id}) \quad} \mathsf{J}[G] \otimes X_{\underline{b}} \tag{11.4.3}$$

$$\text{permute} \downarrow \cong \qquad\qquad \downarrow \lambda_G \left\{ \underline{f}^w \right\}_{w \in G}$$

$$\mathsf{J}[T] \otimes \bigotimes_{j=1}^{n} \left(\mathsf{J}[T_j] \otimes X_{\underline{b}_j} \right)$$

$$\left(\mathrm{Id}, \otimes_j \lambda_{T_j} \left\{ \underline{f}^u \right\}_{u \in T_j} \right) \downarrow$$

$$\mathsf{J}[T] \otimes X_{\underline{c}} \xrightarrow{\quad \lambda_T \left\{ \underline{f}^v \right\}_{v \in T} \quad} X_d$$

is commutative. Here $\pi = \otimes_S 1$ *is the morphism in Lemma 6.2.7 for the grafting* G.

Unity *For each* $c \in \mathfrak{C}$, *the composition*

$$X_c \xrightarrow{\quad \cong \quad} \mathsf{J}[\uparrow_c] \otimes X_c \xrightarrow{\quad \lambda_{\uparrow_c} \{\varnothing\} \quad} X_c \tag{11.4.4}$$

is the identity morphism of X_c.

Equivariance *For each* $T \in \underline{\mathsf{Tree}}^{\mathfrak{C}}\binom{d}{\underline{c}}$, $\{\underline{f}^v\}$ *as above, and permutation* $\sigma \in \Sigma_{|\underline{c}|}$, *the diagram*

$$
\begin{array}{ccc}
\mathsf{J}[T] \otimes X_{\underline{c}} & \xrightarrow{\ \lambda_T\{\underline{f}^v\}_{v \in T}\ } & X_d \\
{\scriptstyle (\mathrm{Id},\sigma^{-1})}\big\downarrow & & \big\| \\
\mathsf{J}[T\sigma] \otimes X_{\underline{c}\sigma} & \xrightarrow{\ \lambda_{T\sigma}\{\underline{f}^v\}_{v \in T\sigma}\ } & X_d
\end{array}
\tag{11.4.5}
$$

is commutative, in which $T\sigma \in \underline{\mathsf{Tree}}^{\mathfrak{C}}\binom{d}{\underline{c}\sigma}$ *is the same as* T *except that its ordering is* $\zeta_T\sigma$ *with* ζ_T *the ordering of* T. *The permutation* $\sigma^{-1} : X_{\underline{c}} \xrightarrow{\cong} X_{\underline{c}\sigma}$ *permutes the factors in* $X_{\underline{c}}$.

Wedge Condition *Suppose* $T \in \underline{\mathsf{Tree}}^{\mathfrak{C}}\binom{d}{\underline{c}}$, $H_v \in \underline{\mathsf{Tree}}^{\mathfrak{C}}(v)$ *for each* $v \in \mathsf{Vt}(T)$, $K = T(H_v)_{v \in T}$ *is the tree substitution, and* $\{\underline{f}^u\} \in \prod_{u \in H_v} \triangle(u)$ *for each* $v \in \mathsf{Vt}(T)$. *Then the diagram*

$$
\begin{array}{ccc}
\mathsf{J}[T] \otimes X_{\underline{c}} & \xrightarrow{\ \lambda_T\{\underline{g}^v\}_{v \in T}\ } & X_d \\
{\scriptstyle (\mathsf{J},\mathrm{Id})}\big\downarrow & & \big\| \\
\mathsf{J}[K] \otimes X_{\underline{c}} & \xrightarrow{\ \lambda_K\{\underline{f}^w\}_{w \in K}\ } & X_d
\end{array}
\tag{11.4.6}
$$

is commutative. Here for each $v \in \mathsf{Vt}(T)$,

$$
\underline{g}^v = \gamma^{\mathsf{O}_{\widehat{\mathfrak{C}}}}_{H_v}\!\left(\{\underline{f}^u\}_{u \in H_v} \right) \in \mathsf{O}_{\widehat{\mathfrak{C}}}(v) = \triangle(v)
$$

with

$$
\mathsf{O}_{\widehat{\mathfrak{C}}}[H_v] = \prod_{u \in H_v} \mathsf{O}_{\widehat{\mathfrak{C}}}(u) \xrightarrow{\ \gamma^{\mathsf{O}_{\widehat{\mathfrak{C}}}}_{H_v}\ } \mathsf{O}_{\widehat{\mathfrak{C}}}(v)
$$

the operadic structure morphism of $\mathsf{O}_{\widehat{\mathfrak{C}}}$ *for* H_v *in* (4.4.10).

A morphism $f : (X, \lambda^X) \longrightarrow (Y, \lambda^Y)$ *of* $\mathsf{WO}^{\mathsf{M}}_{\widehat{\mathfrak{C}}}$-*algebras is a morphism of the underlying* $\mathfrak{C}$-*colored objects that respects the structure morphisms in* (11.4.2) *in the obvious sense.*

Proof. This is the special case of the Coherence Theorem 7.2.1 applied to the $\mathfrak{C}$-colored operad $\mathsf{O}^{\mathsf{M}}_{\widehat{\mathfrak{C}}}$. Indeed, since

$$
\mathsf{O}^{\mathsf{M}}_{\widehat{\mathfrak{C}}}\binom{d}{\underline{c}} = \coprod_{\mathsf{O}_{\widehat{\mathfrak{C}}}\binom{d}{\underline{c}}} \mathbb{1} = \coprod_{\triangle\binom{d}{\underline{c}}} \mathbb{1},
$$

for each $T \in \underline{\mathsf{Tree}}^{\mathfrak{C}}\binom{d}{\underline{c}}$ there is a canonical isomorphism

$$
\mathsf{O}^{\mathsf{M}}_{\widehat{\mathfrak{C}}}[T] = \bigotimes_{v \in T} \mathsf{O}^{\mathsf{M}}_{\widehat{\mathfrak{C}}}(v) = \bigotimes_{v \in T} \left(\coprod_{\triangle(v)} \mathbb{1} \right) \cong \coprod_{\prod_{v \in T} \triangle(v)} \mathbb{1}.
$$

This implies that there is a canonical isomorphism

$$J[T] \otimes O_{\underline{C}}^M[T] \otimes X_{\underline{c}} \cong \coprod_{\prod_{v \in T} \Delta(v)} J[T] \otimes X_{\underline{c}}.$$

Therefore, the structure morphism λ_T in (7.2.2) is uniquely determined by the restricted structure morphisms $\lambda_T \{\underline{f}^v\}_{v \in T}$ in (11.4.2). The above associativity, unity, equivariance, and wedge conditions are those in the Coherence Theorem 7.2.1. □

Interpretation 11.4.7. In a homotopy prefactorization algebra on a configured category $\widehat{C}$, the structure morphism $\lambda_T \{\underline{f}^v\}_{v \in T}$ is specified by (i) first choosing a C-colored tree $T \in \underline{\mathsf{Tree}}^C\binom{d}{\underline{c}}$ and (ii) then choosing a configuration $\underline{f}^v \in \Delta(v)$ for each vertex v in T. In other words, the structure morphisms are parametrized by C-colored trees whose internal edges are decorated by the commutative segment J and whose vertices are decorated by configurations in $\widehat{C}$ with the correct profile. ◇

Example 11.4.8 (Homotopy prefactorization algebras on bounded lattices). For a bounded lattice $(L, \leq)$ with least element 0, consider the configured category $\widehat{L} = (L, \Delta^L)$ in Example 10.2.10. For any two elements $c, d \in L$, there is a morphism $c \longrightarrow d$ in L, which is necessarily unique, if and only if $c \leq d$. Each subset $\Delta^L\binom{d}{\underline{c}}$ of configurations is either empty or a one-element set. If $v \in \mathsf{Vt}(T)$ has profile $\binom{b}{a_1,\dots,a_m}$, then

$$\underline{f}^v \in \Delta^L(v) = \Delta^L\binom{b}{a_1,\dots,a_m}$$

if and only if

- $a_i \leq b$ in L for each $1 \leq i \leq m$ and
- $a_i \wedge a_j = 0$ for all $1 \leq i \neq j \leq m$.

In this case,

$$\underline{f}^v = \{a_i \longrightarrow b\}_{i=1}^m$$

is the unique element in $\prod_{i=1}^m L(a_i, b)$. ◇

Example 11.4.9 (Homotopy Costello-Gwilliam prefactorization algebras). For a topological space X, consider the configured category $\widehat{\mathsf{Open}(X)}$ in Example 10.2.11. Since $\widehat{\mathsf{Open}(X)}$ is an example of $\widehat{L}$ for the bounded lattice $\mathsf{Open}(X)$, everything in Example 11.4.8 applies to $\widehat{\mathsf{Open}(X)}$. Suppose T is an $\mathsf{Open}(X)$-colored tree, and $v \in \mathsf{Vt}(T)$ has profile $\binom{V}{U_1,\dots,U_m}$ with $U_1,\dots,U_m, V \in \mathsf{Open}(X)$. Then

$$\underline{f}^v \in \Delta^X(v) = \Delta^X\binom{V}{U_1,\dots,U_m}$$

if and only if $\{U_i\}_{i=1}^m$ are pairwise disjoint subsets of V. In this case,

$$\underline{f}^v = \{U_i \subset V\}_{i=1}^m$$

is the unique element in $\prod_{i=1}^m \mathsf{Open}(X)(U_i, V)$. ◇

Example 11.4.10 (Homotopy Costello-Gwilliam equivariant prefactorization algebras). For a topological space X equipped with a left action by a group G, consider the configured category $\widehat{\mathsf{Open}(X)}_G$ in Example 10.2.12. The objects in $\mathsf{Open}(X)_G$ are the objects in $\mathsf{Open}(X)$, i.e., open subsets of X, but there are more morphisms in $\mathsf{Open}(X)_G$ than in $\mathsf{Open}(X)$. Suppose T is an $\mathsf{Open}(X)$-colored tree, and $v \in \mathsf{Vt}(T)$ has profile $\binom{V}{U_1,\dots,U_m}$ with $U_1,\dots,U_m, V \in \mathsf{Open}(X)$. Then

$$\underline{f}^v \in \Delta^X_G(v) = \Delta^X_G\binom{V}{U_1,\dots,U_m}$$

if and only if $\underline{f}^v$ has the form

$$\left\{ U_i \xrightarrow{\;g_i\;} g_i U_i \xrightarrow{\;\text{inclusion}\;} V \right\}_{i=1}^m \in \prod_{i=1}^m \mathsf{Open}(X)_G(U_i, V)$$

for some $g_1,\dots,g_m \in G$ such that $\{g_i U_i\}_{i=1}^m$ are pairwise disjoint subsets of V. ◇

11.5 Homotopy Coherent Pointed Diagrams

Using the Coherence Theorem 11.4.1, for the next few sections we will explain the structure that exists in homotopy prefactorization algebras, i.e., in $\mathsf{WO}^{\mathsf{M}}_{\widehat{\mathsf{C}}}$-algebras. In this section, we explain the homotopy coherent pointed diagram structure that exists on each homotopy prefactorization algebra.

Definition 11.5.1. Objects in the category $\mathsf{Alg}_{\mathsf{M}}\big(\mathsf{WO}^{\mathsf{M}}_{\widehat{\mathsf{C}_{\min}}}\big)$ are called *homotopy coherent pointed C-diagrams in* M.

Interpretation 11.5.2. The $\mathfrak{C}$-colored operad $\mathsf{O}^{\mathsf{M}}_{\widehat{\mathsf{C}_{\min}}}$ is the operad for pointed C-diagrams in M, so algebras over its Boardman-Vogt construction are homotopy coherent pointed C-diagrams in M. We saw in Example 11.3.2 that the right adjoint in the change-of-operad adjunction

$$\mathsf{HPFA}(\widehat{\mathsf{C}_{\min}}) = \mathsf{Alg}_{\mathsf{M}}\big(\mathsf{WO}^{\mathsf{M}}_{\widehat{\mathsf{C}_{\min}}}\big) \underset{(\mathsf{WO}^{\mathsf{M}}_{i_0})^*}{\overset{(\mathsf{WO}^{\mathsf{M}}_{i_0})_!}{\rightleftarrows}} \mathsf{Alg}_{\mathsf{M}}\big(\mathsf{WO}^{\mathsf{M}}_{\widehat{\mathsf{C}}}\big) = \mathsf{HPFA}(\widehat{\mathsf{C}})$$

sends each homotopy prefactorization algebra on $\widehat{\mathsf{C}}$ to its underlying homotopy coherent pointed C-diagram in M. ◇

To understand homotopy coherent pointed C-diagrams in M, first we make explicit the colored operad $\mathsf{O}^{\mathsf{M}}_{\widehat{\mathsf{C}_{\min}}}$. The next result is a consequence of the definition of $\widehat{\mathsf{C}_{\min}}$ in Example 10.2.8 and of Definition 10.3.2.

Lemma 11.5.3. *Suppose* C *is a small category with object set* $\mathfrak{C}$*. Then the* $\mathfrak{C}$*-colored operad* $\mathsf{O}^{\mathsf{M}}_{\widehat{\mathsf{C}_{\min}}}$ *has entries*

$$\mathsf{O}^{\mathsf{M}}_{\widehat{\mathsf{C}_{\min}}}\binom{d}{\underline{c}} = \begin{cases} \mathbb{1} & \text{if } \underline{c} = \varnothing, \\ \coprod_{\mathsf{C}(c,d)} \mathbb{1} & \text{if } \underline{c} = c \in \mathfrak{C}, \\ \varnothing & \text{if } |\underline{c}| \geq 2 \end{cases}$$

for $(\underline{c}; d) \in \mathsf{Prof}(\mathfrak{C}) \times \mathfrak{C}$.

Motivation 11.5.4. The following result is the coherence theorem for homotopy coherent pointed diagrams. A pointed C-diagram in M consists of a C-diagram in M and compatible colored units for objects in C. Therefore, a homotopy coherent pointed C-diagram in M should contain a homotopy coherent C-diagram in M along with homotopically compatible homotopy colored units. ◇

We will refer to (i) the Coherence Theorem 7.3.5 for homotopy coherent C-diagrams in M, (ii) the linear graphs $\mathsf{Lin}_?$ in Example 3.1.19, and (iii) the truncated linear graphs $\mathsf{lin}_?$ in Example 3.1.20.

Theorem 11.5.5. *A homotopy coherent pointed C-diagram in M is exactly a triple* (X, λ, θ) *consisting of*

- *a homotopy coherent C-diagram* (X, λ) *in M and*
- *a structure morphism*

$$J[\mathsf{lin}_{\underline{c}}] \xrightarrow{\theta_{\underline{c}}^{\underline{f}}} X_{c_n} \in \mathsf{M} \tag{11.5.6}$$

for

- *each profile* $\underline{c} = (c_1, \ldots, c_n) \in \mathsf{Prof}(\mathfrak{C})$ *with* $n \geq 1$;
- *each sequence of composable C-morphisms* $\underline{f} = (f_2, \ldots, f_n)$ *with* $f_j \in \mathsf{C}(c_{j-1}, c_j)$ *for* $2 \leq j \leq n$

that satisfies the following two conditions.

Associativity *Suppose* $1 \leq n \leq p$, $\underline{c} = (c_1, \ldots, c_n)$, *and* $\underline{c}' = (c_n, \ldots, c_p) \in \mathsf{Prof}(\mathfrak{C})$. *Suppose* $f_j \in \mathsf{C}(c_{j-1}, c_j)$ *for each* $2 \leq j \leq p$ *with* $\underline{f} = (f_2, \ldots, f_n)$ *and* $\underline{f}' = (f_{n+1}, \ldots, f_p)$. *Then the diagram*

$$\begin{array}{ccc} J[\mathsf{Lin}_{\underline{c}'}] \otimes J[\mathsf{lin}_{\underline{c}}] & \xrightarrow{\pi} & J\left[\mathsf{lin}_{(c_1, \ldots, c_p)}\right] \\ \scriptstyle (\mathrm{Id}, \theta_{\underline{c}}^{\underline{f}}) \Big\downarrow & & \Big\downarrow \scriptstyle \theta_{(c_1, \ldots, c_p)}^{(\underline{f}, \underline{f}')} \\ J[\mathsf{Lin}_{\underline{c}'}] \otimes X_{c_n} & \xrightarrow{\lambda_{\underline{c}'}^{\underline{f}'}} & X_{c_p} \end{array} \tag{11.5.7}$$

is commutative. Here the truncated linear graph $\mathsf{lin}_{(c_1, \ldots, c_p)}$ *is regarded as the grafting* (3.3.1) *of the linear graph* $\mathsf{Lin}_{\underline{c}'}$ *and the truncated linear graph* $\mathsf{lin}_{\underline{c}}$ *with* π *the morphism in Lemma 6.2.7.*

Wedge Condition *Suppose* $\underline{c} = (c_1, \ldots, c_n) \in \mathsf{Prof}(\mathfrak{C})$ *with* $n \geq 1$,

$$\underline{b}_j = \begin{cases} \left(b_1^1, \ldots, b_{k_1}^1 = c_1\right) & \text{with } k_1 \geq 1 \text{ for } j = 1, \\ \left(c_{j-1} = b_0^j, b_1^j, \ldots, b_{k_j}^j = c_j\right) & \text{with } k_j \geq 0 \text{ for } 2 \leq j \leq n \end{cases}$$

in $\mathsf{Prof}(\mathfrak{C})$, *and* $\underline{b} = (\underline{b}_1, \ldots, \underline{b}_n)$. *Suppose* $f_i^j \in \mathsf{C}(b_{i-1}^j, b_i^j)$ *for each* $1 \leq j \leq n$ *and* $1 \leq i \leq k_j$ *except for* f_1^1,

$$\underline{f}^j = \begin{cases} (f_2^1, \ldots, f_{k_1}^1) & \text{if } j = 1, \\ (f_1^j, \ldots, f_{k_j}^j) & \text{if } 2 \leq j \leq n, \end{cases}$$

$\underline{f} = (\underline{f}^1, \ldots, \underline{f}^n)$, *and*

$$f^j = \begin{cases} f^1_{k_1} \circ \cdots \circ f^1_2 \in \mathsf{C}(b^1_1, c_1) & \text{if } j = 1, \\ f^j_{k_j} \circ \cdots \circ f^j_1 \in \mathsf{C}(c_{j-1}, c_j) & \text{if } 2 \leq j \leq n. \end{cases}$$

Then the diagram

$$
\begin{array}{ccc}
\mathsf{J}[\mathrm{lin}_{\underline{c}}] & \xrightarrow{\;\theta_{\underline{c}}^{(f^1,\ldots,f^n)}\;} & X_{c_n} \\
{\scriptstyle \mathsf{J}}\downarrow & & \| \\
\mathsf{J}[\mathrm{lin}_{\underline{b}}] & \xrightarrow[\;\theta_{\underline{b}}^f\;]{} & X_{c_n}
\end{array}
\tag{11.5.8}
$$

is commutative. Here the truncated linear graph $\mathrm{lin}_{\underline{b}}$ *is regarded as the tree substitution*

$$\mathrm{lin}_{\underline{b}} = \mathrm{lin}_{\underline{c}}\Big(\mathrm{lin}_{\underline{b}_1}, \mathrm{Lin}_{\underline{b}_2}, \ldots, \mathrm{Lin}_{\underline{b}_n}\Big).$$

Proof. This is the special case of the Coherence Theorem 7.2.1 for the $\mathfrak{C}$-colored operad $\mathsf{O}^{\mathsf{M}}_{\mathfrak{C}_{\min}}$. Indeed, by Lemma 11.5.3 the equivariant structure on $\mathsf{O}^{\mathsf{M}}_{\mathfrak{C}_{\min}}$ is trivial. Since $\mathsf{O}^{\mathsf{M}}_{\mathfrak{C}_{\min}}$ is concentrated in 0-ary and unary entries, if $T \in \underline{\mathsf{Tree}}^{\mathfrak{C}}\binom{d}{\underline{c}}$ is neither a linear graph nor a truncated linear graph, then

$$\mathsf{O}^{\mathsf{M}}_{\mathfrak{C}_{\min}}[T] = \bigotimes_{v \in T} \mathsf{O}^{\mathsf{M}}_{\mathfrak{C}_{\min}}\binom{\mathrm{out}(v)}{\mathrm{in}(v)} = \varnothing.$$

In this case, the structure morphism

$$\mathsf{J}[T] \otimes \mathsf{O}^{\mathsf{M}}_{\mathfrak{C}_{\min}}[T] \otimes X_{\underline{c}} \xrightarrow{\;\lambda_T\;} X_d$$

in (7.2.2) for a $\mathsf{WO}^{\mathsf{M}}_{\mathfrak{C}_{\min}}$-algebra is the trivial morphism $\varnothing \longrightarrow X_d$. In particular, the equivariance condition (7.2.5) is trivial for $\mathsf{WO}^{\mathsf{M}}_{\mathfrak{C}_{\min}}$-algebras. For linear graphs T, we have exactly the structure of a homotopy coherent C-diagram in M as in Theorem 7.3.5.

For $\underline{c} = (c_1, \ldots, c_n) \in \mathsf{Prof}(\mathfrak{C})$ with $n \geq 1$, we have natural isomorphisms

$$\mathsf{O}^{\mathsf{M}}_{\mathfrak{C}_{\min}}[\mathrm{lin}_{\underline{c}}] = \mathsf{O}^{\mathsf{M}}_{\mathfrak{C}_{\min}}\binom{c_1}{\varnothing} \otimes \left[\bigotimes_{j=2}^{n} \mathsf{O}^{\mathsf{M}}_{\mathfrak{C}_{\min}}\binom{c_j}{c_{j-1}}\right] \cong \bigotimes_{j=2}^{n}\left[\coprod_{\mathsf{C}(c_{j-1},c_j)} \mathbb{1}\right] \cong \coprod_{\prod_{j=2}^{n} \mathsf{C}(c_{j-1},c_j)} \mathbb{1}.$$

This implies that there is a natural isomorphism

$$\mathsf{J}[\mathrm{lin}_{\underline{c}}] \otimes \mathsf{O}^{\mathsf{M}}_{\mathfrak{C}_{\min}}[\mathrm{lin}_{\underline{c}}] \cong \coprod_{\prod_{j=2}^{n} \mathsf{C}(c_{j-1},c_j)} \mathsf{J}[\mathrm{lin}_{\underline{c}}].$$

So the structure morphism

$$\mathsf{J}[\mathrm{lin}_{\underline{c}}] \otimes \mathsf{O}^{\mathsf{M}}_{\mathfrak{C}_{\min}}[\mathrm{lin}_{\underline{c}}] \xrightarrow{\;\lambda^{\mathrm{lin}_{\underline{c}}}\;} X_{c_n}$$

in (7.2.2) is uniquely determined by the restrictions $\theta_{\underline{c}}^f$ in (11.5.6). The associativity and wedge conditions are exactly those in the Coherence Theorem 7.2.1 for (truncated) linear graphs. $\qquad\square$

Interpretation 11.5.9. A homotopy coherent pointed C-diagram in M has a homotopy coherent C-diagram in M and additional structure morphisms $\theta_{\underline{c}}^{f}$ for truncated linear graphs. One should think of the structure morphism $\theta_{\underline{c}}^{f}$ in (11.5.6) as determined by the decorated truncated linear graph

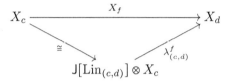

with all but the first vertices decorated by the C-morphisms $f_j \in C(c_{j-1}, c_j)$. Note that if $n = 1$, then $(f_j) = \varnothing$. ◇

Example 11.5.10 (Homotopy colored units). Suppose (X, λ, θ) is a homotopy coherent pointed C-diagram in M. It has structure morphisms

$$
\begin{array}{ccc}
X_c & \xrightarrow{\ \ X_f\ \ } & X_d \\
 & \searrow_{\cong} \qquad \nearrow_{\lambda_{(c,d)}^{f}} & \\
 & J[\mathrm{Lin}_{(c,d)}] \otimes X_c &
\end{array}
$$

for $f \in C(c, d)$ as in Example 7.3.11 and

$$
\mathbb{1} = J[\mathrm{lin}_{(c)}] \xrightarrow{\ \theta_{(c)}^{\varnothing}\ } X_c
$$

for $c \in C$. If X is actually a pointed C-diagram in M as in Definition 10.4.1, then the diagram

$$
\begin{array}{ccc}
\mathbb{1} & =\!=\!= & \mathbb{1} \\
{\scriptstyle \theta_{(c)}^{\varnothing}}\downarrow & & \downarrow{\scriptstyle \theta_{(d)}^{\varnothing}} \\
X_c & \xrightarrow{\ X_f\ } & X_d
\end{array}
$$

is commutative. For a homotopy coherent pointed C-diagram in M, this diagram is homotopy commutative in the following sense.

The diagram

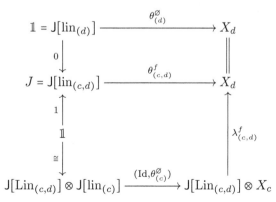

is commutative, where $0, 1 : \mathbb{1} \longrightarrow J$ are part of the commutative segment J.

- The top rectangle is commutative by the wedge condition (11.5.8) for the tree substitution

$$\mathrm{lin}_{(c,d)} = \mathrm{lin}_{(d)}\big(\mathrm{lin}_{(c,d)}\big).$$

Visually $\mathrm{lin}_{(c,d)}$ and $\mathrm{lin}_{(d)}$ are the truncated linear graphs

and similarly for $\mathrm{lin}_{(c)}$.

- The bottom square is commutative by the associativity condition (11.5.7) with $\underline{c} = (c)$ and $\underline{c}' = (c,d)$.

In other words, $\theta^{\varnothing}_{(c)}$ is a homotopy colored unit that is preserved by $\lambda^{f}_{(c,d)}$ up to the homotopy $\theta^{f}_{(c,d)}$ that is also a structure morphism. ◇

11.6   Homotopy Time-Slice Axiom

In this section, we explain a homotopy coherent version of the time-slice axiom in homotopy prefactorization algebras.

Motivation 11.6.1. Suppose $\widehat{\mathsf{C}} = (\mathsf{C}, \triangle)$ is a configured category, and S is a set of morphisms in C. In Example 11.3.3 we saw that there is a change-of-operad adjunction

$$\mathsf{HPFA}(\widehat{\mathsf{C}}) = \mathsf{Alg}_{\mathsf{M}}\big(\mathsf{WO}^{\mathsf{M}}_{\widehat{\mathsf{C}}}\big) \underset{(W\ell^{\mathsf{M}})^{*}}{\overset{W\ell^{\mathsf{M}}_{!}}{\rightleftarrows}} \mathsf{Alg}_{\mathsf{M}}\big(\mathsf{WO}_{\widehat{\mathsf{C}}}[S^{-1}]^{\mathsf{M}}\big) = \mathsf{HPFA}(\widehat{\mathsf{C}}, S) \ .$$

The objects on the right side are homotopy prefactorization algebras on $\widehat{\mathsf{C}}$ satisfying the homotopy time-slice axiom with respect to S. In view of Theorem 10.3.13, for each $s \in S$, we therefore expect the structure morphism $\lambda\{s\}$ to be invertible up to coherent homotopies that are also structure morphisms. We will explain this in the following examples. ◇

Example 11.6.2 (Left homotopy inverses). Suppose (X, λ) is a homotopy prefactorization algebra on $\widehat{\mathsf{C}}$ satisfying the homotopy time-slice axiom with respect to S, i.e., a $\mathsf{WO}_{\widehat{\mathsf{C}}}[S^{-1}]^{\mathsf{M}}$-algebra. Since $\mathsf{O}_{\widehat{\mathsf{C}}}[S^{-1}]^{\mathsf{M}}$ has entries

$$\mathsf{O}_{\widehat{\mathsf{C}}}[S^{-1}]^{\mathsf{M}}\binom{d}{\underline{c}} = \coprod_{\mathsf{O}_{\widehat{\mathsf{C}}}[S^{-1}]\binom{d}{\underline{c}}} \mathbb{1} \quad \text{for} \quad \binom{d}{\underline{c}} \in \mathsf{Prof}(\mathfrak{C}) \times \mathfrak{C},$$

there is a canonical isomorphism

$$\mathsf{O}_{\widehat{\mathsf{C}}}[S^{-1}]^{\mathsf{M}}[T] = \bigotimes_{v \in T} \mathsf{O}_{\widehat{\mathsf{C}}}[S^{-1}]^{\mathsf{M}}(v) = \bigotimes_{v \in T} \left[\coprod_{\mathsf{O}_{\widehat{\mathsf{C}}}[S^{-1}](v)} \mathbb{1} \right] \cong \coprod_{\prod\limits_{v \in T} \mathsf{O}_{\widehat{\mathsf{C}}}[S^{-1}](v)} \mathbb{1}$$

for each $\mathfrak{C}$-colored tree $T \in \underline{\mathsf{Tree}}\binom{d}{\underline{c}}$. Therefore, there is a canonical isomorphism

$$\mathsf{J}[T] \otimes \mathsf{O}_{\widehat{\mathsf{C}}}[S^{-1}]^{\mathsf{M}}[T] \otimes X_{\underline{c}} \cong \coprod_{\prod\limits_{v \in T} \mathsf{O}_{\widehat{\mathsf{C}}}[S^{-1}](v)} \mathsf{J}[T] \otimes X_{\underline{c}}.$$

It follows that the structure morphisms λ_T in (7.2.2) is uniquely determined by the restricted structure morphisms

$$J[T] \otimes X_{\underline{c}} \xrightarrow{\quad \lambda_T\{\underline{f}^v\}_{v \in T} \quad} X_d \in \mathsf{M}$$

for $T \in \underline{\mathsf{Tree}}(\overset{d}{\underline{c}})$ and $\{\underline{f}^v\}_{v \in T} \in \prod_{v \in T} \mathsf{O}_{\widehat{\mathfrak{C}}}[S^{-1}](v)$.

Suppose $f : c \longrightarrow d$ is a morphism in S, so

$$f \in \mathsf{O}_{\widehat{\mathfrak{C}}}[S^{-1}](\overset{d}{c}) \quad \text{and} \quad f^{-1} \in \mathsf{O}_{\widehat{\mathfrak{C}}}[S^{-1}](\overset{c}{d}).$$

Note that

$$\gamma_{\mathrm{Lin}_{(c,d,c)}}^{\mathsf{O}_{\widehat{\mathfrak{C}}}[S^{-1}]}(f, f^{-1}) = 1_c \in \mathsf{O}_{\widehat{\mathfrak{C}}}[S^{-1}](\overset{c}{c}),$$

the c-colored unit in the colored operad $\mathsf{O}_{\widehat{\mathfrak{C}}}[S^{-1}]$. Here

$$\mathsf{O}_{\widehat{\mathfrak{C}}}[S^{-1}][\mathrm{Lin}_{(c,d,c)}] \cong \mathsf{O}_{\widehat{\mathfrak{C}}}[S^{-1}](\overset{d}{c}) \times \mathsf{O}_{\widehat{\mathfrak{C}}}[S^{-1}](\overset{c}{d}) \xrightarrow{\quad \gamma_{\mathrm{Lin}_{(c,d,c)}}^{\mathsf{O}_{\widehat{\mathfrak{C}}}[S^{-1}]} \quad} \mathsf{O}_{\widehat{\mathfrak{C}}}[S^{-1}](\overset{c}{c})$$

is the operadic structure morphism (4.4.10) of $\mathsf{O}_{\widehat{\mathfrak{C}}}[S^{-1}]$ for the linear graph $\mathrm{Lin}_{(c,d,c)}$.
The diagram

$$
\begin{array}{ccc}
\mathbb{1} \otimes X_c = J[\mathrm{Lin}_{(c,c)}] \otimes X_c & \xrightarrow{\lambda_{\mathrm{Lin}_{(c,c)}}\{1_c\}} & X_c \\[2mm]
{\scriptstyle (0,\mathrm{Id})}\downarrow & & \| \\[2mm]
J \otimes X_c = J[\mathrm{Lin}_{(c,d,c)}] \otimes X_c & \xrightarrow{\lambda_{\mathrm{Lin}_{(c,d,c)}}\{f,f^{-1}\}} & X_c \\[2mm]
{\scriptstyle (1,\mathrm{Id})}\uparrow & & \uparrow \\[2mm]
\mathbb{1} \otimes X_c & & \quad{\scriptstyle \lambda_{\mathrm{Lin}_{(d,c)}}\{f^{-1}\}} \\[2mm]
{\scriptstyle \cong}\downarrow & & \\[2mm]
J[\mathrm{Lin}_{(d,c)}] \otimes J[\mathrm{Lin}_{(c,d)}] \otimes X_c & \xrightarrow{(\mathrm{Id},\lambda_{\mathrm{Lin}_{(c,d)}}\{f\})} & J[\mathrm{Lin}_{(d,c)}] \otimes X_d
\end{array}
$$

is commutative:

- The top rectangle is commutative by the wedge condition (7.2.6) for the tree substitution

$$\mathrm{Lin}_{(c,d,c)} = \mathrm{Lin}_{(c,c)}\big(\mathrm{Lin}_{(c,d,c)}\big).$$

- The bottom square is commutative by the associativity condition (7.2.3) for the grafting

$$\mathrm{Lin}_{(c,d,c)} = \mathsf{Graft}\big(\mathrm{Lin}_{(d,c)}; \mathrm{Lin}_{(c,d)}\big).$$

The structure morphism $\lambda_{\mathrm{Lin}_{(c,c)}}\{1_c\}$ is isomorphic to the identity morphism on X_c by Corollary 7.2.8 . Therefore, the above commutative diagram says that the structure morphism $\lambda_{\mathrm{Lin}_{(d,c)}}\{f^{-1}\}$ is a left homotopy inverse of the structure morphism $\lambda_{\mathrm{Lin}_{(c,d)}}\{f\}$ via the homotopy $\lambda_{\mathrm{Lin}_{(c,d,c)}}\{f,f^{-1}\}$ that is also a structure morphism.$\diamond$

Example 11.6.3 (Right homotopy inverses). Similarly, the diagram

$$
\begin{array}{ccc}
\mathbb{1} \otimes X_d = J[\mathrm{Lin}_{(d,d)}] \otimes X_d & \xrightarrow{\lambda_{\mathrm{Lin}_{(d,d)}}\{1_d\}} & X_d \\
{\scriptstyle(0,\mathrm{Id})}\downarrow & & \big\| \\
J \otimes X_d = J[\mathrm{Lin}_{(d,c,d)}] \otimes X_d & \xrightarrow{\lambda_{\mathrm{Lin}_{(d,c,d)}}\{f^{-1},f\}} & X_d \\
{\scriptstyle(1,\mathrm{Id})}\uparrow & & \\
\mathbb{1} \otimes X_d & & \Big\uparrow{\scriptstyle\lambda_{\mathrm{Lin}_{(c,d)}}\{f\}} \\
{\scriptstyle\cong}\downarrow & & \\
J[\mathrm{Lin}_{(c,d)}] \otimes J[\mathrm{Lin}_{(d,c)}] \otimes X_d & \xrightarrow{\left(\mathrm{Id},\lambda_{\mathrm{Lin}_{(d,c)}}\{f^{-1}\}\right)} & J[\mathrm{Lin}_{(c,d)}] \otimes X_c
\end{array}
$$

is commutative, and $\lambda_{\mathrm{Lin}_{(d,d)}}\{1_d\}$ is isomorphic to the identity morphism on X_d. Therefore, the commutative diagram says that the structure morphism $\lambda_{\mathrm{Lin}_{(d,c)}}\{f^{-1}\}$ is a right homotopy inverse of the structure morphism $\lambda_{\mathrm{Lin}_{(c,d)}}\{f\}$ via the homotopy $\lambda_{\mathrm{Lin}_{(d,c,d)}}\{f^{-1},f\}$ that is also a structure morphism. $\diamond$

11.7 E_∞-Algebra Structure

In this section, we explain that some entries in a homotopy prefactorization algebra are E_∞-algebras as in Definition 7.6.2.

Motivation 11.7.1. Suppose $\widehat{C} = (C, \triangle)$ is a configured category with object set $\mathfrak{C}$, and $c \in \mathfrak{C}$ such that $\{\mathrm{Id}_c\}_{i=1}^n \in \triangle\binom{c}{c,\dots,c}$ for all n. For each prefactorization algebra (Y, λ) on $\widehat{C}$, we saw in Proposition 10.5.1 that its c-colored entry Y_c is equipped with the structure of a commutative monoid. For a homotopy prefactorization algebra on $\widehat{C}$, we expect the entry Y_c to be an E_∞-algebra. $\diamond$

The following result is a consequence of Corollary 6.4.9 and Proposition 10.5.1.

Corollary 11.7.2. *Suppose $\widehat{C} = (C, \triangle)$ is a configured category with object set $\mathfrak{C}$, and $c \in \mathfrak{C}$ such that*

$$\{\mathrm{Id}_c\}_{i=1}^n \in \triangle\binom{c}{c,\dots,c}$$

for all n. Then the operad morphism

$$\mathsf{Com} \xrightarrow{\iota_c} \mathsf{O}_{\widehat{C}}^{\mathsf{M}}$$

in Proposition 10.5.1 induces a diagram of change-of-operad adjunctions

$$
\begin{array}{ccc}
\mathsf{Alg_M}(\mathsf{WCom}) & \underset{\mathsf{W}\iota_c^*}{\overset{(\mathsf{W}\iota_c)!}{\rightleftarrows}} & \mathsf{Alg_M}\!\left(\mathsf{WO}_{\widehat{C}}^{\mathsf{M}}\right) = \mathsf{HPFA}(\widehat{C}) \\
{\scriptstyle\eta_!}\uparrow\downarrow{\scriptstyle\eta^*} & & {\scriptstyle\eta_!}\uparrow\downarrow{\scriptstyle\eta^*} \\
\mathsf{Com}(\mathsf{M}) = \mathsf{Alg_M}(\mathsf{Com}) & \underset{\iota_c^*}{\overset{(\iota_c)!}{\rightleftarrows}} & \mathsf{Alg_M}\!\left(\mathsf{O}_{\widehat{C}}^{\mathsf{M}}\right) = \mathsf{PFA}(\widehat{C})
\end{array}
$$

with commuting left adjoint diagram and commuting right adjoint diagram.

Interpretation 11.7.3. The right adjoint $\mathsf{W}\iota_c^*$ sends each homotopy prefactorization algebra (Y, λ) on $\widehat{\mathsf{C}}$ to the E_∞-algebra

$$\mathsf{W}\iota_c^*(Y, \lambda) \in \mathsf{Alg}_\mathsf{M}(\mathsf{WCom}).$$

The underlying object is the entry $Y_c \in \mathsf{M}$. For each one-colored tree $T \in \mathsf{Tree}(n)$, the E_∞-algebra structure morphism

$$J[T] \otimes Y_c^{\otimes n} \xrightarrow{\ \lambda_T\ } Y_c$$

in (7.6.8) is the structure morphism

$$\lambda_{T_c}\left\{\{\mathrm{Id}_c\}_{i=1}^{|\mathrm{in}(v)|}\right\}_{v \in T_c}$$

in (11.4.2). Here $T_c \in \mathsf{Tree}^{\mathfrak{C}}\binom{c}{c,\dots,c}$ is the c-colored tree obtained from T by replacing every edge color by c. For each vertex $v \in T_c$,

$$\{\mathrm{Id}_c\}_{i=1}^{|\mathrm{in}(v)|} \in \Delta\binom{c}{c,\dots,c}$$

is a configuration by assumption. $\diamond$

Example 11.7.4 (Homotopy prefactorization algebras on bounded lattices). For a bounded lattice $(L, \le)$ with least element 0, consider the configured category $\widehat{L} = (L, \Delta^L)$ in Example 10.2.10. The least element $0 \in L$ has the property that

$$\{\mathrm{Id}_0\}_{i=1}^n \in \Delta^L\binom{0}{0,\dots,0} \quad \text{for} \quad n \ge 0.$$

If (Y, λ) is a homotopy prefactorization algebra on $\widehat{L}$, i.e., a $\mathsf{WO}_{\widehat{L}}^\mathsf{M}$-algebra, then the entry $Y_0 \in \mathsf{M}$ is equipped with the structure of an E_∞-algebra by Corollary 11.7.2. $\diamond$

Example 11.7.5 (Homotopy Costello-Gwilliam prefactorization algebras). For a topological space X, consider the configured category

$$\widehat{\mathsf{Open}(X)} = \left(\mathsf{Open}(X), \Delta^X\right)$$

in Example 10.2.11. The category $\mathsf{Open}(X)$ is a bounded lattice with least element $\varnothing_X \subset X$, the empty subset of X. The configured category $\widehat{\mathsf{Open}(X)}$ has the form $\widehat{L}$ in Example 10.2.10. Therefore, as in Example 11.7.4, if (Y, λ) is a homotopy prefactorization algebra on $\widehat{\mathsf{Open}(X)}$, i.e., a $\mathsf{WO}_{\widehat{\mathsf{Open}(X)}}^\mathsf{M}$-algebra, then the entry $Y_{\varnothing_X} \in \mathsf{M}$ is equipped with the structure of an E_∞-algebra by Corollary 11.7.2. A similar statement holds for the configured category

$$\widehat{\mathsf{Open}(X)}_G = \left(\mathsf{Open}(X)_G, \Delta_G^X\right)$$

in Example 10.2.12. $\diamond$

11.8   Objectwise E_∞-Module

Suppose $(L, \leq)$ is an arbitrary but fixed bounded lattice with least element 0. Consider the configured category $\widehat{L} = (L, \triangle^L)$ in Example 10.2.10. We saw in Example 11.7.4 that for each homotopy prefactorization algebra on $\widehat{L}$, the 0-entry is an E_∞-algebra. In this section, we explain that every other entry has the structure of an E_∞-module over the E_∞-algebra at the 0-entry.

Motivation 11.8.1. For a monoid A in M, recall from Definition 2.6.6 that a left A-module is an object X equipped with a left action $m : A \otimes X \longrightarrow X$ that satisfies the associativity and unity axioms. These two axioms can be read off from those of a monoid in Proposition 2.6.2 by replacing the last A-entry by X. We can define modules over an A_∞-algebra or an E_∞-algebra in the same way, by replacing the last entry with the module in each structure morphism and each axiom. For our current objective of understanding homotopy prefactorization algebras, we will need the following concept of an E_∞-module over an E_∞-algebra. To define E_∞-modules, we will need two colors, one color 0 for the E_∞-algebra and one color d for the E_∞-modules on which it acts.   ◇

Recall the concept of a directed path in Definition 3.1.11. We first define the colored trees that parametrize the structure morphisms of an E_∞-module. In the next definition, 0 and d are two distinct symbols, not necessarily elements in a bounded lattice. In practice, 0 is the least element in a bounded lattice L, and $0 \neq d \in L$.

Definition 11.8.2. A $\{0, d\}$-*tree* is a $\{0, d\}$-colored tree

$$T \in \underline{\mathsf{Tree}}^{\{0,d\}} \binom{d}{0,\ldots,0,d}$$

in which $(0, \ldots, 0)$ is a possibly empty profile of copies of 0. It is required that the following conditions be satisfied.

- T has at least one input, the last of which and the output are colored by d. All other inputs of T are colored by 0.
- Each $v \in \mathsf{Vt}(T)$ has at least one input.
- If T does not have any vertices, then T is the d-colored exceptional edge $\uparrow_d$.
- If T has a non-empty set of vertices, consider the unique directed path P_T^d in T whose initial vertex contains the last input of T and whose terminal vertex contains the output of T.
  - Each vertex $v \in P_T^d$ has profile $\binom{d}{0,\ldots,0,d}$, where $(0, \ldots, 0)$ has $|\mathsf{in}(v)| - 1$ copies of 0.
  - Each vertex $v \notin P_T^d$ has profile $\binom{0}{0,\ldots,0}$, where $(0, \ldots, 0)$ has $|\mathsf{in}(v)|$ copies of 0.

Interpretation 11.8.3. In a $\{0, d\}$-tree T, the directed path P_T^d from the last input to the output is d-colored, where we abbreviate the singleton $\{d\}$ to d. All other edges in T are colored by 0. For the vertex v that contains the last input of T, the last input of v is also the last input of T, both of which are d-colored. ◇

Example 11.8.4. Every d-colored linear graph $\mathsf{Lin}_{(d,d,\dots,d)}$ as in Example 3.1.19 is a $\{0,d\}$-tree. On the other hand, a truncated linear graph as in Example 3.1.20 cannot be a $\{0,d\}$-tree because it does not have any inputs. ◇

Example 11.8.5. The $(\underline{c};d)$-corolla in Example 3.1.21 is a $\{0,d\}$-tree if and only if $\underline{c}$ has the form $(0,\dots,0,d)$, where $(0,\dots,0)$ is a possibly empty profile of copies of 0. ◇

Example 11.8.6. The 2-level tree $T(\{\underline{b}_j\};\underline{c};d)$ in Example 3.1.23 is a $\{0,d\}$-tree if and only if the following conditions hold:

- Every $k_j \geq 1$.
- $c_i = 0$ for $1 \leq i \leq m-1$, and $c_m = d$.
- $b_{j,l} = 0$ for all $1 \leq j \leq m$ and $1 \leq l \leq k_j$, except for $b_{m,k_m} = d$.

For example, the 2-level tree on the right

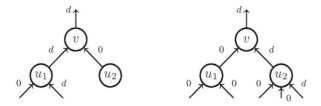

is a $\{0,d\}$-tree, but the one on the left is not a $\{0,d\}$-tree because u_2 does not have any inputs. ◇

Example 11.8.7. Suppose T is a $\{0,d\}$-tree with $n \geq 1$ inputs, and T_n is also a $\{0,d\}$-tree. For each $1 \leq i \leq n-1$, suppose $T_i \in \underline{\mathsf{Tree}}^{\{0\}}$ in which every vertex has at least one input. Then the grafting

$$G = \mathsf{Graft}(T;T_1,\dots,T_n)$$

is also a $\{0,d\}$-tree. Indeed, the last input of T_n becomes the last input of the grafting G, and the output of T becomes the output of G. The output of T_n and the last input of T are both d-colored. The d-colored edges in T and T_n yield the required d-colored directed path from the last input of the grafting to its output. All other edges are colored by 0. ◇

Example 11.8.8. Suppose T is a $\{0,d\}$-tree with n inputs, and $H_v \in \underline{\mathsf{Tree}}^{\{0,d\}}(v)$ is a $\{0,d\}$-tree if $v \in P_T^d$. Otherwise, $H_v \in \underline{\mathsf{Tree}}^{\{0\}}(v)$ in which every vertex has at least one input. Then the tree substitution $K = T(H_v)_{v \in T}$ is also a $\{0,d\}$-tree. Indeed, the unique d-colored directed paths in H_v for $v \in P_T^d$ together form the unique d-colored directed path in K from its last input to the output. All other edges in K are 0-colored. ◇

The following definition of an E_∞-module is modeled after the Coherence Theorem 7.6.7 for E_∞-algebras.

Definition 11.8.9. Suppose (A, λ^A) is an E_∞-algebra. An E_∞-*module over* (A, λ) is a pair (X, λ^X) consisting of

- an object $X \in \mathsf{M}$ and
- a structure morphism

$$\mathsf{J}[T] \otimes A^{\otimes n-1} \otimes X \xrightarrow{\ \lambda^X_T\ } X \ \in \mathsf{M} \qquad (11.8.10)$$

for each $\{0, d\}$-tree T with $n \geq 1$ inputs

that satisfies the following four conditions.

Associativity Suppose T and T_n are $\{0, d\}$-trees in which T has $n \geq 1$ inputs, and $T_j \in \underline{\mathsf{Tree}}^{\{0\}}$ for $1 \leq j \leq n-1$ in which every vertex has at least one input. Suppose T_j has k_j inputs for $1 \leq j \leq n$, and $k = k_1 + \cdots + k_n$. Suppose $G = \mathsf{Graft}(T; T_1, \ldots, T_n)$ is the grafting. Then the diagram

$$\mathsf{J}[T] \otimes \left(\bigotimes_{j=1}^{n} \mathsf{J}[T_j] \right) \otimes A^{\otimes k-1} \otimes X \xrightarrow{(\pi, \mathrm{Id})} \mathsf{J}[G] \otimes A^{\otimes k-1} \otimes X$$

$$\downarrow{\scriptstyle \mathrm{permute}}\ {\cong}$$

$$\mathsf{J}[T] \otimes \left[\bigotimes_{j=1}^{n-1} \left(\mathsf{J}[T_j] \otimes A^{\otimes k_j} \right) \right] \otimes \left(\mathsf{J}[T_n] \otimes A^{\otimes k_n - 1} \otimes X \right) \qquad \lambda^X_G$$

$$\downarrow{\left(\mathrm{Id}, \bigotimes_{j=1}^{n-1} \lambda^A_{T_j}, \lambda^X_{T_n} \right)}$$

$$\mathsf{J}[T] \otimes A^{\otimes n-1} \otimes X \xrightarrow{\qquad\qquad \lambda^X_T \qquad\qquad} X$$

$$(11.8.11)$$

is commutative. Here $\pi = \otimes_S 1$ is the morphism in Lemma 6.2.7 for the grafting G.

Unity The composition

$$X \xrightarrow{\ \cong\ } \mathsf{J}[\uparrow_d] \otimes X \xrightarrow{\ \lambda^X_\uparrow\ } X \qquad (11.8.12)$$

is the identity morphism of X, where $\uparrow_d$ is the d-colored exceptional edge.

Equivariance For a $\{0, d\}$-tree T with $n \geq 1$ inputs and $\sigma \in \Sigma_{n-1}$, the diagram

$$\begin{array}{ccc} \mathsf{J}[T] \otimes A^{\otimes n-1} \otimes X & \xrightarrow{\ \lambda^X_T\ } & X \\ {\scriptstyle (\mathrm{Id}, \sigma^{-1}, \mathrm{Id})} \downarrow & & \| \\ \mathsf{J}[T\sigma] \otimes A^{\otimes n-1} \otimes X & \xrightarrow{\ \lambda^X_{T\sigma}\ } & X \end{array} \qquad (11.8.13)$$

is commutative. Here $T\sigma$ is the $\{0, d\}$-tree obtained from T by replacing its ordering ζ_T by $\zeta_T \circ (\sigma \oplus \mathrm{id}_1)$.

Wedge Condition Suppose T is a $\{0,d\}$-tree with n inputs, and $H_v \in \underline{\mathsf{Tree}}^{\{0,d\}}(v)$ is a $\{0,d\}$-tree if $v \in P_T^d$. For $v \in \mathsf{Vt}(T)$ with $v \notin P_T^d$, $H_v \in \underline{\mathsf{Tree}}^{\{0\}}(v)$ in which every vertex has at least one input. Suppose $K = T(H_v)_{v \in T}$ is the tree substitution. Then the diagram

$$
\begin{array}{ccc}
\mathsf{J}[T] \otimes A^{\otimes n-1} \otimes X & \xrightarrow{\;\lambda_T^X\;} & X \\
{\scriptstyle (\mathsf{J},\mathrm{Id})}\Big\downarrow & & \Big\| \\
\mathsf{J}[K] \otimes A^{\otimes n-1} \otimes X & \xrightarrow{\;\lambda_K^X\;} & X
\end{array}
\tag{11.8.14}
$$

is commutative.

In a bounded lattice $(L, \le)$ with least element 0, regarded as a small category, the unique morphism $0 \longrightarrow d$ is denoted by 0_d for $d \in L$. In particular, $0_0 = \mathrm{Id}_0$. The next result is the main observation in this section.

Corollary 11.8.15. *Suppose $(L, \le)$ is a bounded lattice with least element 0, and (Y, λ) is a homotopy prefactorization algebra on $\widehat{L}$, i.e., a $\mathsf{WO}_{\widehat{L}}^{\mathsf{M}}$-algebra. For each $d \in L$, the entry Y_d is an E_∞-module over the E_∞-algebra Y_0 when equipped with the structure morphisms*

$$
\mathsf{J}[T] \otimes Y_0^{\otimes n-1} \otimes Y_d \xrightarrow{\;\lambda_T\{\underline{f}^v\}_{v \in T}\;} Y_d \in \mathsf{M}
$$

in (11.4.2) for $\{0,d\}$-trees T with $n \ge 1$ inputs. Here for $v \in \mathsf{Vt}(T)$,

$$
\underline{f}^v = \begin{cases} \{0_d, \dots, 0_d, \mathrm{Id}_d\} \in \Delta^L\binom{d}{0,\dots,0,d} & \text{if } v \in P_T^d, \\ \{\mathrm{Id}_0\}_{i=1}^{|\mathrm{in}(v)|} \in \Delta^L\binom{0}{0,\dots,0} & \text{if } v \notin P_T^d. \end{cases}
\tag{11.8.16}
$$

Proof. This is a special case of the Coherence Theorem 11.4.1 applied to the configured category $\widehat{L}$. To check that the E_∞-module associativity (11.8.11) is a special case of the associativity condition (11.4.3), we use the fact, which we explained in Interpretation 11.7.3, that the E_∞-algebra structure morphisms of Y_0 are the structure morphisms

$$
\lambda_{T_0}\left\{\{\mathrm{Id}_0\}_{i=1}^{|\mathrm{in}(v)|}\right\}_{v \in T_0}
$$

with $T_0 \in \underline{\mathsf{Tree}}^{\{0\}}$.

To check that the E_∞-module wedge condition (11.8.14) is a special case of the wedge condition (11.4.6), we use the fact that, for $v \in P_T^d$, $H_v \in \underline{\mathsf{Tree}}^{\{0,d\}}(v)$ is a $\{0,d\}$-tree. For $u \in \mathsf{Vt}(H_v)$, $\underline{f}^u \in \Delta^L(u)$ is defined as in (11.8.16). So we have

$$
\gamma_{H_v}^{\mathsf{O}_{\widehat{L}}}\left(\{\underline{f}^u\}_{u \in H_v}\right) = \{0_d, \dots, 0_d, \mathrm{Id}_d\} \in \Delta^L\binom{d}{0,\dots,0,d} = \Delta^L(v).
$$

On the other hand, for $v \in \mathsf{Vt}(T)$ with $v \notin P_T^d$, we have that $H_v \in \underline{\mathsf{Tree}}^{\{0\}}(v)$ and that

$$
\gamma_{H_v}^{\mathsf{O}_{\widehat{L}}}\left(\left\{\{\mathrm{Id}_0\}_{i=1}^{|\mathrm{in}(u)|}\right\}_{u \in H_v}\right) = \{\mathrm{Id}_0\}_{i=1}^{|\mathrm{in}(v)|} \in \Delta^L\binom{0}{0,\dots,0} = \Delta^L(v).
$$

$\square$

Example 11.8.17 (Homotopy Costello-Gwilliam prefactorization algebras). For a topological space X, consider the configured category

$$\widetilde{\mathsf{Open}(X)} = \left(\mathsf{Open}(X), \triangle^X\right)$$

in Example 10.2.11. The category $\mathsf{Open}(X)$ is a bounded lattice with least element $\varnothing_X \subset X$, the empty subset of X. The configured category $\widetilde{\mathsf{Open}(X)}$ has the form $\widehat{L}$ in Example 10.2.10. Therefore, by Corollary 11.8.15, if (Y,λ) is a homotopy prefactorization algebra on $\widetilde{\mathsf{Open}(X)}$, i.e., a $\mathsf{WO}^{\mathsf{M}}_{\widetilde{\mathsf{Open}(X)}}$-algebra, then each entry Y_U with $U \in \mathsf{Open}(X)$ is equipped with the structure of an E_∞-module over the E_∞-algebra $Y_{\varnothing_X}$ in Example 11.7.5.   ◇

Example 11.8.18 (Homotopy Costello-Gwilliam equivariant prefactorization algebras). Suppose G is a group, and X is a topological space in which G acts on the left by homeomorphisms. Consider the configured category

$$\widetilde{\mathsf{Open}(X)}_G = \left(\mathsf{Open}(X)_G, \triangle^X_G\right)$$

in Example 10.2.12. By the change-of-operad adjunction

$$\mathsf{Alg}_{\mathsf{M}}\left(\mathsf{WO}^{\mathsf{M}}_{\widetilde{\mathsf{Open}(X)}}\right) \underset{(\mathsf{WO}^{\mathsf{M}}_\iota)^*}{\overset{(\mathsf{WO}^{\mathsf{M}}_\iota)_!}{\rightleftarrows}} \mathsf{Alg}_{\mathsf{M}}\left(\mathsf{WO}^{\mathsf{M}}_{\widetilde{\mathsf{Open}(X)}_G}\right)$$

in Example 11.3.6, every homotopy prefactorization algebra (Y,λ) on $\widetilde{\mathsf{Open}(X)}_G$ has an underlying homotopy prefactorization algebra on $\widetilde{\mathsf{Open}(X)}$. Therefore, by Example 11.8.17, each entry Y_U with $U \in \mathsf{Open}(X)$ is equipped with the structure of an E_∞-module over the E_∞-algebra $Y_{\varnothing_X}$.   ◇

11.9   Homotopy Coherent Diagrams of E_∞-Modules

Suppose $(L,\leq)$ is an arbitrary but fixed bounded lattice with least element 0. Consider the configured category $\widehat{L} = (L, \triangle^L)$ in Example 10.2.10. In this section, we explain that, for each homotopy prefactorization algebra on $\widehat{L}$, the objectwise E_∞-module structure over the E_∞-algebra at the 0-entry in Section 11.8 is compatible with the homotopy coherent diagram structure in Section 11.5.

Motivation 11.9.1. In Corollary 10.6.3 we saw that, for a prefactorization algebra (Y,λ) on $\widehat{L}$, the objectwise left Y_0-module structure is compatible with the L-diagram structure. For a homotopy prefactorization algebra (Y,λ) on $\widehat{L}$, the entry Y_0 is equipped with the structure of an E_∞-algebra by Example 11.7.4. Furthermore, by Corollary 11.8.15, every other entry Y_d with $d \in L$ is equipped with the structure of an E_∞-module over the E_∞-algebra Y_0. We expect the homotopy coherent L-diagram structure in Y in Section 11.5 to be homotopically compatible with the entrywise E_∞-module structure. To explain precisely how they are compatible, we need the following notation.   ◇

Assumption 11.9.2. Suppose

- $\underline{c} = \left(c = c_0, \ldots, c_m = d\right) \in \mathsf{Prof}(L)$ with $m \geq 1$ and $c = c_0 \leq \cdots \leq c_m = d$ in L.
- $L_{\underline{c}} = \mathsf{Lin}_{\underline{c}} \in \mathsf{Linear}^L\binom{d}{c}$ is the corresponding linear graph in Example 3.1.19.
- $L_{cd} = \mathsf{Lin}_{(c,d)} \in \mathsf{Linear}^L\binom{d}{c}$.
- $g_i \in L(c_{i-1}, c_i)$ is the unique element for $1 \leq i \leq m$, and $g = g_m \cdots g_1 \in L(c, d)$.
- $T_d \in \mathsf{Tree}^{\{0,d\}}\binom{d}{0,\ldots,0,d}$ is a $\{0, d\}$-tree as in Definition 11.8.2, in which $(0, \ldots, 0)$ has $n - 1$ copies of 0 for some $n \geq 1$.
- $T_c \in \mathsf{Tree}^{\{0,c\}}\binom{c}{0,\ldots,0,c}$ is the $\{0, c\}$-tree obtained from T_d by replacing every d-colored edge by a c-colored edge.

Define the $\{0, c, d\}$-colored trees

$$T^1 = \mathsf{Graft}\left(L_{\underline{c}}; T_c\right), \quad T^2 = \mathsf{Graft}\Big(T_d; \underbrace{\uparrow_0, \ldots, \uparrow_0}_{n-1 \text{ copies}}, L_{cd}\Big), \quad \text{and} \quad C = \mathsf{Cor}_{\left((0,\ldots,0,c);d\right)}$$

in $\mathsf{Tree}^{\{0,c,d\}}\binom{d}{0,\ldots,0,c}$. Here $\uparrow_0$ is the $\{0\}$-colored exceptional edge in Example 3.1.18, and $\mathsf{Cor}_?$ is the corolla in Example 3.1.21, with Graft the grafting in Definition 3.3.1.

We can visualize T^1 (on the left) and T^2 (on the right) as follows.

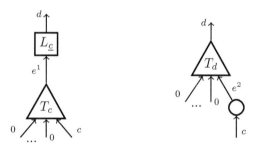

In T^1 the c-colored internal edge connecting T_c to $L_{\underline{c}}$ is denoted by e^1. In T^2 the d-colored internal edge connecting the linear graph L_{cd} to T_d is denoted by e^2.

For $v \in \mathsf{Vt}(T_d)$, the configuration $\underline{f}^v \in \Delta^L$ in (11.8.16) will be denoted by $\underline{f}_d^v$. The corresponding c-colored version, with d replaced by c everywhere, is denoted by $\underline{f}_c^v$.

The following is the main result of this section. A copy of the morphism $1 : \mathbb{1} \longrightarrow J$ indexed by an internal edge e will be denoted by 1_e. To simplify the notation, we will omit writing some of the identity morphisms below. We will use the notation in the Coherence Theorem 11.4.1 for homotopy prefactorization algebras.

Theorem 11.9.3. *Suppose (Y, λ) is a homotopy prefactorization algebra on $\widehat{L}$, i.e.,* a $\mathsf{WO}_{\overline{L}}^{\mathsf{M}}$-*algebra. Under Assumption 11.9.2, the diagram*

$$
\begin{array}{ccc}
\mathsf{J}[L_{\underline{c}}] \otimes \mathsf{J}[T_c] \otimes Y_0^{\otimes n-1} \otimes Y_c & \xrightarrow{\ \lambda_{T_c}\{\underline{f}_c^v\}_{v\in T_c}\ } & \mathsf{J}[L_{\underline{c}}] \otimes Y_c \\
\end{array}
$$

$$
\begin{array}{ccc}
(1_{e^1})(\cong)\ \Big\downarrow & & \Big\downarrow\ \lambda_{L_{\underline{c}}}\{g_i\}_{i=1}^m \\
\mathsf{J}[T^1] \otimes Y_0^{\otimes n-1} \otimes Y_c & \xrightarrow{\ \lambda_{T_1}\big\{\{\underline{f}_c^v\}_{v\in T_c},\,\{g_i\}_{i=1}^n\big\}\ } & Y_d \\
\end{array}
$$

$$
\begin{array}{ccc}
(0^{\otimes|T^1|})(\cong)\ \Big\uparrow & & \Big\| \\
\mathsf{J}[C] \otimes Y_0^{\otimes n-1} \otimes Y_c & \xrightarrow{\ \lambda_C\{0_d,\dots,0_d,g\}\ } & Y_d \\
\end{array}
$$

$$
\begin{array}{ccc}
(0^{\otimes|T^2|})(\cong)\ \Big\downarrow & & \Big\| \\
\mathsf{J}[T^2] \otimes Y_0^{\otimes n-1} \otimes Y_c & \xrightarrow{\ \lambda_{T_2}\big\{g,\{\underline{f}_d^v\}_{v\in T_d}\big\}\ } & Y_d \\
\end{array}
$$

$$
\begin{array}{ccc}
(1_{e^2})(\cong)\ \Big\uparrow & & \Big\uparrow\ \lambda_{T_d}\{\underline{f}_d^v\}_{v\in T_d} \\
\mathsf{J}[T_d] \otimes Y_0^{\otimes n-1} \otimes \mathsf{J}[L_{cd}] \otimes Y_c & \xrightarrow{\ \lambda_{L_{cd}}\{g\}\ } & \mathsf{J}[T_d] \otimes Y_0^{\otimes n-1} \otimes Y_d \\
\end{array}
$$

is commutative, where $0, 1 : \mathbb{1} \longrightarrow J$ *are part of the commutative segment* J.

Proof. This follows from the Coherence Theorem 11.4.1. Indeed, in the above diagram from top to bottom:

- The first rectangle is commutative by the associativity condition (11.4.3) and the grafting definition of T^1.
- The second rectangle is commutative by the wedge condition (11.4.6) applied to the tree substitution $T^1 = C(T^1)$.
- The third rectangle is commutative by the same wedge condition applied to the tree substitution $T^2 = C(T^2)$.
- The bottom rectangle is commutative by the associativity condition (11.4.3), the grafting definition of T^2, and the unity condition (11.4.4).

In the second and the third rectangles, one observes that the set $\Delta^L\binom{d}{0,\dots,0,c}$ contains only the configuration

$$
\big\{\underbrace{0_d,\dots,0_d}_{n-1\ \text{copies}},g\big\} = \gamma_{T^1}^{O_{\overline{L}}}\Big(\{\underline{f}_c^v\}_{v\in T_c},\,\{g_i\}_{i=1}^n\Big) = \gamma_{T^2}^{O_{\overline{L}}}\Big(g,\{\underline{f}_d^v\}_{v\in T_d}\Big).
$$

Along the left side of the diagram, the top and the bottom isomorphisms are of the form $? \cong \mathbb{1}\otimes?$. The middle two isomorphisms are of the form $? \cong \mathbb{1}^{\otimes|T^r|}\otimes?$ with $r = 1, 2$. $\qquad\square$

Interpretation 11.9.4. In Theorem 11.9.3, the structure morphisms

$$
\lambda_{T_c}\Big\{\underline{f}_c^v\Big\}_{v\in T_c} \quad \text{and} \quad \lambda_{T_d}\Big\{\underline{f}_d^v\Big\}_{v\in T_d}
$$

are E_∞-module structure morphisms of Y_c and Y_d, respectively, as in Corollary 11.8.15. The structure morphisms

$$
\lambda_{L_{\underline{c}}}\{g_i\}_{i=1}^m \quad \text{and} \quad \lambda_{L_{cd}}\{g\}
$$

are part of the underlying homotopy coherent L-diagram of (Y, λ). Therefore, the commutative diagram says that the homotopy coherent L-diagram structure of (Y, λ) commutes with the objectwise E_∞-module structure over the E_∞-algebra Y_0 up to specified homotopies that are also structure morphisms. ◇

Example 11.9.5. In Assumption 11.9.2, suppose $m = 1$, so $\underline{c} = (c, d)$ and $L_{\underline{c}} = L_{cd} = \mathrm{Lin}_{(c,d)}$. In this case, the commutative diagram in Theorem 11.9.3 says that the diagram

$$
\begin{array}{ccc}
\mathsf{J}[L_{cd}] \otimes \mathsf{J}[T_c] \otimes Y_0^{\otimes n-1} \otimes Y_c & \xrightarrow{\lambda_{T_c}\{\underline{f}_c^v\}_{v \in T_c}} & \mathsf{J}[L_{cd}] \otimes Y_c \\
{\scriptstyle \cong} \downarrow & & \downarrow {\scriptstyle \lambda_{L_{cd}}\{g\}} \\
\mathsf{J}[T_d] \otimes Y_0^{\otimes n-1} \otimes \mathsf{J}[L_{cd}] \otimes Y_c & & \\
{\scriptstyle \lambda_{L_{cd}}\{g\}} \downarrow & & \\
\mathsf{J}[T_d] \otimes Y_0^{\otimes n-1} \otimes Y_d & \xrightarrow{\lambda_{T_d}\{\underline{f}_d^v\}_{v \in T_d}} & Y_d
\end{array}
$$

is commutative up to specified homotopies that are also structure morphisms. This is the homotopy coherent analogue of the commutative diagram

$$
\begin{array}{ccc}
Y_0 \otimes Y_c & \xrightarrow{\lambda\{0_c, \mathrm{Id}_c\}} & Y_c \\
{\scriptstyle (\mathrm{Id}, \lambda\{g\})} \downarrow & & \downarrow {\scriptstyle \lambda\{g\}} \\
Y_0 \otimes Y_d & \xrightarrow{\lambda\{0_d, \mathrm{Id}_d\}} & Y_d
\end{array}
$$

in Corollary 10.6.3 for a prefactorization algebra (Y, λ) on $\widehat{L}$. ◇

Example 11.9.6 (Homotopy Costello-Gwilliam prefactorization algebras). For a topological space X, the configured category

$$
\widehat{\mathsf{Open}(X)} = \left(\mathsf{Open}(X), \triangle^X\right)
$$

in Example 10.2.11 has the form $\widehat{L}$ for the bounded lattice $\mathsf{Open}(X)$ with least element $\varnothing_X \subset X$. Therefore, Theorem 11.9.3 applies to every homotopy prefactorization algebra (Y, λ) on $\widehat{\mathsf{Open}(X)}$. So the homotopy coherent $\mathsf{Open}(X)$-diagram structure in Y is homotopically compatible with the objectwise E_∞-module structure over the E_∞-algebra $Y_{\varnothing_X}$. ◇

Example 11.9.7 (Homotopy Costello-Gwilliam equivariant prefactorization algebras). Suppose G is a group, and X is a topological space in which G acts on the left by homeomorphisms. The homotopy prefactorization algebras on the configured category

$$
\widehat{\mathsf{Open}(X)}_G = \left(\mathsf{Open}(X)_G, \triangle_G^X\right)
$$

in Example 10.2.12 are related to those on the configured category $\widetilde{\mathsf{Open}(X)}$ via the change-of-operad adjunction

$$\mathsf{Alg}_\mathsf{M}\big(\mathsf{WO}^\mathsf{M}_{\overline{\mathsf{Open}(X)}}\big) \xrightleftharpoons[\;(\mathsf{WO}^\mathsf{M}_\iota)^*\;]{\;(\mathsf{WO}^\mathsf{M}_\iota)_!\;} \mathsf{Alg}_\mathsf{M}\big(\mathsf{WO}^\mathsf{M}_{\overline{\mathsf{Open}(X)}_G}\big)$$

$$\|\qquad\qquad\qquad\qquad\qquad\|$$

$$\mathsf{HPFA}\big(\widetilde{\mathsf{Open}(X)}\big) \qquad\qquad \mathsf{HPFA}\big(\widetilde{\mathsf{Open}(X)}_G\big)$$

in Example 11.3.6. In particular, every homotopy prefactorization algebra (Y, λ) on $\widetilde{\mathsf{Open}(X)}_G$ has an underlying homotopy prefactorization algebra on $\widetilde{\mathsf{Open}(X)}$. Therefore, by Example 11.9.6, for each homotopy prefactorization algebra (Y, λ) on $\widetilde{\mathsf{Open}(X)}_G$, the homotopy coherent $\mathsf{Open}(X)$-diagram structure in Y is homotopically compatible with the objectwise E_∞-module structure over the E_∞-algebra $Y_{\varnothing_X}$. $\qquad\qquad\diamond$

11.10 Homotopy Coherent Diagrams of E_∞-Algebras

Suppose C is a small category. In this section, we explain that homotopy prefactorization algebras on the maximal configured category $\overline{\mathsf{C}_{\mathsf{max}}} = (\mathsf{C}, \triangle^\mathsf{C}_{\mathsf{max}})$ in Example 10.2.9 are homotopy coherent C-diagrams of E_∞-algebras. Recall the colored operads Com^C in Example 4.5.23 and $\mathsf{O}_{\overline{\mathsf{C}}}$ in Definition 10.3.2.

Lemma 11.10.1. *Suppose C is a small category with object set $\mathfrak{C}$. There is an equality*

$$\mathsf{O}^\mathsf{M}_{\overline{\mathsf{C}_{\mathsf{max}}}} = \mathsf{Com}^\mathsf{C}$$

of $\mathfrak{C}$-colored operads in M.

Proof. Both $\mathfrak{C}$-colored operads have entries

$$\mathsf{O}^\mathsf{M}_{\overline{\mathsf{C}_{\mathsf{max}}}}\big({}^d_{\underline{c}}\big) = \coprod_{\triangle^\mathsf{C}_{\mathsf{max}}\big({}^d_{\underline{c}}\big)} \mathbb{1} = \coprod_{\prod_{j=1}^n \mathsf{C}(c_j, d)} \mathbb{1} = \mathsf{Com}^\mathsf{C}\big({}^d_{\underline{c}}\big)$$

for $\big({}^d_{\underline{c}}\big) = \big({}^{\quad d}_{c_1,\dots,c_n}\big) \in \mathsf{Prof}(\mathfrak{C}) \times \mathfrak{C}$. From Example 4.5.23 and Definition 10.3.2, their operad structures also coincide. $\qquad\square$

Recall from Definition 7.8.2 that WCom^C-algebras are called homotopy coherent C-diagrams of E_∞-algebras in M.

Corollary 11.10.2. *There is an equality*

$$\mathsf{Alg}_\mathsf{M}\big(\mathsf{WO}^\mathsf{M}_{\overline{\mathsf{C}_{\mathsf{max}}}}\big) = \mathsf{Alg}_\mathsf{M}\big(\mathsf{WCom}^\mathsf{C}\big)$$

between the category of homotopy prefactorization algebras on $\overline{\mathsf{C}_{\mathsf{max}}}$, i.e., $\mathsf{WO}^\mathsf{M}_{\overline{\mathsf{C}_{\mathsf{max}}}}$-algebras and the category of homotopy coherent C-diagrams of E_∞-algebras in M.

Proof. We first apply the Boardman-Vogt construction in Theorem 6.3.11 to the equality of colored operads in Lemma 11.10.1 and then take the category of algebras.
□

Example 11.10.3 (Homotopy Costello-Gwilliam prefactorization algebras on manifolds). Suppose Emb^n is a small category equivalent to the category of smooth n-manifolds with open embeddings as morphisms. Recall from Example 10.7.4 that symmetric monoidal functors $\mathsf{Emb}^n \longrightarrow \mathsf{M}$ are called prefactorization algebras on n-manifolds with values in M. The category of such objects is isomorphic to the category $\mathsf{PFA}(\widehat{\mathsf{Emb}^n_{\mathsf{max}}})$ of prefactorization algebras on $\widehat{\mathsf{Emb}^n_{\mathsf{max}}}$. By Corollary 11.10.2 there is an equality

$$\mathsf{Alg}_{\mathsf{M}}\left(\mathsf{WO}^{\mathsf{M}}_{\widehat{\mathsf{Emb}^n_{\mathsf{max}}}}\right) = \mathsf{Alg}_{\mathsf{M}}\left(\mathsf{WCom}^{\mathsf{Emb}^n}\right)$$

between the category of homotopy prefactorization algebras on the maximal configured category $\widehat{\mathsf{Emb}^n_{\mathsf{max}}}$, i.e., $\mathsf{WO}^{\mathsf{M}}_{\widehat{\mathsf{Emb}^n_{\mathsf{max}}}}$-algebras, and the category of homotopy coherent Emb^n-diagrams of E_∞-algebras in M. There is a similar equality for Hol^n, which is a small category equivalent to the category of complex n-manifolds with open holomorphic embeddings as morphisms. ◇

Chapter 12

Comparing Prefactorization Algebras and AQFT

In this chapter, we compare (homotopy) prefactorization algebras and (homotopy) algebraic quantum field theories.

Recall from Chapter 8 and Chapter 9 that algebraic quantum field theories and their homotopy analogues are defined as algebras over the colored operads $\mathsf{O}_{\overline{\mathsf{C}}}^{\mathsf{M}}$ and $\mathsf{WO}_{\overline{\mathsf{C}}}^{\mathsf{M}}$ for an orthogonal category $\overline{\mathsf{C}}$. In Chapter 10 and Chapter 11, prefactorization algebras and their homotopy analogues are defined as algebras over the colored operads $\mathsf{O}_{\widehat{\mathsf{C}}}^{\mathsf{M}}$ and $\mathsf{WO}_{\widehat{\mathsf{C}}}^{\mathsf{M}}$ for a configured category $\widehat{\mathsf{C}}$. To compare these objects, in Section 12.1 we first observe that every orthogonal category yields a configured category in which the configurations are the finite sequences of pairwise orthogonal morphisms.

In Section 12.2 we observe that we can also go backward, from configured categories to orthogonal categories, by restricting to binary configurations. More formally, the category of orthogonal categories embeds in the category of configured categories as a full reflective subcategory. We will show by examples that this is not an adjoint equivalence, so the two categories are genuinely different.

In Section 12.3 we show that, for each configured category, there is a comparison morphism from the colored operad defining prefactorization algebras to the colored operad defining algebraic quantum field theories. This comparison morphism is well-behaved with respect to configured functors and the time-slice axiom. As a consequence, we have various comparison adjunctions between (homotopy) prefactorization algebras and (homotopy) algebraic quantum field theories.

In Section 12.4 we illustrate the comparison adjunctions with many examples. In Section 12.5 we identify precisely the prefactorization algebras that come from algebraic quantum field theories.

As in previous chapters, $(\mathsf{M}, \otimes, \mathbb{1})$ is a cocomplete symmetric monoidal closed category, such as $\mathsf{Chain}_{\mathbb{K}}$, with a commutative segment $(J, \mu, 0, 1, \epsilon)$ as in Definition 6.2.1. For a small category C, its object set is denoted by $\mathfrak{C}$.

12.1 Orthogonal Categories as Configured Categories

In this section, we show that orthogonal categories as in Definition 8.2.1 yield configured categories as in Definition 10.2.1.

Definition 12.1.1. Suppose $\overline{\mathsf{C}} = (\mathsf{C}, \perp^{\mathsf{C}})$ is an orthogonal category. For objects $c_1, \ldots, c_n, d \in \mathsf{C}$ with $n \geq 0$ and $\underline{c} = (c_1, \ldots, c_n)$, define $\Delta^{\mathsf{C}}\binom{d}{\underline{c}}$ as the set of pairs $\left(d; \{f_i\}_{i=1}^n\right)$ such that

- $\{f_i\}_{i=1}^n \in \prod_{i=1}^n \mathsf{C}(c_i, d)$ and
- if $1 \leq i \neq j \leq n$, then $f_i \perp^{\mathsf{C}} f_j$.

Interpretation 12.1.2. Configurations in Δ^{C} are those $\{f_i\} \in \prod \mathsf{C}(c_i, d)$ such that the f_i's are pairwise orthogonal. As in Notation 10.2.7 we will usually write $(d; \{f_i\})$ as $\{f_i\}$. $\qquad\diamond$

Recall from Definitions 8.2.1 and 10.2.5 that OrthCat is the category of orthogonal categories and that ConfCat is the category of configured categories.

Proposition 12.1.3. *Suppose* $\overline{\mathsf{C}} = (\mathsf{C}, \perp^{\mathsf{C}})$ *is an orthogonal category.*

(1) With Δ^{C} as in Definition 12.1.1, $\Psi\overline{\mathsf{C}} = (\mathsf{C}, \Delta^{\mathsf{C}})$ is a configured category.
(2) This construction defines a functor

$$\Psi : \mathsf{OrthCat} \longrightarrow \mathsf{ConfCat}$$

that leaves the underlying categories and functors unchanged.
(3) Ψ sends each orthogonal equivalence to a configured equivalence.

Proof. For the first assertion, the subset axiom and the inclusivity axiom follow directly from the definition of Δ^{C}. The symmetry axiom follows from that of the orthogonality relation $\perp^{\mathsf{C}}$. The composition axiom follows from the fact that the orthogonality relation is closed under both post-compositions and pre-compositions. Indeed, using the notation in (10.2.2), we need to show that the $f_i g_{ij}$'s are pairwise orthogonal.

- If $1 \leq j \neq j' \leq k_i$ for some $1 \leq i \leq n$, then

$$f_i g_{ij} \perp^{\mathsf{C}} f_i g_{ij'}$$

  by the post-composition axiom of $\perp^{\mathsf{C}}$ because $g_{ij} \perp^{\mathsf{C}} g_{ij'}$ by assumption.
- If $1 \leq i \neq i' \leq n$, $1 \leq j \leq k_i$, and $1 \leq j' \leq k_{i'}$, then

$$f_i g_{ij} \perp^{\mathsf{C}} f_{i'} g_{i'j'}$$

  by the pre-composition axiom of $\perp^{\mathsf{C}}$ because $f_i \perp^{\mathsf{C}} f_{i'}$ by assumption.

Therefore, $\widehat{\mathsf{C}}$ is a configured category.

For the functoriality of this construction, observe that an orthogonal functor F sends a configuration $\{f_i\}$ to $\{Ff_i\}$, where the Ff_i's are pairwise orthogonal because the f_i's are. The last assertion follows immediately from the definition. $\qquad\square$

In particular, every orthogonal category in Section 8.4 is sent by the functor Ψ to a configured category.

Example 12.1.4 (Minimal and maximal orthogonal categories). For each small category C, there are equalities

$$\Psi\overline{\mathsf{C}_{\mathsf{min}}} = (\mathsf{C}, \triangle_{\mathsf{min}}^{\mathsf{C}}) = \widehat{\mathsf{C}_{\mathsf{min}}} \quad \text{and} \quad \Psi\overline{\mathsf{C}_{\mathsf{max}}} = (\mathsf{C}, \triangle_{\mathsf{max}}^{\mathsf{C}}) = \widehat{\mathsf{C}_{\mathsf{max}}}.$$

Here $\overline{\mathsf{C}_{\mathsf{min}}}$ and $\overline{\mathsf{C}_{\mathsf{max}}}$ are the minimal and maximal orthogonal categories in Examples 8.4.1 and 8.4.2. On the other hand, $\widehat{\mathsf{C}_{\mathsf{min}}}$ and $\widehat{\mathsf{C}_{\mathsf{max}}}$ are the minimal and maximal configured categories in Examples 10.2.8 and 10.2.9. ⬦

Example 12.1.5 (Bounded lattices). Suppose $(L, \leq)$ is a bounded lattice with least element 0 as in Example 2.2.12. There is an equality

$$\Psi(L, \perp) = (L, \triangle^L) = \widehat{L}$$

with $(L, \perp)$ the orthogonal category in Example 8.4.4 and $\widehat{L}$ the configured category in Example 10.2.10. ⬦

Example 12.1.6 (Topological spaces). For each topological space X, there is an equality

$$\Psi\overline{\mathsf{Open}(X)} = \widehat{\mathsf{Open}(X)}$$

with $\overline{\mathsf{Open}(X)}$ the orthogonal category in Example 8.4.5 and $\widehat{\mathsf{Open}(X)}$ the configured category in Example 10.2.11. ⬦

Example 12.1.7 (Equivariant topological spaces). Suppose G is a group, and X is a topological space in which G acts on the left by homeomorphisms. There is an equality

$$\Psi\overline{\mathsf{Open}(X)_G} = \widehat{\mathsf{Open}(X)}_G$$

with $\overline{\mathsf{Open}(X)_G}$ the orthogonal category in Example 8.4.6 and $\widehat{\mathsf{Open}(X)}_G$ the configured category in Example 10.2.12. ⬦

Example 12.1.8 (Oriented manifolds). Recall from Example 8.4.7 the orthogonal category $(\mathsf{Man}^d, \perp)$, where Man^d is the category of d-dimensional oriented manifolds with orientation-preserving open embeddings as morphisms in Example 2.2.15. Two morphisms $g_1 : X_1 \longrightarrow X$ and $g_2 : X_2 \longrightarrow X$ in Man^d are orthogonal if and only if their images are disjoint subsets in X. In the configured category

$$\Psi(\mathsf{Man}^d, \perp) = (\mathsf{Man}^d, \triangle),$$

a finite sequence of morphisms $\{g_i : X_i \to X\}_{i=1}^n$ is a configuration if and only if the images $g_i X_i$ are pairwise disjoint in X. Similar statements hold for the orthogonal categories in Examples 8.4.8 and 8.4.10–8.4.17. ⬦

12.2   Configured Categories to Orthogonal Categories

In this section, we observe that each configured category yields an orthogonal category in which an orthogonal pair is exactly a binary configuration. Moreover, the category of orthogonal categories embeds as a full reflective subcategory of the category of configured categories.

Definition 12.2.1. Suppose $\widehat{\mathsf{C}} = (\mathsf{C}, \triangle^{\mathsf{C}})$ is a configured category. Define $\perp^{\mathsf{C}}$ as the set of pairs $\{g_1, g_2\}$ in $\triangle^{\mathsf{C}}$.

Proposition 12.2.2. *Suppose $\widehat{\mathsf{C}} = (\mathsf{C}, \triangle^{\mathsf{C}})$ is a configured category.*

(1) With $\perp^{\mathsf{C}}$ as in Definition 12.2.1, $\Phi\widehat{\mathsf{C}} = (\mathsf{C}, \perp^{\mathsf{C}})$ is an orthogonal category.
(2) This construction defines a functor

$$\Phi : \mathsf{ConfCat} \longrightarrow \mathsf{OrthCat}$$

that leaves the underlying categories and functors unchanged.
(3) Φ sends each configured equivalence to an orthogonal equivalence.

Proof. The symmetry of $\perp^{\mathsf{C}}$ follows from that of $\triangle^{\mathsf{C}}$. By the inclusivity axiom of $\triangle^{\mathsf{C}}$, each morphism f in C yields a configuration $\{f\}$. So the composition axiom of $\triangle^{\mathsf{C}}$ implies both the post-composition axiom and the pre-composition axiom of $\perp^{\mathsf{C}}$. For the second assertion, observe that a configured functor preserves all the configurations, in particular the binary configurations $\{g_1, g_2\}$. The last assertion follows immediately from the definition. $\square$

Definition 12.2.3. For a configured category $\widehat{\mathsf{C}}$, we call $\Phi\widehat{\mathsf{C}}$ the *associated orthogonal category*.

The next observation says that the category $\mathsf{OrthCat}$ of orthogonal categories embeds in the category $\mathsf{ConfCat}$ of configured categories via the functor Ψ in Proposition 12.1.3 as a full reflective subcategory.

Theorem 12.2.4. *There is an adjunction*

$$\mathsf{ConfCat} \overset{\Phi}{\underset{\Psi}{\rightleftarrows}} \mathsf{OrthCat}$$

with left adjoint Φ such that the counit $\Phi\Psi \longrightarrow \mathrm{Id}_{\mathsf{OrthCat}}$ is the identity natural transformation. In particular, every orthogonal category is the Φ-image of some configured category.

Proof. Since both functors Φ and Ψ leave the underlying categories and functors unchanged, to establish the adjunction, it suffices to prove the following statement. Suppose $\widehat{\mathsf{C}} = (\mathsf{C}, \triangle^{\mathsf{C}})$ is a configured category and $\overline{\mathsf{D}} = (\mathsf{D}, \perp^{\mathsf{D}})$ is an orthogonal category. For each functor $F : \mathsf{C} \longrightarrow \mathsf{D}$, the following statements are equivalent:

(1) F sends each binary configuration in $\widehat{\mathsf{C}}$ to an orthogonal pair in $\overline{\mathsf{D}}$.

(2) The image of each configuration in $\widehat{\mathsf{C}}$ under F is pairwise orthogonal in $\overline{\mathsf{D}}$.

To see that (2) implies (1), simply note that each binary configuration is a configuration. To see that (1) implies (2), suppose $\{g_i\}_{i=1}^n$ is a configuration in $\widehat{\mathsf{C}}$ with $1 \le i \ne j \le n$. By the subset axiom of a configured category, $\{g_i, g_j\}$ is also a configuration in $\widehat{\mathsf{C}}$. So by (1) their images $\{Fg_i, Fg_j\}$ are orthogonal in $\overline{\mathsf{D}}$.

The equality

$$\Phi\Psi = \mathrm{Id}_{\mathsf{OrthCat}}$$

follows directly from the definition of the functors Φ and Ψ. $\qquad\square$

In Theorem 12.2.4 we observed that the counit $\Phi\Psi \longrightarrow \mathrm{Id}_{\mathsf{OrthCat}}$ is the identity natural transformation, so each orthogonal category $\overline{\mathsf{C}}$ is equal to $\Phi\Psi\overline{\mathsf{C}}$. In particular, this is true for all the orthogonal categories in Section 8.4. Below are some examples.

Example 12.2.5 (Empty orthogonality and minimal configuration). For each small category C, there are equalities

$$\Psi(\mathsf{C}, \varnothing) = (\mathsf{C}, \triangle_{\min}^{\mathsf{C}}) \quad \text{and} \quad \Phi(\mathsf{C}, \triangle_{\min}^{\mathsf{C}}) = (\mathsf{C}, \varnothing).$$

Here $\varnothing$ is the empty orthogonality relation in Example 8.4.1, and $(\mathsf{C}, \triangle_{\min}^{\mathsf{C}})$ is the minimal configured category on C in Example 10.2.8. $\qquad\diamond$

Example 12.2.6 (Maximal orthogonality and maximal configuration). For each small category C, there are equalities

$$\Psi(\mathsf{C}, \perp_{\max}) = (\mathsf{C}, \triangle_{\max}^{\mathsf{C}}) \quad \text{and} \quad \Phi(\mathsf{C}, \triangle_{\max}^{\mathsf{C}}) = (\mathsf{C}, \perp_{\max}).$$

Here $\perp_{\max}$ is the orthogonality relation in Example 8.4.2, and $(\mathsf{C}, \triangle_{\max}^{\mathsf{C}})$ is the maximal configured category on C in Example 10.2.9. $\qquad\diamond$

Example 12.2.7 (Orthogonality and configuration of bounded lattices). For each bounded lattice $(L, \le)$, there are equalities

$$\Psi(L, \perp) = (L, \triangle^L) \quad \text{and} \quad \Phi(L, \triangle^L) = (L, \perp).$$

Here $\perp$ is the orthogonality relation in Example 8.4.4, and $(L, \triangle^L)$ is the configured category on L in Example 10.2.10. $\qquad\diamond$

Example 12.2.8 (Orthogonality and configuration of topological spaces). For each topological space X, there are equalities

$$\Psi\big(\overline{\mathsf{Open}(X)}\big) = \widehat{\mathsf{Open}(X)} \quad \text{and} \quad \Phi\big(\widehat{\mathsf{Open}(X)}\big) = \overline{\mathsf{Open}(X)}.$$

Here $\overline{\mathsf{Open}(X)}$ is the orthogonal category in Example 8.4.5, and $\widehat{\mathsf{Open}(X)}$ is the configured category in Example 10.2.11. $\qquad\diamond$

Example 12.2.9 (Orthogonality and configuration of equivariant topological spaces). For each topological space X with a left action by a group G, there are equalities

$$\Psi\left(\overline{\mathsf{Open}(X)_G}\right) = \widehat{\mathsf{Open}(X)}_G \quad\text{and}\quad \Phi\left(\widehat{\mathsf{Open}(X)}_G\right) = \overline{\mathsf{Open}(X)_G}.$$

Here $\overline{\mathsf{Open}(X)_G}$ is the orthogonal category in Example 8.4.6, and $\widehat{\mathsf{Open}(X)}_G$ is the configured category in Example 10.2.12.   ◇

The next two examples show that the unit of the adjunction $\mathrm{Id}_{\mathsf{ConfCat}} \longrightarrow \Psi\Phi$ is not a natural isomorphism. Therefore, the adjunction $\Phi \dashv \Psi$ is not an adjoint equivalence.

Example 12.2.10 (Ψ is not essentially surjective). Consider the category C with four objects $\mathsf{Ob}(\mathsf{C}) = \{a, b_1, b_2, b_3\}$ and only three non-identity morphisms $f_i : b_i \longrightarrow a$ for $i = 1, 2, 3$. We may visualize the category C as follows.

$$
\begin{array}{ccc}
 & b_2 & \\
 & \Big\downarrow {\scriptstyle f_2} & \\
b_1 \xrightarrow{\ f_1\ } & a & \xleftarrow{\ f_3\ } b_3
\end{array}
$$

For any two objects c, d in C, define the sets:

$$\Delta^{\mathsf{C}}\binom{c}{\varnothing} = *,$$

$$\Delta^{\mathsf{C}}\binom{d}{c} = \mathsf{C}(c,d),$$

$$\Delta^{\mathsf{C}}\binom{a}{c,d} = \mathsf{C}(c,a) \times \mathsf{C}(d,a) \quad \text{if } c \neq d \text{ and } c, d \in \{b_1, b_2, b_3\}$$

All other sets $\Delta^{\mathsf{C}}\binom{d}{c}$ are empty. Then $\widehat{\mathsf{C}} = (\mathsf{C}, \Delta^{\mathsf{C}})$ is a configured category.

However, the configured category $\widehat{\mathsf{C}}$ is not in the essential image of the functor Ψ. Indeed, if it is in the essential image of Ψ, then Δ^{C} is as in Definition 12.1.1 for some orthogonality relation $\perp^{\mathsf{C}}$ on C. By the definition of Δ^{C}, we have $f_i \perp^{\mathsf{C}} f_j$ for $1 \leq i \neq j \leq 3$. But then $\{f_1, f_2, f_3\}$ is also pairwise orthogonal, so it forms a configuration, which contradicts the definition of Δ^{C}.   ◇

Example 12.2.11 (Φ is not an embedding). Suppose C is the category in Example 12.2.10. Define Δ^{C}_0 to be the same as Δ^{C} except that

$$\Delta^{\mathsf{C}}_0\binom{a}{x,y,z} = \mathsf{C}(x,a) \times \mathsf{C}(y,a) \times \mathsf{C}(z,a) \quad \text{if } \{x,y,z\} = \{b_1, b_2, b_3\}$$

as sets. Then $(\mathsf{C}, \Delta^{\mathsf{C}}_0)$ is also a configured category, which is not isomorphic to $(\mathsf{C}, \Delta^{\mathsf{C}})$ since the latter has no triple configurations. However, the images $\Phi(\mathsf{C}, \Delta^{\mathsf{C}})$ and $\Phi(\mathsf{C}, \Delta^{\mathsf{C}}_0)$ are equal as orthogonal categories because Δ^{C} and Δ^{C}_0 have the same binary configurations. So the functor Φ is not an embedding.   ◇

12.3 Comparison Adjunctions

In this section, we show that for each configured category, there is a comparison morphism from the colored operad for prefactorization algebras to the colored operad for algebraic quantum field theories. This comparison morphism induces a comparison adjunction between the category of (homotopy) prefactorization algebras and the category of (homotopy) algebraic quantum field theories on the associated orthogonal category. The comparison morphism is also compatible with changing the configured category and with the time-slice axiom.

Recall from Definition 8.3.2 the colored operad $O_{\overline{C}}$ for an orthogonal category $\overline{C}$ and from Definition 10.3.2 the colored operad $O_{\widehat{C}}$ for a configured category $\widehat{C}$. Also recall from Example 5.3.3 the change-of-category functor

$$(-)^M : \mathsf{Operad}^{\mathfrak{C}}(\mathsf{Set}) \longrightarrow \mathsf{Operad}^{\mathfrak{C}}(M).$$

We will use the augmentation $\eta : W \longrightarrow \mathrm{Id}$ of the Boardman-Vogt construction in Theorem 6.4.4.

Theorem 12.3.1. *Suppose $\widehat{C} = (C, \triangle)$ is a configured category with object set $\mathfrak{C}$, and $\Phi\widehat{C} = \overline{C}$ is the associated orthogonal category.*

(1) There is a morphism

$$O_{\widehat{C}} \xrightarrow{\;\delta\;} O_{\overline{C}}$$

of $\mathfrak{C}$-operads in Set *that is entrywise defined as*

$$O_{\widehat{C}}\binom{d}{\underline{c}} \xrightarrow{\;\delta\;} O_{\overline{C}}\binom{d}{\underline{c}}, \qquad \{f_i\}_{i=1}^n \longmapsto [\mathrm{id}_n, \{f_i\}_{i=1}^n\,]$$

for $\binom{d}{\underline{c}} = \binom{d}{c_1,\dots,c_n} \in \mathsf{Prof}(\mathfrak{C}) \times \mathfrak{C}$ and $\{f_i\}_{i=1}^n \in O_{\widehat{C}}\binom{d}{\underline{c}} = \triangle\binom{d}{\underline{c}}$.

(2) There is a commutative diagram

$$
\begin{array}{ccc}
\mathsf{WO}_{\widehat{C}}^M & \xrightarrow{\;\mathsf{W}\delta^M\;} & \mathsf{WO}_{\overline{C}}^M \\
{\scriptstyle\eta}\downarrow & & \downarrow{\scriptstyle\eta} \\
O_{\widehat{C}}^M & \xrightarrow{\;\delta^M\;} & O_{\overline{C}}^M
\end{array}
$$

of $\mathfrak{C}$-colored operads in M*, in which $\eta : \mathsf{WO}_{\widehat{C}}^M \longrightarrow O_{\widehat{C}}^M$ and $\eta : \mathsf{WO}_{\overline{C}}^M \longrightarrow O_{\overline{C}}^M$ are the augmentations of $O_{\widehat{C}}^M$ and $O_{\overline{C}}^M$.*

(3) There is a diagram of change-of-operad adjunctions

$$
\begin{array}{ccc}
\mathsf{HPFA}(\widehat{C}) = \mathsf{Alg}_M\big(\mathsf{WO}_{\widehat{C}}^M\big) & \underset{(\mathsf{W}\delta^M)^*}{\overset{\mathsf{W}\delta_!^M}{\rightleftarrows}} & \mathsf{Alg}_M\big(\mathsf{WO}_{\overline{C}}^M\big) = \mathsf{HQFT}(\overline{C}) \\[4pt]
{\scriptstyle\eta_!}\big\uparrow\big\downarrow{\scriptstyle\eta^*} & & {\scriptstyle\eta_!}\big\uparrow\big\downarrow{\scriptstyle\eta^*} \\[4pt]
\mathsf{PFA}(\widehat{C}) = \mathsf{Alg}_M\big(O_{\widehat{C}}^M\big) & \underset{(\delta^M)^*}{\overset{\delta_!^M}{\rightleftarrows}} & \mathsf{Alg}_M\big(O_{\overline{C}}^M\big) \cong \mathsf{QFT}(\overline{C})
\end{array}
$$

with commuting left adjoint diagram and commuting right adjoint diagram.

Proof. For the first assertion, first observe that the morphism δ is entrywise a well-defined function. That δ preserves the colored units and the operadic compositions follows immediately from the definition. To see that δ preserves the equivariant structures, suppose given a configuration $\{f_i\}_{i=1}^n \in \Delta\binom{d}{c}$ and a permutation $\sigma \in \Sigma_n$. We must show that the middle equality in

$$\delta\Big(\{f_i\}_{i=1}^n \sigma\Big) = \Big[\mathrm{id}_n, \{f_{\sigma(i)}\}_{i=1}^n\Big] = \Big[\sigma, \{f_{\sigma(i)}\}_{i=1}^n\Big] = \Big(\delta\{f_i\}_{i=1}^n\Big)\sigma$$

holds in $\mathsf{O}_{\overline{\mathsf{C}}}\binom{d}{c}$. For any $1 \le i \ne j \le n$, by the subset axiom of a configured category, we have that $\{f_i, f_j\} \in \Delta$. It follows that $f_i \perp f_j$ in $(\mathsf{C}, \perp) = \overline{\mathsf{C}}$, since $\perp$ is defined as the set of binary configurations. In other words, the f_i's are pairwise orthogonal. So the middle equality above holds by the definition of the equivalence relation $\sim$ that defines $\mathsf{O}_{\overline{\mathsf{C}}}$.

The second assertion follows from the first assertion, the change-of-category functor $(-)^\mathsf{M}$, and the naturality of the Boardman-Vogt construction. The third assertion follows from the second assertion and Corollary 6.4.9. $\qquad\square$

Definition 12.3.2. In the setting of Theorem 12.3.1:

(1) The morphism

$$\delta : \mathsf{O}_{\widehat{\mathsf{C}}} \longrightarrow \mathsf{O}_{\overline{\mathsf{C}}}$$

of $\mathfrak{C}$-colored operads and its image δ^M in M are called the *comparison morphisms*.

(2) The morphism

$$\mathsf{W}\delta^\mathsf{M} : \mathsf{WO}_{\widehat{\mathsf{C}}}^\mathsf{M} \longrightarrow \mathsf{WO}_{\overline{\mathsf{C}}}^\mathsf{M}$$

of $\mathfrak{C}$-colored operads in M is called the *homotopy comparison morphism*.

(3) The adjunction $\delta_!^\mathsf{M} \dashv (\delta^\mathsf{M})^*$ is called the *comparison adjunction*.

(4) The adjunction $\mathsf{W}\delta_!^\mathsf{M} \dashv (\mathsf{W}\delta^\mathsf{M})^*$ is called the *homotopy comparison adjunction*.

Interpretation 12.3.3. The comparison adjunction

$$\mathsf{PFA}(\widehat{\mathsf{C}}) = \mathsf{Alg}_\mathsf{M}\big(\mathsf{O}_{\widehat{\mathsf{C}}}^\mathsf{M}\big) \xrightleftharpoons[(\delta^\mathsf{M})^*]{\delta_!^\mathsf{M}} \mathsf{Alg}_\mathsf{M}\big(\mathsf{O}_{\overline{\mathsf{C}}}^\mathsf{M}\big) \cong \mathsf{QFT}(\overline{\mathsf{C}})$$

compares prefactorization algebras on $\widehat{\mathsf{C}}$ with algebraic quantum field theories on the associated orthogonal category $\overline{\mathsf{C}} = \Phi\widehat{\mathsf{C}}$. The homotopy comparison adjunction

$$\mathsf{HPFA}(\widehat{\mathsf{C}}) = \mathsf{Alg}_\mathsf{M}\big(\mathsf{WO}_{\widehat{\mathsf{C}}}^\mathsf{M}\big) \xrightleftharpoons[(\mathsf{W}\delta^\mathsf{M})^*]{\mathsf{W}\delta_!^\mathsf{M}} \mathsf{Alg}_\mathsf{M}\big(\mathsf{WO}_{\overline{\mathsf{C}}}^\mathsf{M}\big) = \mathsf{HQFT}(\overline{\mathsf{C}})$$

compares homotopy prefactorization algebras on $\widehat{\mathsf{C}}$ with homotopy algebraic quantum field theories on the associated orthogonal category $\overline{\mathsf{C}}$. $\qquad\diamond$

Remark 12.3.4. There is an analogue of the comparison adjunction in [Gwilliam and Rejzner (2017)] (Theorems 1 and 2) for the free scalar field that compares Costello-Gwilliam factorization algebra with the perturbative algebraic quantum field theory in [Fredenhagen and Rejzner (2012,b)]. A factorization algebra [Costello and Gwilliam (2017)] is a prefactorization algebra that satisfies a local-to-global gluing condition. The comparison in [Gwilliam and Rejzner (2017)] is a quasi-isomorphism from a restriction of a factorization algebra to the composition of an algebraic quantum field theory followed by a forgetful functor to cochain complexes. As far as the author can tell, the comparison in [Gwilliam and Rejzner (2017)] cannot be obtained as a special case of our comparison adjunction because we only deal with prefactorization algebras here.   ◇

The following observation is the relative version of Theorem 12.3.1.

Corollary 12.3.5. *Suppose* $F : \widehat{\mathsf{C}} \longrightarrow \widehat{\mathsf{D}}$ *is a configured functor with* $\Phi\widehat{\mathsf{C}} = \overline{\mathsf{C}}$ *and* $\Phi\widehat{\mathsf{D}} = \overline{\mathsf{D}}$.

(1) There is a commutative diagram

$$
\begin{array}{ccc}
\mathsf{O}_{\widehat{\mathsf{C}}} & \xrightarrow{\ \mathsf{O}_F\ } & \mathsf{O}_{\widehat{\mathsf{D}}} \\
{\scriptstyle \delta}\downarrow & & \downarrow{\scriptstyle \delta} \\
\mathsf{O}_{\overline{\mathsf{C}}} & \xrightarrow{\ \mathsf{O}_{\Phi F}\ } & \mathsf{O}_{\overline{\mathsf{D}}}
\end{array}
$$

of colored operads in Set *with both morphisms* δ *as in Theorem 12.3.1.*

(2) There is a commutative cube

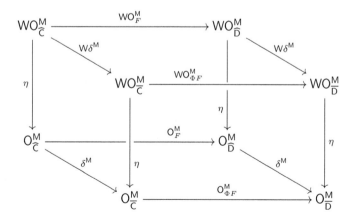

of colored operads in M, *in which every morphism* η *is an augmentation.*

(3) There is a diagram of change-of-operad adjunctions

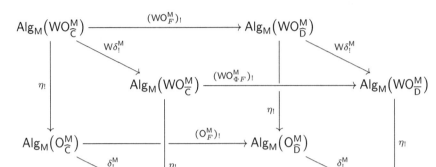

in which only the left adjoints are displayed. In each face, the left adjoint diagram and the right adjoint diagram are commutative.

Proof. The first assertion follows from Theorem 12.3.1(1) and the fact that Φ leaves the underlying functor F unchanged. The second assertion follows from the first assertion, the change-of-category functor $(-)^M$, and the naturality of the Boardman-Vogt construction in Theorem 6.4.4. The last assertion follows from assertion (2) and Theorem 5.1.8. $\qquad\square$

Interpretation 12.3.6. In the commutative cube in Corollary 12.3.5(3), the left and the right faces are the diagrams in Theorem 12.3.1(3) for $\widehat{\mathsf{C}}$ and $\widehat{\mathsf{D}}$, respectively. The front face is the diagram in Corollary 9.2.7, and the back face is the diagram in Corollary 11.2.6. The bottom face compares prefactorization algebras on $\widehat{\mathsf{C}}$ and $\widehat{\mathsf{D}}$ and algebraic quantum field theories on $\overline{\mathsf{C}}$ and $\overline{\mathsf{D}}$. The top face compares homotopy prefactorization algebras on $\widehat{\mathsf{C}}$ and $\widehat{\mathsf{D}}$ and homotopy algebraic quantum field theories on $\overline{\mathsf{C}}$ and $\overline{\mathsf{D}}$. $\qquad\diamond$

Next we consider the situation when the time-slice axiom is present. Recall the time-slice axiom from Definition 8.2.3 for algebraic quantum field theories and from Definition 10.3.4 for prefactorization algebras. Also recall localization of a category $\ell : \mathsf{C} \longrightarrow \mathsf{C}[S^{-1}]$ from Definition 2.8.1 and localization of a colored operad $\ell : \mathsf{O} \longrightarrow \mathsf{O}[S^{-1}]$ from Definition 5.4.3.

Corollary 12.3.7. *Suppose $\widehat{\mathsf{C}} = (\mathsf{C}, \triangle)$ is a configured category with object set $\mathfrak{C}$, and $\Phi\widehat{\mathsf{C}} = \overline{\mathsf{C}}$ is the associated orthogonal category. Suppose S is a set of morphisms in C.*

(1) There exists a unique morphism

$$\delta_S : \mathsf{O}_{\widehat{\mathsf{C}}}[S^{-1}] \longrightarrow \mathsf{O}_{\overline{\mathsf{C}[S^{-1}]}}$$

of $\mathfrak{C}$-colored operads such that the diagram

$$
\begin{array}{ccc}
\mathsf{O}_{\widehat{\mathsf{C}}} & \xrightarrow{\;\ell\;} & \mathsf{O}_{\widehat{\mathsf{C}}}[S^{-1}] \\
\delta\downarrow & & \downarrow\delta_S \\
\mathsf{O}_{\widehat{\mathsf{C}}} & \xrightarrow{\;\mathsf{O}_\ell\;} & \mathsf{O}_{\overline{\mathsf{C}[S^{-1}]}}
\end{array}
$$

of $\mathfrak{C}$-colored operads in Set *is commutative, in which the morphism δ is from Theorem 12.3.1(1).*

(2) There is a commutative cube

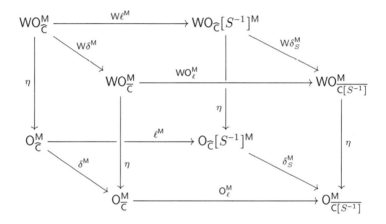

of $\mathfrak{C}$-colored operads in M, *in which every morphism η is an augmentation.*

(3) There is a diagram of change-of-operad adjunctions

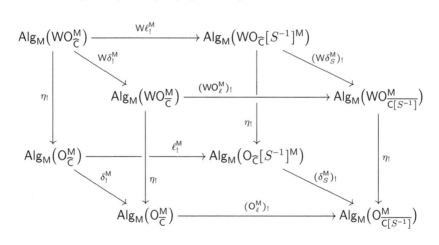

in which only the left adjoints are displayed. In each face, the left adjoint diagram and the right adjoint diagram are commutative.

Proof. For the first assertion, consider the solid-arrow diagram

$$
\begin{array}{ccc}
\mathsf{O}_{\widehat{\mathsf{C}}} & \xrightarrow{\quad \ell \quad} & \mathsf{O}_{\widehat{\mathsf{C}}}[S^{-1}] \\
\delta \downarrow & & \vdots \, \delta_S \\
\mathsf{O}_{\overline{\mathsf{C}}} & \xrightarrow{\quad \mathsf{O}_\ell \quad} & \mathsf{O}_{\overline{\mathsf{C}[S^{-1}]}}
\end{array}
$$

of $\mathfrak{C}$-colored operads in Set. For each morphism $f \in S$, regarded as a unary element in the $\mathfrak{C}$-colored operad $\mathsf{O}_{\widehat{\mathsf{C}}}$, the unary element

$$
\mathsf{O}_\ell \delta(\{f\}) = [\mathrm{id}_1, f]
$$

is invertible in $\mathsf{O}_{\overline{\mathsf{C}[S^{-1}]}}$ with inverse $[\mathrm{id}_1, f^{-1}]$. This is true because the morphism $f \in \mathsf{C}[S^{-1}]$ is invertible. Therefore, by the universal property of an S-localization of $\mathsf{O}_{\widehat{\mathsf{C}}}$, there exists a unique morphism δ_S that makes the diagram commutative.

The second assertion follows from the first assertion, the change-of-category functor $(-)^{\mathsf{M}}$, and the naturality of the Boardman-Vogt construction in Theorem 6.4.4. The last assertion follows from assertion (2) and Theorem 5.1.8. □

Interpretation 12.3.8. The adjunction

$$
\mathsf{PFA}(\widehat{\mathsf{C}}, S) = \mathsf{Alg}_{\mathsf{M}}\big(\mathsf{O}_{\widehat{\mathsf{C}}}[S^{-1}]^{\mathsf{M}}\big) \xrightleftharpoons[(\delta^{\mathsf{M}}_S)^*]{(\delta^{\mathsf{M}}_S)_!} \mathsf{Alg}_{\mathsf{M}}\big(\mathsf{O}^{\mathsf{M}}_{\overline{\mathsf{C}[S^{-1}]}}\big) \cong \mathsf{QFT}(\overline{\mathsf{C}}, S)
$$

compares (i) prefactorization algebras on the configured category $\widehat{\mathsf{C}}$ satisfying the time-slice axiom with respect to S and (ii) algebraic quantum field theories on the associated orthogonal category $\overline{\mathsf{C}}$ satisfying the time-slice axiom with respect to S. The adjunction

$$
\mathsf{HPFA}(\widehat{\mathsf{C}}, S) = \mathsf{Alg}_{\mathsf{M}}\big(\mathsf{WO}_{\widehat{\mathsf{C}}}[S^{-1}]^{\mathsf{M}}\big) \xrightleftharpoons[(\mathsf{W}\delta^{\mathsf{M}}_S)^*]{(\mathsf{W}\delta^{\mathsf{M}}_S)_!} \mathsf{Alg}_{\mathsf{M}}\big(\mathsf{WO}^{\mathsf{M}}_{\overline{\mathsf{C}[S^{-1}]}}\big) = \mathsf{HQFT}(\overline{\mathsf{C}[S^{-1}]})
$$

compares (i) homotopy prefactorization algebras on $\widehat{\mathsf{C}}$ satisfying the homotopy time-slice axiom with respect to S and (ii) homotopy algebraic quantum field theories on $\overline{\mathsf{C}[S^{-1}]}$.   ◇

12.4   Examples of Comparison

In this section, we provide examples that illustrate the comparison adjunctions in Section 12.3 between (homotopy) prefactorization algebras and (homotopy) algebraic quantum field theories.

Example 12.4.1 (The minimal case). Suppose C is a small category. In Example 12.2.5 we noted that

$$
\Phi(\mathsf{C}, \triangle^{\mathsf{C}}_{\min}) = \overline{\mathsf{C}_{\min}},
$$

with $\widehat{C_{min}} = (C, \triangle^C_{min})$ the minimal configured category on C in Example 10.2.8 and $\overline{C_{min}} = (C, \varnothing)$ the orthogonal category with the empty orthogonality relation in Example 8.4.1. The comparison morphism

$$O^M_{\widehat{C_{min}}} \xrightarrow{\delta^M_{min}} O^M_{\overline{C_{min}}} = O^M_C$$

was described in Proposition 10.4.7, in which the equality comes from Example 4.5.22. In Proposition 10.4.3 we noted that $O^M_{\widehat{C_{min}}}$ is the $\mathfrak{C}$-colored operad whose algebras are pointed C-diagrams in M. In Corollary 10.4.8 we observed that the comparison adjunction is the free-forgetful adjunction

$$M^C_* \cong PFA(\widehat{C_{min}}) = Alg_M\left(O^M_{\widehat{C_{min}}}\right) \underset{(\delta^M_{min})^*}{\overset{(\delta^M_{min})_!}{\rightleftarrows}} Alg_M\left(O^M_{\overline{C_{min}}}\right) \cong QFT(\overline{C_{min}}) = Mon(M)^C$$

between pointed C-diagrams in M and C-diagrams of monoids in M.

Furthermore, the homotopy comparison adjunction

$$HPFA(\widehat{C_{min}}) = Alg_M\left(WO^M_{\widehat{C_{min}}}\right) \underset{(W\delta^M_{min})^*}{\overset{(W\delta^M_{min})_!}{\rightleftarrows}} Alg_M\left(WO^M_C\right) = Alg_M\left(WO^M_{\overline{C_{min}}}\right) = HQFT(\overline{C_{min}})$$

is the free-forgetful adjunction between the category of homotopy coherent pointed C-diagrams in M in Definition 11.5.1 and the category of homotopy coherent C-diagrams of A_∞-algebras in M in Definition 7.7.2. $\diamond$

Example 12.4.2 (The classical case). Suppose C is a small category. In Example 12.2.6 we noted that $\Phi\overline{C_{max}} = \widehat{C_{max}}$. The isomorphism

$$O_{\widehat{C_{max}}} \xrightarrow[\cong]{\delta_{max}} O_{\overline{C_{max}}}$$

in Proposition 10.7.1 coincides with the comparison morphism δ in Theorem 12.3.1. In Example 10.7.2 we noted that the induced functor $(\delta^M)^* = (\delta^M_{max})^*$ is an isomorphism

$$PFA(\widehat{C_{max}}) = Alg_M\left(O^M_{\widehat{C_{max}}}\right) \xleftarrow[\cong]{(\delta^M_{max})^*} Alg_M\left(O^M_{\overline{C_{max}}}\right) \cong QFT(\overline{C_{max}}) = Com(M)^C .$$

In this case, the category of prefactorization algebras and the category of algebraic quantum field theories are both isomorphic to the category $Com(M)^C$ of C-diagrams of commutative monoids in M. We interpreted this situation as saying that the two mathematical approaches to quantum field theory coincide in the classical case.

Furthermore, since the comparison morphism

$$Com^C = O^M_{\widehat{C_{max}}} \xrightarrow[\cong]{\delta^M} O^M_{\overline{C_{max}}}$$

is an isomorphism of $\mathfrak{C}$-colored operads with the first equality from Lemma 11.10.1, the homotopy comparison morphism

$$WCom^C = WO^M_{\widehat{C_{max}}} \xrightarrow[\cong]{W\delta^M} WO^M_{\overline{C_{max}}}$$

is also an isomorphism of $\mathfrak{C}$-colored operads. Therefore, the induced functor is an isomorphism

$$\mathsf{HPFA}(\widehat{\mathsf{C}_{\max}}) = \mathsf{Alg}_\mathsf{M}(\mathsf{WCom}^\mathsf{C}) = \mathsf{Alg}_\mathsf{M}(\mathsf{WO}^\mathsf{M}_{\widehat{\mathsf{C}_{\max}}}) \xleftarrow[\cong]{(\mathsf{W}\delta^\mathsf{M})^*} \mathsf{Alg}_\mathsf{M}(\mathsf{WO}^\mathsf{M}_{\overline{\mathsf{C}_{\max}}}) = \mathsf{HQFT}(\overline{\mathsf{C}_{\max}}) .$$

In this case, both the category of homotopy prefactorization algebras and the category of homotopy algebraic quantum field theories are isomorphic to the category of homotopy coherent C-diagrams of E_∞-algebras in M in Definition 7.8.2. $\qquad \diamond$

Example 12.4.3 (Homotopy Costello-Gwilliam PFA and QFT on manifolds). For any small category Emb^n equivalent to the category of smooth n-manifolds with open embeddings as morphisms, we observed in Example 11.3.8 that the category of prefactorization algebras on n-manifolds with values in M in the sense of Costello-Gwilliam is isomorphic to the categories

$$\mathsf{Com}(\mathsf{M})^{\mathsf{Emb}^n} = \mathsf{PFA}(\widehat{\mathsf{Emb}^n_{\max}}) \cong \mathsf{QFT}(\overline{\mathsf{Emb}^n_{\max}}).$$

By Example 12.4.2 there is also an isomorphism

$$\mathsf{Alg}_\mathsf{M}(\mathsf{WCom}^{\mathsf{Emb}^n}) = \mathsf{Alg}_\mathsf{M}(\mathsf{WO}^\mathsf{M}_{\widehat{\mathsf{Emb}^n_{\max}}}) \xleftarrow[\cong]{(\mathsf{W}\delta^\mathsf{M})^*} \mathsf{Alg}_\mathsf{M}(\mathsf{WO}^\mathsf{M}_{\overline{\mathsf{Emb}^n_{\max}}}) .$$
$$\| \qquad\qquad\qquad\qquad\qquad\qquad\qquad\qquad \|$$
$$\mathsf{HPFA}(\widehat{\mathsf{Emb}^n_{\max}}) \qquad\qquad\qquad\qquad\qquad \mathsf{HQFT}(\overline{\mathsf{Emb}^n_{\max}})$$

So the category of homotopy prefactorization algebras on n-manifolds and the category of homotopy algebraic quantum field theories on $\overline{\mathsf{Emb}^n_{\max}}$ are both isomorphic to the category of homotopy coherent Emb^n-diagrams of E_∞-algebras in M. Analogous to Emb^n, we can also consider a small category Hol^n equivalent to the category of complex n-manifolds with open holomorphic embeddings as morphisms. $\qquad \diamond$

Example 12.4.4 (Homotopy PFA and QFT on bounded lattices and topological spaces). Suppose $(L, \leq)$ is a bounded lattice as in Example 2.2.12. Consider the configured category $\widehat{L}$ in Example 10.2.10. Then

$$\Phi\widehat{L} = \overline{L} = (L, \perp)$$

is the orthogonal category in Example 8.4.4. By Theorem 12.3.1 there is a diagram of change-of-operad adjunctions

$$\mathsf{HPFA}(\widehat{L}) = \mathsf{Alg}_\mathsf{M}(\mathsf{WO}^\mathsf{M}_{\widehat{L}}) \underset{(\mathsf{W}\delta^\mathsf{M})^*}{\overset{\mathsf{W}\delta^\mathsf{M}_!}{\rightleftarrows}} \mathsf{Alg}_\mathsf{M}(\mathsf{WO}^\mathsf{M}_{\overline{L}}) = \mathsf{HQFT}(\overline{L})$$

$$\eta_! \Big\downarrow\Big\uparrow \eta^* \qquad\qquad\qquad\qquad\qquad \eta_! \Big\downarrow\Big\uparrow \eta^*$$

$$\mathsf{PFA}(\widehat{L}) = \mathsf{Alg}_\mathsf{M}(\mathsf{O}^\mathsf{M}_{\widehat{L}}) \underset{(\delta^\mathsf{M})^*}{\overset{\delta^\mathsf{M}_!}{\rightleftarrows}} \mathsf{Alg}_\mathsf{M}(\mathsf{O}^\mathsf{M}_{\overline{L}}) \cong \mathsf{QFT}(\overline{L})$$

with commuting left/right adjoint diagrams that compares (homotopy) prefactorization algebras on $\widehat{L}$ and (homotopy) algebraic quantum field theories on $\overline{L}$. In particular, this works when:

- $\widehat{L}$ is the configured category $\widehat{\mathsf{Open}(X)}$ in Example 10.2.11 for some topological space X, and $\overline{L}$ is the orthogonal category $\overline{\mathsf{Open}(X)}$ in Example 8.4.5.
- $\widehat{L}$ is the configured category $\widehat{\mathsf{Open}(X)}_G$ in Example 10.2.12 with X a topological space on which a group G acts on the left by homeomorphisms, and $\overline{L}$ is the orthogonal category

$$\Phi\widehat{\mathsf{Open}(X)}_G = \overline{\mathsf{Open}(X)_G}$$

in Example 8.4.6. ◇

Example 12.4.5 (Homotopy chiral conformal PFA and QFT). Consider the category Man^d of d-dimensional oriented manifolds with orientation-preserving open embeddings as morphisms in Example 2.2.15. Define $\triangle$ to be the set of finite sequences of morphisms $\{g_i : X_i \to X\}_{i=1}^n$ in Man^d such that the images $g_i X_i$ are pairwise disjoint subsets of X. Then

$$\widehat{\mathsf{Man}^d} = (\mathsf{Man}^d, \triangle)$$

is the configured category in Example 12.1.8 such that

$$\Phi\widehat{\mathsf{Man}^d} = \overline{\mathsf{Man}^d}$$

is the orthogonal category in Example 8.4.7. Algebras over the colored operads $\mathsf{O}_{\widehat{\mathsf{Man}^d}}^{\mathsf{M}}$ and $\mathsf{WO}_{\widehat{\mathsf{Man}^d}}^{\mathsf{M}}$ are called *chiral conformal prefactorization algebras* and *homotopy chiral conformal prefactorization algebras*, respectively. By Theorem 12.3.1 there is a diagram of change-of-operad adjunctions

$$
\begin{array}{ccc}
\mathsf{HPFA}(\widehat{\mathsf{Man}^d}) = \mathsf{Alg}_\mathsf{M}\left(\mathsf{WO}_{\widehat{\mathsf{Man}^d}}^{\mathsf{M}}\right) & \underset{(\mathsf{W}\delta^{\mathsf{M}})^*}{\overset{\mathsf{W}\delta^{\mathsf{M}}_!}{\rightleftarrows}} & \mathsf{Alg}_\mathsf{M}\left(\mathsf{WO}_{\overline{\mathsf{Man}^d}}^{\mathsf{M}}\right) = \mathsf{HQFT}(\overline{\mathsf{Man}^d}) \\[2mm]
\eta_! \downarrow\uparrow \eta^* & & \eta_! \downarrow\uparrow \eta^* \\[2mm]
\mathsf{PFA}(\widehat{\mathsf{Man}^d}) = \mathsf{Alg}_\mathsf{M}\left(\mathsf{O}_{\widehat{\mathsf{Man}^d}}^{\mathsf{M}}\right) & \underset{(\delta^{\mathsf{M}})^*}{\overset{\delta^{\mathsf{M}}_!}{\rightleftarrows}} & \mathsf{Alg}_\mathsf{M}\left(\mathsf{O}_{\overline{\mathsf{Man}^d}}^{\mathsf{M}}\right) \cong \mathsf{QFT}(\overline{\mathsf{Man}^d})
\end{array}
$$

with commuting left/right adjoint diagrams that compares (homotopy) chiral conformal prefactorization algebras and (homotopy) chiral conformal quantum field theories. There are similar comparison adjunctions for the categories in Examples 2.2.16–2.2.19 and 2.2.23. ◇

Example 12.4.6 (Homotopy PFA and QFT on structured spacetimes). As in Example 8.4.16, suppose

$$\pi : \mathsf{Str} \longrightarrow \mathsf{Loc}^d$$

is a functor between small categories, and $S_\pi \subset \mathsf{Mor}(\mathsf{Str})$ is the π-pre-image of the set S of Cauchy morphisms in Loc^d. Define $\triangle$ to be the set of finite sequences of morphisms $\{g_i : X_i \to X\}_{i=1}^n$ in Str such that the images $(\pi g_i)(\pi X_i)$ are pairwise causally disjoint subsets of πX. Then the configured category

$$\widehat{\mathsf{Str}} = (\mathsf{Str}, \triangle)$$

satisfies

$$\Phi\widehat{\mathsf{Str}} = \overline{\mathsf{Str}},$$

which is the orthogonal category in Example 8.4.16. Algebras over the colored operads $\mathsf{O}_{\widehat{\mathsf{Str}}}[S_\pi^{-1}]^{\mathsf{M}}$ and $\mathsf{WO}_{\widehat{\mathsf{Str}}}[S_\pi^{-1}]^{\mathsf{M}}$ are called *prefactorization algebras on π* and *homotopy prefactorization algebras on π*, respectively.

By Corollary 12.3.7 there is a diagram of change-of-operad adjunctions

$$
\begin{array}{ccc}
\mathsf{HPFA}(\widehat{\mathsf{Str}}, S_\pi) = \mathsf{Alg}_{\mathsf{M}}\big(\mathsf{WO}_{\widehat{\mathsf{Str}}}[S_\pi^{-1}]^{\mathsf{M}}\big) & \underset{(\mathsf{W}\delta_{S_\pi}^{\mathsf{M}})^*}{\overset{(\mathsf{W}\delta_{S_\pi}^{\mathsf{M}})_!}{\rightleftarrows}} & \mathsf{Alg}_{\mathsf{M}}\big(\mathsf{WO}_{\mathsf{Str}[S_\pi^{-1}]}^{\mathsf{M}}\big) = \mathsf{HQFT}\big(\overline{\mathsf{Str}[S_\pi^{-1}]}\big) \\
\eta_! \Big\uparrow\Big\downarrow \eta^* & & \eta_! \Big\uparrow\Big\downarrow \eta^* \\
\mathsf{PFA}(\widehat{\mathsf{Str}}, S_\pi) = \mathsf{Alg}_{\mathsf{M}}\big(\mathsf{O}_{\widehat{\mathsf{Str}}}[S_\pi^{-1}]^{\mathsf{M}}\big) & \underset{(\delta_{S_\pi}^{\mathsf{M}})^*}{\overset{(\delta_{S_\pi}^{\mathsf{M}})_!}{\rightleftarrows}} & \mathsf{Alg}_{\mathsf{M}}\big(\mathsf{O}_{\mathsf{Str}[S_\pi^{-1}]}^{\mathsf{M}}\big) \cong \mathsf{QFT}(\widehat{\mathsf{Str}}, S_\pi)
\end{array}
$$

with commuting left/right adjoint diagrams that compares (homotopy) prefactorization algebras on π and (homotopy) quantum field theories on π. For example, this applies to the following functors.

- The forgetful functor

$$\pi : \mathsf{Loc}_G^d \longrightarrow \mathsf{Loc}^d$$

in Example 8.4.13, where Loc_G^d is the category of d-dimensional oriented, time-oriented, and globally hyperbolic Lorentzian manifolds equipped with a principal G-bundle.

- The forgetful functor

$$\pi p : \mathsf{Loc}_{G,\mathsf{con}}^d \longrightarrow \mathsf{Loc}^d$$

in Example 8.4.14, where $\mathsf{Loc}_{G,\mathsf{con}}^d$ is the category of triples (X, P, C) with $(X, P) \in \mathsf{Loc}_G^d$ and C a connection on P.

- The forgetful functor

$$\pi : \mathsf{SLoc}^d \longrightarrow \mathsf{Loc}^d$$

in Example 8.4.15, where SLoc^d is the category of d-dimensional oriented, time-oriented, and globally hyperbolic Lorentzian spin manifolds.   ◇

12.5   Prefactorization Algebras from AQFT

In this section, we identify the essential image of the right adjoint

$$\mathsf{PFA}(\overline{\mathsf{C}}) = \mathsf{Alg}_{\mathsf{M}}\big(\mathsf{O}_{\overline{\mathsf{C}}}^{\mathsf{M}}\big) \xleftarrow{\;(\delta^{\mathsf{M}})^*\;} \mathsf{Alg}_{\mathsf{M}}\big(\mathsf{O}_{\mathsf{C}}^{\mathsf{M}}\big) \cong \mathsf{QFT}(\overline{\mathsf{C}})$$

in the comparison adjunction in Theorem 12.3.1. In other words, we characterize the prefactorization algebras that come from algebraic quantum field theories. We will use the notation in the Coherence Theorem 10.3.6 for prefactorization algebras.

Theorem 12.5.1. *Suppose* $\widehat{\mathsf{C}} = (\mathsf{C}, \triangle)$ *is a configured category with object set* $\mathfrak{C}$, *and* $\Phi\widehat{\mathsf{C}} = \overline{\mathsf{C}}$ *is the associated orthogonal category. Suppose* $(X, \lambda) \in \mathsf{Alg}_{\mathsf{M}}(O^{\mathsf{M}}_{\overline{\mathsf{C}}})$. *Then the following two statements are equivalent.*

(1) There exist $B \in \mathsf{Alg}_{\mathsf{M}}(O^{\mathsf{M}}_{\mathsf{C}})$ *and an isomorphism*

$$(X, \lambda) \cong (\delta^M)^* B.$$

(2) For each $c \in \mathfrak{C}$, *the object* $X_c \in \mathsf{M}$ *can be equipped with a monoid structure* $(X_c, \mu_c, 1_c)$ *such that the following three conditions are satisfied.*

   (a) The structure morphism

$$\mathbb{1} \xrightarrow{\ \lambda\{(d;\varnothing)\}\ } X_d \in \mathsf{M}$$

   is the monoid unit 1_d *of* X_d *for each* $d \in \mathfrak{C}$.

   (b) For each morphism $f : c \longrightarrow d$ *in* C, *the structure morphism*

$$X_c \xrightarrow{\ \lambda\{f\}\ } X_d \in \mathsf{M}$$

   respects the monoid structure.

   (c) For each configuration $\{f_i : c_i \to d\}^n_{i=1} \in \triangle\binom{d}{c}$ *with* $n \geq 2$, *the diagram*

$$
\begin{array}{ccc}
\displaystyle\bigotimes_{i=1}^{n} X_{c_i} & \xrightarrow{\ \lambda\{f_i\}^n_{i=1}\ } & X_d \\[2mm]
{\scriptstyle\bigotimes_{i=1}^{n} \lambda\{f_i\}}\Big\downarrow & & \Big\| \\[2mm]
\displaystyle\bigotimes_{i=1}^{n} X_d & \xrightarrow{\ \mu_d\ } & X_d
\end{array}
$$

   is commutative, in which μ_d *is the* $(n-1)$-*fold iterate of the monoid multiplication on* X_d.

Proof. The implication (1) $\Longrightarrow$ (2) follows from the definition of the operad morphism $\delta : O_{\overline{\mathsf{C}}} \longrightarrow O_{\widehat{\mathsf{C}}}$ and the fact that each $O^{\mathsf{M}}_{\overline{\mathsf{C}}}$-algebra, i.e., algebraic quantum field theory on $\overline{\mathsf{C}}$, is a functor $\mathsf{C} \longrightarrow \mathsf{Mon}(\mathsf{M})$. Every $O^{\mathsf{M}}_{\overline{\mathsf{C}}}$-algebra of the form $(\delta^M)^* B$ for some $B \in \mathsf{Alg}_{\mathsf{M}}(O^{\mathsf{M}}_{\mathsf{C}})$ satisfies the three conditions in (2). Therefore, so does any $O^{\mathsf{M}}_{\overline{\mathsf{C}}}$-algebra in the essential image of $(\delta^M)^*$.

For (2) $\Longrightarrow$ (1), suppose (X, λ) satisfies the conditions in (2). Define a functor $B : \mathsf{C} \longrightarrow \mathsf{Mon}(\mathsf{M})$ by setting

$$B(c) = (X_c, \mu_c, 1_c) \quad \text{for} \quad c \in \mathfrak{C},$$

$$B(c) \xrightarrow{\ B(f) = \lambda\{f\}\ } B(d) \quad \text{for} \quad f \in \mathsf{C}(c, d).$$

The functoriality of B follows from the inclusivity axiom of a configured category and the associativity condition (10.3.8) of (X, λ).

To check that B satisfies the causality axiom (8.2.4), suppose given an orthogonal pair $(f : a \to c) \perp (g : b \to c)$. By the definition of Φ, this means that $\{f, g\}$ is a configuration. We must show that the outermost diagram in

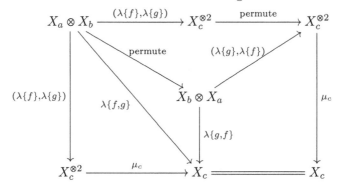

is commutative.

- The top triangle is commutative by the naturality of the symmetry isomorphism in M.
- The left triangle and the right trapezoid are commutative by assumption (2)(c), since $\{f, g\}$ and $\{g, f\}$ are both configurations.
- The middle triangle is commutative by the equivariance condition (10.3.9) of (X, λ).

Therefore, B is an algebraic quantum field theory on $\overline{\mathsf{C}}$. By the assumed conditions (2)(a)-(2)(c), we also have that $(X, \lambda) = (\delta^{\mathsf{M}})^* B$. $\qquad\qquad\square$

Interpretation 12.5.2. Physically, Theorem 12.5.1 tells us which prefactorization algebras on $\widehat{\mathsf{C}}$ arise from algebraic quantum field theories on the associated orthogonal category $\overline{\mathsf{C}}$. We will see below examples of both kinds, i.e., prefactorization algebras that arise from algebraic quantum field theories and those that do not. $\diamond$

Recall from Example 10.2.11 the configured category $\widehat{\mathsf{Open}}(X)$ for a topological space X.

Example 12.5.3 (Costello-Gwilliam associative prefactorization algebras on $\mathbb{R}$). Here we provide examples of prefactorization algebras that do not come from algebraic quantum field theories.

Suppose (A, μ, ε) is a monoid in M. Define a prefactorization algebra $(A^{\mathsf{fact}}, \lambda)$ on the configured category $\widehat{\mathsf{Open}}(\mathbb{R})$ as follows. For the empty subset of $\mathbb{R}$, we define $A^{\mathsf{fact}}_{\varnothing} = \mathbb{1}$. For each open interval $(a, b) \subset \mathbb{R}$, we define $A^{\mathsf{fact}}_{(a,b)} = A$. For a finite disjoint union of open intervals $V = \coprod_{j=1}^{n}(a_j, b_j)$ with $b_j \leq a_{j+1}$ for $1 \leq j \leq n-1$, we define
$$A^{\mathsf{fact}}_V = A^{\mathsf{fact}}_{(a_1, b_1)} \otimes \cdots \otimes A^{\mathsf{fact}}_{(a_n, b_n)} = A^{\otimes n}.$$
For a general disjoint union of open intervals $U = \coprod_{i \in I}(a_i, b_i)$, we define
$$A^{\mathsf{fact}}_U = \operatorname*{colim}_{J \subset I} A^{\mathsf{fact}}_{\coprod_{j \in J}(a_j, b_j)} \in \mathsf{M}$$

with the colimit indexed by the partially ordered set of finite subsets $J \subset I$ under inclusion. If $J \subset J'$ are finite subsets of I, then the morphism

$$A^{\text{fact}}_{\coprod_{j \in J} (a_j, b_j)} \longrightarrow A^{\text{fact}}_{\coprod_{j' \in J'} (a_{j'}, b_{j'})}$$

is induced by the unit $\varepsilon : \mathbb{1} \longrightarrow A$ for each element in $J' \smallsetminus J$.

Consider a configuration $\{f_i : U_i \subset V\}_{i=1}^n$ in $\widehat{\text{Open}(\mathbb{R})}$; i.e., the U_i's are pairwise disjoint open subsets in V. The structure morphism

$$\bigotimes_{i=1}^n A^{\text{fact}}_{U_i} \xrightarrow{\ \lambda\{f_i\}_{i=1}^n\ } A^{\text{fact}}_V \in \mathsf{M}$$

in (10.3.7) is defined by the colimits involved, the equivariance condition (10.3.9), and the following special cases on open intervals.

- If $n = 0$, then

$$\lambda\{\varnothing\} : A^{\text{fact}}_\varnothing = \mathbb{1} \longrightarrow A = A^{\text{fact}}_{(a,b)}$$

  is the unit $\varepsilon : \mathbb{1} \longrightarrow A$.
- If $n = 1$ and if $f : (a,b) \subset (c,d)$, then

$$\lambda\{f\} : A^{\text{fact}}_{(a,b)} = A \longrightarrow A = A^{\text{fact}}_{(c,d)}$$

  is the identity morphism.
- If $n \geq 2$ and if $f_i : (a_i, b_i) \subset (c,d)$ are pairwise disjoint in (c,d) for $1 \leq i \leq n$ with $b_i \leq a_{i+1}$ for $1 \leq i \leq n-1$, then

$$\bigotimes_{i=1}^n A^{\text{fact}}_{(a_i,b_i)} = \bigotimes_{i=1}^n A \xrightarrow{\ \lambda\{f_i\}_{i=1}^n\ } A = A^{\text{fact}}_{(c,d)}$$

  is the $(n-1)$-fold iterate of the multiplication μ.

For instance, consider the inclusions

$$f_1 : (2,3) \subset (1,8) \quad \text{and} \quad f_2 : (4,6) \subset (1,8).$$

- The structure morphisms

$$A^{\text{fact}}_{(2,3)} = A \xrightarrow{\ \lambda\{f_1\}\ } A = A^{\text{fact}}_{(1,8)} \quad \text{and} \quad A^{\text{fact}}_{(4,6)} = A \xrightarrow{\ \lambda\{f_2\}\ } A = A^{\text{fact}}_{(1,8)}$$

  are both the identity morphism.
- The structure morphism

$$A^{\text{fact}}_{(2,3)} \otimes A^{\text{fact}}_{(4,6)} = A^{\otimes 2} \xrightarrow{\ \lambda\{f_1, f_2\}\ } A = A^{\text{fact}}_{(1,8)}$$

  is the multiplication μ. So for open intervals already in the correct order in $\mathbb{R}$, the structure morphism is just the multiplication.
- On the other hand, for the configuration $\{f_2, f_1\}$, the structure morphism

$$A^{\text{fact}}_{(4,6)} \otimes A^{\text{fact}}_{(2,3)} = A^{\otimes 2} \xrightarrow{\ \lambda\{f_2, f_1\}\ } A$$

is the opposite multiplication $\mu \circ (1\ 2)$; i.e., permute the two domain factors before multiplying. So for open intervals not in the correct order in $\mathbb{R}$, we must first permute the domain factors back to the correct order before multiplying.

One can check that A^{fact} is actually a prefactorization algebra on $\widetilde{\mathsf{Open}(\mathbb{R})}$ using the Coherence Theorem 10.3.6. This prefactorization algebra is a key example in [Costello and Gwilliam (2017)] Section 3.1.1. Moreover, by Theorem 12.5.1, A^{fact} is *not* in the essential image of the right adjoint $(\delta^{\mathsf{M}})^*$ because it does not satisfy condition (2)(c) there. Indeed, if it satisfies condition (2)(c), then the structure morphism $\lambda\{f_2, f_1\}$ in the previous paragraph would just be the multiplication μ, which is not true. ◇

Example 12.5.4 (Costello-Gwilliam commutative prefactorization algebras on $\mathbb{R}$). Suppose (A, μ, ε) is a commutative monoid. Then the prefactorization algebra $(A^{\mathsf{fact}}, \lambda)$ on the configured category $\widetilde{\mathsf{Open}(\mathbb{R})}$ in Example 12.5.3 is in the image of the right adjoint $(\delta^{\mathsf{M}})^*$. In other words, it arises from an algebraic quantum field theory on the associated orthogonal category $\overline{\mathsf{Open}(\mathbb{R})}$. Indeed, since A is a commutative monoid, each entry A^{fact}_U inherits from A the structure of a commutative monoid. So this construction defines a functor

$$A^{\mathsf{fact}}_? : \mathsf{Open}(\mathbb{R}) \longrightarrow \mathsf{Com}(\mathsf{M}),$$

and the causality axiom (8.2.4) is satisfied. In other words, $A^{\mathsf{fact}}_?$ is an algebraic quantum field theory on $\overline{\mathsf{Open}(\mathbb{R})}$. Applying the right adjoint $(\delta^{\mathsf{M}})^*$, it becomes the prefactorization algebra A^{fact} on $\widetilde{\mathsf{Open}(\mathbb{R})}$. ◇

Example 12.5.5 (Costello-Gwilliam symmetric prefactorization algebras). Here we provide examples of prefactorization algebras that come from algebraic quantum field theories. Suppose

$$F : \mathsf{Open}(\mathbb{R}) \longrightarrow \mathsf{M}$$

is any functor, and suppose

$$\mathsf{Com} : \mathsf{M} \longrightarrow \mathsf{Com}(\mathsf{M})$$

is the free commutative monoid functor, which is left adjoint to the forgetful functor. Then their composition

$$\mathsf{Com} \circ F : \mathsf{Open}(\mathbb{R}) \longrightarrow \mathsf{Com}(\mathsf{M})$$

defines an algebraic quantum field theory on $\overline{\mathsf{Open}(\mathbb{R})}$. Applying the right adjoint, it becomes a prefactorization algebra on $\widetilde{\mathsf{Open}(\mathbb{R})}$. A prefactorization algebra of the form $\mathsf{Com} \circ F$ is another example from [Costello and Gwilliam (2017)] Section 3.1.1. ◇

Example 12.5.6 (The right adjoint is not injective). In this example, we illustrate that the right adjoint $(\delta^{\mathsf{M}})^*$, from algebraic quantum field theories to prefactorization algebras, is in general not injective on objects. Suppose X is an indiscrete topological space; i.e.,

$$\mathsf{Open}(X) = \{\varnothing \subset X\}$$

is a category with only two objects and one non-identity morphism. Suppose (A, μ, ε) is a monoid in M. Define a functor

$$F^A : \mathsf{Open}(X) \longrightarrow \mathsf{Mon}(\mathsf{M})$$

by setting

$$F^A(\varnothing) = (\mathbb{1}, \mathbb{1} \otimes \mathbb{1} \cong \mathbb{1}, \mathrm{Id}_\mathbb{1}),$$
$$F^A(X) = (A, \mu, \varepsilon),$$
$$F^A(\varnothing \subset X) = \varepsilon : \mathbb{1} \longrightarrow A.$$

In the orthogonal category $\overline{\mathsf{Open}(X)}$, there are only four orthogonal pairs:

$$\varnothing \subset \varnothing \supset \varnothing, \quad \varnothing \subset X \supset \varnothing, \quad \varnothing \subset X \supset X, \quad \text{and} \quad X \subset X \supset \varnothing.$$

It follows that the functor F^A satisfies the causality axiom (8.2.4) because the multiplication μ on A is not involved. So F^A is an algebraic quantum field theory on $\overline{\mathsf{Open}(X)}$.

In the prefactorization algebra $(\delta^\mathsf{M})^* F^A$ on $\widetilde{\mathsf{Open}(X)}$, the only structure morphisms are of the forms

$$\mathbb{1}^{\otimes n} \cong \mathbb{1}, \quad \mathbb{1}^{\otimes n} \cong \mathbb{1} \xrightarrow{\ \varepsilon\ } A, \quad \text{and} \quad \overbrace{\mathbb{1} \otimes \cdots \otimes A \otimes \cdots \otimes \mathbb{1}}^{\text{only one } A} \xrightarrow{\ \cong\ } A.$$

They correspond to the unique configurations in

$$\triangle^X \left(\begin{smallmatrix} \varnothing \\ \varnothing, \ldots, \varnothing \end{smallmatrix}\right), \quad \triangle^X \left(\begin{smallmatrix} X \\ \varnothing, \ldots, \varnothing \end{smallmatrix}\right), \quad \text{and} \quad \triangle^X \left(\begin{smallmatrix} X \\ \varnothing, \ldots, X, \ldots, \varnothing \end{smallmatrix}\right).$$

In particular, the multiplication μ is not needed to specify this prefactorization algebra. Therefore, if we change (A, μ, ε) to the monoid $A^\mathsf{op} = (A, \mu \circ (1\ 2), \varepsilon)$ with the opposite multiplication, then there is an equality

$$(\delta^\mathsf{M})^* F^A = (\delta^\mathsf{M})^* F^{A^\mathsf{op}}$$

of prefactorization algebras on $\widetilde{\mathsf{Open}(X)}$. However, the algebraic quantum field theories F^A and F^{A^op} are different because their values at X are different. $\diamond$

List of Notations

Notation	Page	Description
Chapter 2		
$\mathsf{Ob}(\mathsf{C})$	11	class of objects in a category C
$\mathsf{C}(a,b)$	11	set of morphisms from a to b in C
Id_a	11	identity morphism of a
$\mathsf{Mor}(\mathsf{C})$	12	class of all morphisms in C
C^{op}	12	opposite category of C
$\cong$	12	isomorphism
$\mathsf{Fun}(\mathsf{C},\mathsf{D})$, D^{C}	14	category of functors from C to D
$a \downarrow \mathsf{C}$	14	under category
Set	15	category of sets
$\mathbb{K}$	15	a field
$\mathsf{Vect}_{\mathbb{K}}$	15	category of $\mathbb{K}$-vector spaces
$\mathsf{Chain}_{\mathbb{K}}$	15	category of chain complexes of $\mathbb{K}$-vector spaces
Top	15	category of compactly generated weak Hausdorff spaces
Δ	15	simplex category
SSet	15	category of simplicial sets
Cat	15	category of small categories
$(S, \le)$	15	partially ordered set
$a \vee b$	15	least upper bound
$a \wedge b$	15	greatest lower bound
$\mathsf{Open}(X)$	15	category of open subsets in X
$\mathsf{Open}(X)_G$	16	category of open subsets in X with a G-action
Man^d	16	d-dimensional oriented manifolds
$\mathbb{R}$	16	field of real numbers
Disc^d	16	oriented manifolds diffeomorphic to $\mathbb{R}^d$
Riem^d	16	d-dimensional oriented Riemannian manifolds
Loc^d	17	d-dimensional Lorentzian manifolds
$\mathsf{Gh}(X)$	17	globally hyperbolic open subsets of X

Chapter 8

Chapter 9

Chapter 10

Bibliography

Bär, C. and Ginoux, N. (2011). Classical and quantum fields on Lorentzian manifolds, in: *Global Differential Geometry*, Springer Proceedings in Math. **17** (Springer-Verlag), pp.359–400.

Bär, C., Ginoux, N. and Pfäffle, F. (2007). *Wave Equations on Lorentzian Manifolds and Quantization* (Eur. Math. Soc., Zürich).

Bartels, A., Douglas, C. L. and Henriques, A. (2015). Conformal nets I: Coordinate-free nets, *Int. Math. Res. Not.* **13**, pp. 4975–5052.

Batanin, M. A. and Berger, C. (2017). Homotopy theory for algebras over polynomial monads, *Theory Appl. Categ.* **32**, pp. 148–253.

Becker, C., Benini, M., Schenkel, A. and Szabo, R. J. (2017). Abelian duality on globally hyperbolic spacetimes, *Comm. Math. Phys.* **349**, pp. 361–392.

Becker, C., Schenkel, A. and Szabo, R. J. (2017). Differential cohomology and locally covariant quantum field theory, *Rev. Math. Phys.* **29**, 1750003.

Beem, J. K., Ehrlich, P. E. and Easley, K. L. (1996). *Global Lorentzian Geometry*, 2nd edn. (Marcel Dekker).

Benini, M., Dappiaggi, C., Hack, T. P. and Schenkel, A. (2014). A C^*-algebra for quantized principal $U(1)$-connections on globally hyperbolic Lorentzian manifolds, *Comm. Math. Phys.* **332**, pp. 477–504.

Benini, M., Dappiaggi, C. and Schenkel, A. (2014). Quantized Abelian principal connections on Lorentzian manifolds, *Comm. Math. Phys.* **330**, pp. 123–152.

Benini, M., Dappiaggi, C. and Schenkel, A. (2018). Algebraic quantum field theory on spacetimes with timelike boundary, *Ann. Henri Poincaré* **19**, 2401.

Benini, M. and Schenkel, A. (2017). Quantum field theories on categories fibered in groupoids, *Comm. Math. Phys.* **356**, pp. 19–64.

Benini, M., Schenkel, A. and Schreiber, U. (2018). The stack of Yang-Mills fields on Lorentzian manifolds, *Comm. Math. Phys.* **359**, 765.

Benini, M., Schenkel, A. and Szabo, R. (2015). Homotopy colimits and global observables in Abelian gauge theory, *Lett. Math. Physics* **105**, pp. 1193–1222.

Benini, M., Schenkel, A. and Woike, L. (2017). Operads for algebraic quantum field theory, https://arxiv.org/abs/1709.08657.

Benini, M., Schenkel, A. and Woike, L. (2019). Involutive categories, colored *-operads and quantum field theory, *Theory Appl. Categories* **34**, pp. 13–57.

Benini, M., Schenkel, A. and Woike, L. (2019). Homotopy theory of algebraic quantum field theories, *Lett. Math. Physics* **109**, pp. 1487–1532.

Berger, C. and Moerdijk, I. (2013). Axiomatic homotopy theory for operads, *Comm. Math. Helv.* **78**, pp. 805–831.

Berger, C. and Moerdijk, I. (2006). The Boardman-Vogt resolution of operads in monoidal model categories, *Topology* **45**, pp. 807–849.

Berger, C. and Moerdijk, I. (2007). Resolution of coloured operads and rectification of homotopy algebras, *Contemporary Math.* **431**, pp. 31–58.

Boardman, J. M. and Vogt, R. M. (1972). *Homotopy Invariant Algebraic Structures on Topological Spaces*, Lecture Notes in Math. **347** (Springer-Verlag, Berlin).

Borceux, F. (1994). *Handbook of Categorical Algebra 1, Basic Category Theory* (Cambridge Univ. Press, Cambridge).

Borceux, F. (1994). *Handbook of Categorical Algebra 2, Categories and Structures* (Cambridge Univ. Press, Cambridge).

Bruinsma, S. and Schenkel, A. (2018). Algebraic field theory operads and linear quantization, `https://arxiv.org/abs/1809.05319`.

Brunetti, R., Fredenhagen, K. and Verch, R. (2003). The generally covariant locality principle: A new paradigm for local quantum field theory, *Comm. Math. Phys.* **237**, pp. 31–68.

Cordier, J. M. and Porter, T. (1986). Vogt's theorem on categories of homotopy coherent diagrams, *Math. Proc. Camb. Phil. Soc.* **100**, pp. 65–90.

Cordier, J. M. and Porter, T. (1997). Homotopy coherent category theory, *Trans. Amer. Math. Soc.* **349**, pp. 1–54.

Costello, K. and Gwilliam, O. (2017). *Factorization Algebras in Quantum Field Theory, Vol. 1*, New Math. Monographs **31** (Cambridge).

Dappiaggi, C., Hack, T.-P. and Pinamonti, N. (2009). The extended algebra of observables for Dirac fields and the trace anomaly of their stress-energy tensor, *Rev. Math. Phys.* **21**, pp. 1241–1312.

Dappiaggi, C. and Lang, B. (2012). Quantization of Maxwells Equations on Curved Backgrounds and General Local Covariance, *Lett. Math. Phys.* **101**, pp. 265–287.

Dzhunushaliev, V. D. (1994). Particle scattering in nonassociative quantum field theory, *Theoret. Math. Phys.* **100**, pp. 1082–1085.

Eilenberg, S. and Mac Lane, S. (1945). General theory of natural equivalences, *Trans. Amer. Math. Soc.* **58**, pp. 231–294.

Fewster, C. J. (2013). Endomorphisms and automorphisms of locally covariant quantum field theories, *Rev. Math. Phys.* **25**, 1350008.

Fewster, C. J. and Verch, R. (2015). Algebraic quantum field theory in curved spacetimes, in R. Brunetti et. al. (eds.), *Adv. in algebraic quantum field theory* (Springer Verlag, Heidelberg), pp. 125–189.

Fredenhagen, K. and Rejzner, K. (2012). Batalin-Vilkovisky formalism in the functional approach to classical field theory, *Comm. Math. Physics* **314**, pp. 93–127.

Fredenhagen, K. and Rejzner, K. (2012). Batalin-Vilkovisky formalism in perturbative algebraic quantum field theory, *Comm. Math. Physics* **317**, pp. 697–725.

Fresse, B. (2009). *Modules over Operads and Functors*, Lecture Notes in Math. **1967** (Springer-Verlag, Berlin).

Gwilliam, O. and Rejzner, K. (2017). Relating nets and factorization algebras of observables: free field theories, `https://arxiv.org/abs/1711.06674v2`.

Haag, R. and Kastler, D. (1964). An algebraic approach to quantum field theory, *J. Math. Phys.* **5**, pp. 848–861.

Hirschhorn, P. S. (2003). *Model Categories and Their Localizations*, Math. Surveys and Monographs **99** (Amer. Math. Soc. Providence, RI).

Hollander, S. (2007). Descent for quasi-coherent sheaves on stacks, *Alg. Geom. Topology* **7**, pp. 411–437.

Hollander, S. (2007). A homotopy theory for stacks, *Israel J. Math.* **163**, pp. 93–124.

Hollander, S. (2008). Characterizing algebraic stacks, *Proc. Amer. Math. Soc.* **136**, pp. 1465–1476.

Hovey, M. (1999). *Model Categories*, Math. Surveys and Monographs **63** (Amer. Math. Soc., Providence, RI).

Kan, D. M. (1958). Adjoint functors, *Trans. Amer. Math. Soc.* **87**, pp. 294–329.

Kawahigashi, Y. (2015). Conformal field theory, tensor categories and operator algebras, *J. Phys. A* **48**, 303001.

Kelly, G. M. (2005). On the operads of J. P. May, *Theory Appl. Categ.* **13**, pp. 1–13.

Lambek, J. (1969). Deductive systems and categories. II. Standard constructions and closed categories, in *1969 Category Theory, Homology Theory and their Applications, I (Battelle Institute Conference, Seattle, Wash.)* (Springer, Berlin), pp. 76–122.

Lee, J. M. (2013). *Introduction to smooth manifolds* (Springer-Verlag, New York).

Loregian, F. (2015). *This is the (co)end, my only (co)friend*, `https://arxiv.org/abs/1501.02503`.

Mac Lane, S. (1998). *Categories for the Working Mathematician*, Grad. Texts in Math. **5**, 2nd edn. (Springer-Verlag, New York).

Majid, S. (2005). Gauge theory on nonassociative spaces, *J. Math. Phys.* **46**, 103519.

Markl, M., Shnider, S. and Stasheff, J. (2002). *Operads in Algebra, Topology and Physics*, Math. Surveys and Monographs **96** (Amer. Math. Soc., Providence, RI).

Massey, W. S. (1991). *A Basic Course in Algebraic Topology*, Grad. Texts in Math. **127** (Springer, New York).

May, J. P. (1972). *The geometry of iterated loop spaces*, Lecture Notes in Math. **271** (Springer-Verlag, New York).

May, J. P. and Ponto, K. (2012). *More Concise Algebraic Topology*, Chicago Lectures in Math. Series (U. Chicago Press, Chicago).

de Medeiros, P. and Ramgoolam, S. (2005). Non-associative gauge theory and higher spin interactions, *J. High Energy Physics* **03**, 072.

Okubo, S. (1995). *Introduction to Octonion and Other Non-Associative Algebras in Physics* (Cambridge).

O'Neill, B. (1983). *Semi-Riemannian Geometry: With Applications to Relativity* (Academic Press, San Diego).

Prasolov, A. V. (2016). Cosheafification, *Theory Appl. Categ.* **31**, pp. 1134–1175.

Quillen, D. G. (1967). *Homotopical Algebra*, Lecture Notes in Math. **43** (Springer-Verlag, Berlin).

Ramgoolam, S. (2004). Towards gauge theory for a class of commutative and non-associative fuzzy spaces, *J. High Energy Physics* **03**, 034.

Rehren, K. H. (2015). Algebraic conformal quantum field theory in perspective, in R. Brunetti et. al. (eds.), *Adv. in algebraic quantum field theory* (Springer Verlag, Heidelberg), pp. 331–364.

Rezk, C. (2000). A model category for categories, `https://faculty.math.illinois.edu/~rezk/papers.html`.

Sanders, K. (2010). The locally covariant Dirac field, *Rev. Math. Phys.* **22**, pp. 381–430.

Sanders, K., Dappiaggi, C. and Hack, T. P. (2014). Electromagnetism, local covariance, the Aharonov-Bohm effect and Gauss' law, *Comm. Math. Phys.* **328**, pp. 625–667.

Schenkel, A. and Zahn, J. (2017). Global anomalies on Lorentzian space-times, *Ann. Henri Poincaré* **18**, pp. 2693–2714.

Schlingemann, D. (1999). From Euclidean field theory to quantum field theory, *Rev. Math. Phys.* **11**, pp. 1151–1178.

Schwede, S. and Shipley, B. (2000). Algebras and modules in monoidal model categories, *Proc. London Math. Soc.* **80**, pp. 491–511.

Stasheff, J. D. (1963). Homotopy associativity of H-spaces I, II, *Trans. Amer. Math. Soc.* **108**, pp. 275–312.

Verch, R. (2001). A spin statistics theorem for quantum fields on curved space-time manifolds in a generally covariant framework, *Comm. Math. Phys.* **223**, pp. 261–288.

Vogt, R. M. (1973). Homotopy limits and colimits, *Math. Z.* **134**, pp. 11–52.

Vogt, R. M. (2003). Cofibrant operads and universal E_∞ operads, *Top. Appl.* **133**, pp. 69–87.

Weibel, C. A. (1997). *An Introduction to Homological Algebra*, Cambridge Studies in Adv. Math. **38** (Cambridge).

White, D. and Yau, D. (2018). Bousfield localization and algebras over colored operads, *Appl. Categ. Structures* **26**, pp. 153–203.

White, D. and Yau, D. (2019). Homotopical Adjoint Lifting Theorem, *Appl. Categ. Structures*, `https://doi.org/10.1007/s10485-019-09560-2`.

Yau, D. (2016). *Colored Operads*, AMS Graduate Studies in Math. **170** (Amer. Math. Soc., Providence, RI).

Yau, D. and Johnson, M. W. (2015). *A Foundation for PROPs, Algebras, and Modules*, Math. Surveys and Monographs **203** (Amer. Math. Soc., Providence, RI).

Yau, D. and Johnson, M. W. (2017). *Boardman-Vogt Resolutions of Generalized Props*, `http://u.osu.edu/yau.22/main`.

Yoneda, N. (1960). On ext and exact sequences, *J. Fac. Sci. Tokyo, Sec. I* **8**, pp. 507–526.

Zahn, J. (2014). The renormalized locally covariant Dirac field, *Rev. Math. Phys.* **26**, 1330012.

Index

CPSIA information can be obtained
at www.ICGtesting.com
Printed in the USA
JSHW011023201119
2436JS00004BA/9